Gender Differences in Science Archievement in 10 Countries
1970/71 to 1983/84

Empirische Schul- und Unterrichtsforschung

Herausgegeben von Manfred Herbig

Band 9

PETER LANG
Frankfurt am Main · Berlin · Bern · New York · Paris · Wien

Dieter Kotte

Gender Differences in Science Achievement in 10 Countries

1970/71 to 1983/84

PETER LANG
Frankfurt am Main · Berlin · Bern · New York · Paris · Wien

Die Deutsche Bibliothek - CIP-Einheitsaufnahme

Kotte, Dieter:

Gender differences in science achievement in 10 countries : 1970/71 to 1983/84 / Dieter Kotte. - Frankfurt am Main ; Berlin ; Bern ; New York ; Paris ; Wien : Lang, 1992
(Empirische Schul- und Unterrichtsforschung ; Bd. 9)
Zugl.: Hamburg, Univ., Diss., 1992
ISBN 3-631-44918-6

NE: GT

D 18
ISSN 0724-4460
ISBN 3-631-44918-6

Printed in Germany 1 2 3 4 6 7

Table of Contents

List of Appendices

List of Figures

List of Tables

List of Abbreviations

ACER	Australian Council for Educational Research
FIMS	First International Mathematics Study (IEA)
FISS	First International Science Study (IEA)
HLM	hierarchical linear modelling
HS&B	High School and Beyond Study (United States)
LV	latent variable
MV	manifest variable
NAEP	National Assessment in Educational Progress
PLS	partial least square
SIMS	Second International Mathematics Study (IEA)
SISS	Second International Science Study (IEA)

HomeBack	Home background of student
Aptitude	Student's aptitude
Gender	Sex of student
SciValue	Science values of student
Effort	Effort of student
SciAtt	Science attitudes of student
Achieve	Science achievement,
ExpEd	Expected education of student
ExpOcc	Expected occupation of student
FutEnrol	Future enrolment in science courses
TchBack	Teacher's educational background
TchGend	Sex of teacher
SchFacil	Facilities available for science teaching at school
TchEmph	Emphasis on science content taught
Practice	Practical and experimental work in class
Region	Region school was located in
SchType	Type of school (Australia only)
RegType	Combination of region and type of school (England only)

α	Cronbach's alpha
a	factor loading
$\bar{a}$	mean factor loading
β	beta, standardized regression coefficient, path coefficient
$\bar{\beta}$	mean beta
d	effect size
$\bar{d}$	mean effect size
$f()$	function of...
i	indirect inner model effect
I	interaction effect in HLM analyses
N	number of cases, sample size
p	path coefficient giving the direct effect
$\bar{p}$	mean direct effect
r	product-moment correlation coefficient
sd	standard deviation
t	total effect
$\bar{t}$	mean total effect
Δ	difference
$\bar{x}$	mean
R^2	explained variance of path model, squared multiple correlation
r^2	squared regression coefficient
s^2	variance
γ^*	gamma (*), HLM path coefficient equivalent to β and p
χ^2	chi-square

Acknowledgements

This thesis would not have been possible without the help and assistance, encouragement, and dedication of several people.

First and foremost, I am indebted to my parents and my aunt without whom I would not have had the chance for a good education. Furthermore, I am indebted to my brother, whose critical remarks were always challenging.

Heidrun and Petra deserve special thanks as they had always provided me with a lot of motivation and had helped me in every aspect.

But this work would not have been completed without the dedication of three professors. First of all, I would like to thank Dr. Rainer H. Lehmann at the University of Hamburg. It was he who had raised my interest in educational statistics.

It was, however, the idea of Dr. John P. Keeves from Australia who suggested the topic of this thesis. His wide knowledge both in the field of science education and statistics helped me both conceptually and methodologically. I shall always be indebted to him for his guidance and efforts.

Finally, my very special thanks go to Dr. T. Neville Postlethwaite for introducing me to the world of international comparative research and for provoking me to continue with this thesis - even in moments of despair.

Thus, I thank 'my' three professors for all their patience, endurance, and even their sometimes 'pushy' behaviour. Perhaps it was not all in vain.

Break on thru to the other side…

Jim Morrison, Lyrics and Vocals, The Doors

1. Introduction

Until the latter half of the twentieth century the different roles of men and women in society in the western world remained largely unquestioned, although since 1920 women gradually had gained the right to vote, and general access to education at all levels.

As many women claimed the right to be treated as equals alongside men in all aspects of social, political and cultural life, the demand for further societal changes was evident. Many of these legitimate claims of women began to be constitutionalized and institutionalized in numerous countries in the late 1960s. Nonetheless, the masked imbalance between the sexes in many fields of employment was not overcome completely. There are relatively few female scientists and engineers at the professional level, and even fewer technicians and tradeswomen at the skilled worker level (Kelly, 1978; Kann, 1981; Moss, 1982; Keeves and Kotte, 1991).

The origins of such differences can be traced back to participation in studying science at school, from the earliest grades onwards (Eccles, 1984; Collis, 1985; Duncan, 1989). It cannot be ruled out that such differences are generated at an even earlier stage in the socialization process taking place at home. However, there seems little doubt that these differences between the sexes are established and consolidated during formal schooling (Keeves, 1991).

It was in 1970/71 when the International Association for the Evaluation for Educational Achievement (IEA) conducted the First IEA Science Study (FISS) with the data collection for the Second IEA Science Study (SISS) following in 1983/84. Data from these two cross-national science studies are believed to provide a unique source for the analysis of gender differences in science achievement.

It must, however, be acknowledged that the secondary analyses of data from the FISS and SISS studies imply a number of restrictions on this thesis. In particular, certain variables useful for the analysis of gender differences in science achievement had not been collected in 1970/71 or 1983/84 as the research design of FISS and SISS focussed on different aims and research questions. Also, shortcomings were encountered due to the data cleaning and selection of variables for country data sets intended to be used for subsequent multivariate analyses. These problems will be dealt with in greater detail in the relevant chapters.

This study is concerned with how the school engenders such differences through science education programs and with the changes that have been recorded over a period of 14 years from 1970 until 1984.

1.1 Gender and Science Education

When assessing students at the age of 14, an age which for many young boys and girls is the last year of formal schooling, international comparative studies have shown the existence of gender differences in writing abilities (Gorman, Purves, and Degenhart, 1989), mathematics (Marsh, 1989; Robitaille and Garden, 1989), English as a foreign language (Lewis and Massad, 1975), and computer literacy (Lockheed et al., 1983) to name just a few important subjects.

Yet, there is rarely such a clear-cut picture to be found as with science education when examining gender disparities across countries. Boys outperform girls on science achievement tests at the 10-year-old and 14-year-old levels in all countries (Comber and Keeves, 1973; Postlethwaite and Wiley, 1991).

Moreover, boys and girls have shown different attitudes to science subjects at different age and grade levels (Ormerod and Duckworth, 1975; Humrich, 1988 [1]; Duncan, 1989). Despite many attempts to liberalize education such that equal opportunities exist for both genders in a school system boys' participation in science courses at the terminal secondary level is generally substantially higher. Furthermore, boys favour studying science at the university, and girls are more hesitant to opt for a science career (Keeves, 1991; Keeves and Kotte, 1991).

A few important questions can be asked:

- ☞ What differences, if any, are found between boys and girls at different age levels in terms of achievement in science?
- ☞ What changes, if any, occurred in the differences between boys and girls at the different grade levels in terms of achievement in science between 1970 and 1984?
- ☞ In which way are attitudes towards science of boys and girls different?
- ☞ What factors contribute to gender differences in science achievement and how do they operate?
- ☞ What factors contribute to gender differences in participation rates in science and how do they operate?
- ☞ How can the participation rate of students in science courses be increased?
- ☞ Is coeducational schooling more effective than single-sex schooling for the achievement level of boys and girls in science?
- ☞ What measures can be recommended to help balance gender differences in science achievement?

Data from the FISS and SISS studies are accessed, which may, in part, help to find answers to these questions which themselves should be of major importance to educators and policy-makers. If a nation requires larger numbers of qualified school-leavers to proceed to education at the tertiary level, or to fill skilled vacancies in the labour market in order to advance the technological development of the nation, then these can only come from those who studied science with some success at school. Under such circumstances it is frequently asked, what curricular and organizational changes are prerequisites in order to obtain these goals.

In all Western countries science education is being called on to play an even more important role in the future. More students are needed to become competent in the key science subjects of physics, chemistry, and biology. This demands planning decisions with respect to updated curricula, changed classroom environments, and new teaching strategies. However, increased numbers of qualified students could be achieved by motivating more girls to enrol in science subjects at higher grade levels (see Husén and Keeves, 1991).

The objective of this study is to identify those factors which influence science achievement and participation in science courses. This should be of help in finding ways to increase a country's science-trained output of students and to provide more and more equal opportunities for studying science to both boys and girls. It is, however, recognized that societal changes in the roles assigned to men and women might not be wished to be obtained in every country.

1.1.1 *Significance of Research Topic*

In the economic competitive environment of the industrialized countries each educational system is expected to 'produce' an optimum number of technologically qualified personnel who are needed by the labour market. This has implications for the planning of the educational system of each country. Not only are more science-trained students expected to graduate from high school, but there is also a proportionately higher demand for female students as societies become more responsive to women in science careers. In the past, many of the more prestigious and more highly paid jobs in Western societies have gone to men who have been trained in science based programs, such as medicine, engineering and technology. Since girls have not studied science courses at school to the same extent as have boys, such occupations have been filled by men rather than women. Only by increasing the participation of girls in such science courses during the years of upper secondary schooling, at grade levels beyond which science is mandatory, is it possible for greater numbers of women to enter science-based occupations. Consequently, encouragement is given by parents and teachers in many countries to a greater emphasis for girls to enrol in science courses at the upper secondary school level (Kann, 1981; Riis et al., 1988; Duncan, 1989; Keeves, 1991; Keeves and Kotte, 1991).

The theme of this research study is timely. Detailed information is needed on how to achieve the goal of having more girls enrol in science courses, how to reduce gender differences in science achievement, and how to improve the achievement level of students in science. To answer these important questions, several data are to be gathered from students, teachers, and schools.

Examining the questions raised form a necessary precondition for obtaining a higher proportion of science-trained and science-oriented girls on the completion of secondary schooling. Optimizing science achievement and at the same time reducing differences in performance levels between boys and girls may eventually lead to greater economic efficiency within a system. In this process, gender differences can be reduced as increased opportunities become available to girls (Duncan, 1989; Husén and Keeves, 1991; Keeves and Kotte, 1991).

1.1.2 *The Need for Research*

During the past 30 years most educational systems have undergone administrative and curricular changes by introducing coeducational schooling, which was a challenging novelty in its early stages, with its pros and cons hotly debated by politicians, educators, and parents. While in theory the intent of such a radical change from the traditional sex-segregated educational structure into an alternative form of schooling in terms of administration, organization, curriculum, and teaching style seemed reasonable, since it followed a general trend of liberalization and permissiveness within the society. In practice, it did not necessarily succeed in removing gender differences in achievement completely (Ormerod, 1975; Kelly, 1978).

With respect to science the FISS study identified gender differences of varying degree across the science subjects of biology, chemistry, and physics and across the different age groups tested (Comber and Keeves, 1973). In the same study gender differences in students' attitudes towards science were detected. While FISS revealed gender differences for the first time cross-nationally it did not attempt to provide causes.

But more research is also needed to answer questions, such as the ones raised in Section 1.1 above, to obtain more reliable information about gender differences in science achievement and the factors promoting them. This justifies the need for further research.

1.2 The Advantage of International Comparisons

Both data from FISS and SISS offer the unique possibility of examining important research questions across different educational systems. International comparisons allow the identification of factors which typically influence attitudes and achievement as well as participation rates in science across countries, using the natural variation that exists within and between countries for the investigation of those effects.

This permits differences and similarities in the factors influencing outcomes to be analyzed.

Differences are useful in that systems with low levels of science achievement or a large variability of achievement between boys and girls can be compared with those systems having high achievement levels and less variability between the sexes. Differences also show the extent of variation observed. It is this variation across countries, school types, levels of schooling, and also across classrooms which is be used to investigate the causes for such gender differences.

Similarities between countries in science education and achievement may serve as a basis for identifying 'typical' mechanisms for promoting science achievement at the various grade levels. Relevant factors can be more easily detected and suggestions made to science educators and educational planners on how to improve the achievement levels in science while at the same time controlling for possible gender differences. Similarities can also lead to the development of a general theory of gender influences on science achievement and attitudes.

Furthermore, from the data sets available cross-national comparisons can be undertaken by examining the data across two points in time, since the data had been obtained for a set of 10 countries which took part in both science surveys (FISS in 1970/71 and SISS in 1983/84). The data also allow for a longitudinal comparison of countries that should serve to uncover changes and identify trends in science education.

1.3 Limitations of the Study

This study is restricted to data from 10 countries which were collected at two points in time: in 1970/71 and in 1983/84.

While data for three levels of schooling are available, this report concentrates on the grade level at which most of the 14-year-olds are enrolled. It is envisaged that at this age many students have to decide about their future education and training as many of them will leave school at the end of the period of compulsory schooling. Thus, the 14-year-old level is a critical stage both for students in making decisions to proceed to further education or to seek employment. It is also a critical stage for educators and educational planners in their task of improving educational provisions to obtain a well-trained and a well-educated output of students from the educational system.

Nevertheless, it should be emphasized that whenever data for the 10-year-old level as well as for the terminal secondary level are presented, it must be remembered that even at the 10-year-old level differences in achievement and participation in science occur (Comber and Keeves, 1973). Obtaining data from the terminal secondary school level permits an examination of gender differences in the course of their development and their effects on the enrolment rates in science courses.

A major limitation in analyzing the factors leading to gender differences is that from the 1983/84 data collection only two countries, namely Australia and England, still had sizeable numbers of single-sex schools in operation although in the 1970/71 data collection there were greater numbers of single-sex schools in some other countries.

This shift over the 14-year period in the amount of co-education in some countries in preference to single-sex schools is also of considerable interest and importance. However, it restricts the number of countries from which data sets can be analyzed to examine the effects of single-sex schools on the outcomes of science teaching in schools to 10. These countries are Australia, England, Finland, Hungary, Italy, Japan, the Netherlands, Sweden, Thailand, and the United States.

2. The Research Question

Before attempting to conduct analyses aimed at identifying factors affecting differences between boys and girls in science achievement across various educational systems, a review of the current state-of-the-art research on gender differences in science education will be given in this chapter. Most of the results reported are based on country- or school-type-specific rather than cross-national studies.

This is especially true for the Federal Republic of Germany where educational policy is part of each state legislature rather than the federal administration as was shown by Weinbach (1988) and Srocke (1989). Further to the absence of national curricula for science subjects in Germany, there is little empirical evidence on gender differences in science achievement due to a widespread preference for qualitative methods among researchers in what is called 'Frauenforschung', or 'Women Research', which in many instances is confined to 'soft' data, such as anecdotal case studies and speculative reasoning rather than 'hard' empirical data (see Roloff et al., 1987; Brehmer et al., 1989; Enders-Dragesser and Fuchs, 1990).

The results of a first major attempt to measure educational achievements across countries were published by UNESCO in the early 1960s (Foshay et al., 1962). In the following, it was the IEA, under which auspices international large-scale surveys in various school subjects have been carried out. In fact, the IEA First International Science Study (FISS) for which the data collection took place in 1970/71 was the first major international comparative study ever undertaken in the field of school science. In addition, the Second International Science Study (SISS) with its data collection in 1983/84 was the other major cross-national research project aimed at analyzing science achievement at different levels of formal schooling. Numerous reports and papers have since been published which deal with particular aspects of these two studies. Recently, a summary of these publications was edited by Degenhart (1990).

However, due to the design of these large-scale assessments little is known about the factors 'producing', increasing or decreasing, gender differences in science achievement at different age levels. Often research projects were confined to limited samples which did not permit generalizations across other classrooms, schools, or countries. This makes it desirable to review also research results from related subject areas where this seems appropriate, such as mathematics, or computer sciences. Likewise, findings about gender differences in the general cognitive ability levels of students, where these are available, have to be taken into consideration.

It is, of course, recognized that wherever gender differences are found home background factors as well as teacher and school factors also play a critical role. Similarly, gender differences between students, regardless of the subject one examines, cannot be understood in their full complexity if the students' individual psychological characteristics such as expectations, interests, attitudes, and values are left aside. This implies that in the following research review due attention has to be paid to these background and individual aspects.

After the summary of the research on gender differences in science achievement, the general and specific hypotheses which this thesis will examine are derived and formulated.

2.1 Review of Research

This review is structured according to those components, which as will be shown, are believed to contribute in one way or another to the manifestation of gender differences in science achievement among students. First, the effects of students' individual characteristics, home background, and socio-cultural factors are discussed. Secondly, the role of teacher and school factors on achievement differences between the sexes will be investigated before research findings on the effects of students' attitudes, values, and expectations are described.

A classification of the factors influencing gender differences in science achievement in terms of their relative strength would not be sound as grave differences exist between and within countries which can be seen later in this chapter.

It is, therefore, believed to be more appropriate to review the current research on gender differences in science achievement by structuring possible influential factors from a more chronological perspective as the effects operating on gender differences between students during their growing up are of primary importance.

2.1.1 Individual Characteristics

Even if learning opportunities and teaching strategies would be equally effective in science instruction for every boy or girl in class a formal test given at the end of a certain curricular sequence would still yield marked differences between boys, between girls, and also between boys and girls (Husén, 1989). Thus, if all external influences could be kept constant while teaching science to students at a certain age level, which, of course, is only a theoretical possibility, the question is what are the individual characteristics of students that lead to differences in achievement? Without running the risk of delving into the 'old' discussion of what intelligence is and how it is measured and without putting too much emphasis onto biological factors possibly responsible for differences between the sexes in terms of science achievement a few selected findings are worthy of presentation.

Biological and Genetical Determinants Until today no clear picture has emerged as to which degree biological or genetical factors are responsible, if at all, for gender differences in science achievement (Humrich, 1988 [1], pp. 12-13).

Sperry and his colleagues, based on their 'split-brain' neuro-psychological experiments, proposed differential functions to each of the two hemispheres of the cortex (Sperry, 1961; Levy-Agresti and Sperry, 1968). While he concluded that the left hemisphere would rather perform tasks of 'sequential information processing' and the right hemisphere a more 'simultaneous information processing' of neural impulses individual preferences of higher neural activities in either hemisphere would be likely. These, however, were not related to the genders (Sperry, 1972).

Witelson (1977), in her meta-analysis of functional asymmetry of the cerebral regions found that a specialization of either hemisphere for certain motoric and cognitive activities was observed in both sexes at as early an age which could be experimentally studied.

To some extent, these findings were in conflict with research from Gullahorn (1979 [1], 1979 [2]), who stated a different brain lateralization for males and females. Using experimental data from patients with brain-damage, the degree of language impairment from left-hemispheric lesions was related to visuo-spatial disabilities in women but not in men. In patients with lesions of the right hemisphere, the performance of visuo-spatial tasks was less affected for women than men. This suggested a different psycho-motoric organisation of cerebral regions for males and females though caution has to be used not to over-interpret these findings because of the very restricted number of cases.

Furthermore, right hemispheric functions of the brain seem to mature at an earlier age in boys than in girls. Levy and Levy (1978) attributed this to the possible role steroid hormones play in developing the relative maturational rate of the hemispheres.

There are some indications for a gender-specifity caused by different hormonal and metabolic activities in the development process of girls and boys. Following the analysis of some psycho-endocrinological research in the United States, Baker (1980) presented data indicating that a certain configuration of pre-natal hormones can influence some gender role behaviour throughout life, a finding underlined by reports from Kolata (1980) who, based on clinical research, saw particular pre-natal steroids being responsible for variation in achievement and behaviour.

Languis and Naour (1985) suggested a vector model of sex differences. They assumed that for the individual, gender difference fell along a feminine-masculine continuum with strong neuro-developmental influences at various points throughout the life-span. These neuro-developmental influences can be conceptualized in a vector model of sex differences. They see the various vector attributes, direction and magnitude, as influenced initially by differences in levels of sex hormones at developmentally sensitive points during an individual's cognitive development. According to their model hormonally induced variation would be secondarily reinforced by gender role influences of society and culture. A hormonally induced and culturally reinforced vector model of gender differences would be appropriate to account for the large range of variation in cognitive and learning patterns of individuals within female and male populations, as well as the mean difference between the gender groups.

Similarly, this opinion was expressed by Gullahorn (1979 [1]; 1979 [2]), who claimed that parental expectations tended to reinforce differential role behaviour which initially was a function of genetical, hormonal, and neuro-physiological factors.

Given these findings, gender differences may be viewed in the broader sense as differences of intelligence as proposed, for example, by Guilford (1967), Jensen (1970; 1973), or Eysenck (1967; 1970; 1979). Though a discussion of intelligence being a function of 'nature' and 'nurture' could be certainly interesting in itself, it is considered to be of marginal help investigating gender differences in science achievement in this review. While the results of Gullahorn or Languis and Naour reinforced the belief that achievement differences are largely attributable to individual ability differences being a function of genetical, hormonal, and neuro-physiological factors ('nature') other research findings seem to suggest that gender differences are more likely the effect of enculturation and socialization ('nurture').

Reports from a British study presented by Dickson (1986) provided no conclusive evidence that low mathematic performance of girls could be attributed to a possible innate ability or disability. Much more it was found that observed gender differences were the result of attitudinal and expectational differences of peers, such as parents, teachers, and examination boards who jointly viewed mathematics as 'masculine'. This stereotyped projection begins at an early age and seems to be definitely present already in primary schooling. The notion of boys achieving higher in mathematics will, consequently, lead to biassed teaching styles for boys and girls as will be also seen later in this review.

Though biological, hormonal, and psycho-physiological attributes vary between boys and girls even at pre-natal stages of development today's evidence is inconclusive as to which degree these factors themselves account for gender differences related to achievement (Petersen et al., 1982). It is more likely to assume socialization processes and special learning and developmental environments as the actual cause of biological or cognitive differences found to exist between males and female students.

Visual and Spatial Skills Biological premises, specifically genes, hormones, brain structure and function, as well as the timing of pubertal changes, have been used to support gender differences in spatial and visual ability. However, under closer scrutiny only steroid hormones, and then only as they relate to the levels present in the body, have been correlated with gender differences in cognitive performance (Wittig and Petersen, 1979; Petersen, 1983).

In analyzing gender differences in academic achievement over the past decade, three areas (visual and spatial ability, mathematics, and verbal ability) would appear to establish consistent gender differences. However, upon closer examination of spatial ability research, only one distinct area within this cognitive domain, namely mental rotations, showed slight gender differences favouring male students (Amthauer, 1970; Smith and Litman, 1979). Yet, these spatial differences between the sexes could not be traced among younger children.

Other areas, such as Piagetian horizontality (water level) and verticality (plumb line) tasks, were believed not to measure truly spatial ability and therefore should not confound the research and results as was stressed by Petersen (1983).

In an experimental study by Golbeck (1986), sex-related differences on horizontality and verticality tasks were examined by testing a selected sample of college students. Results showed that college-aged males and females generally did not differ in spatial competence although they might have been differentially influenced by task content.

In another study from the United States, Mitchell (1981) reported differences between the sexes on a set of spatial ability items administered to various age groups. While adolescent males scored higher than females, young and adolescent girls achieved a higher test score on items of verbal reasoning. The latter finding could also be supported by results from an earlier German study (Amthauer, 1970).

It is, however, interesting to note that if young students (10-year-olds) received special training in mastering items which required spatial thinking, both boys and girls in the experimental group outperformed those in the control group (Smith and Schroeder, 1979). Offering the same instruction to older students only resulted in improved test scores for male students, a finding which casts doubts on many science curricula for coeducational schooling according to which boys and girls of all age groups are supposed to be taught the same way.

This seemed to be supported by results from a study conducted by Kelly and Smail (1986; see also Smail, 1984 [1]). They gave special instruction to 11-year-old students in technical crafts doing technical drawings and hands-on work with metal and wooden materials. It could be shown that boys and girls from the experimental group succeeded in increasing their scores on a pre-post spatial visualization test significantly as compared to students from the control group.

Thus, differences in spatial abilities of lower secondary students found between the sexes could be associated with lower science achievement scores on particular practical or theoretical test items which demand cognitive processes of spatial intelligence, but these differences appear not to be innate but are much more the result of other effects such as the instruction given to the students (Smail, 1984 [2]).

Moreover, those gender differences found in spatial abilities of students were small. In a meta-analysis of cognitive gender differences Hyde (1981) reported that the biggest proportion of any differences occurred in visuo-spatial and quantifying abilities.

Mathematical and Verbal Ability The classification of mental or cognitive abilities generally assumes the structuring of these abilities into subsets of different, and largely independent factors. Next to visual and spatial abilities two other important intellectual factors which, to a large extent, determine the general achievement level and, in particular, the achievement in science are mathematical and verbal abilities (Thurstone, 1938; Tyler, 1965; Guilford, 1967; Hilgard and Bower, 1966; Scandura, 1973).

IEA had conducted two major cross-national studies on mathematics achievement (Husén, 1967; Robitaille and Garden, 1989; Travers and Westbury, 1989). Gender differences were identified in both the First and Second International Mathematics Studies, 1964 (FIMS) and 1982 (SIMS). These were found to be mainly in favour of boys at all age levels. Typically, differences occurred between the grade levels tested and between countries.

However, in order to solve science or mathematics test items a variety of abilities have to be used. Agreement has been reached on differentiating the general mathematical ability into numerical, computational, spatial, and verbal components (Husén, 1967; Travers and Westbury, 1989) even if the terms used for these traits sometimes seem to vary between the research fields of educational psychology, cognitive psychology, education, or neuro-physiology. Different theories also exist as to how these mathematical abilities are acquired, shaped, and related to other cognitive abilities (Li, 1986; Cobb, 1987; Silver, 1987).

By analyzing data obtained from large samples from United States students, Maccoby and Jacklin (1974) found gender differences at various age levels in verbal skills, commonly in favour of the girls. But the authors concluded also that the general intelligence level was the same for both sexes at all age levels tested.

On the other hand, the findings reported by Linn et al. (1987), who reviewed results from the large-scale National Assessment of Educational Progress (NAEP) study conducted in 1976/77 for 17-years-old students in the United States stated no significant gender differences for mathematics test items which required visual and verbal abilities.

Interesting results were presented by Marsh (1989) who, by reanalyzing data from the American High School and Beyond Study (HS&B), could show that mathematics and verbal attitudes were basically uncorrelated, better mathematics skills being associated with poorer verbal attitudes and better verbal skills being associated with poorer attitudes towards mathematics.

From a meta-analysis of research findings on mathematics and science achievement Linn and Hyde (1989) concluded that gender differences in cognitive tasks were small and declining. In their opinion, this was similarly true for mathematical ('quantitative' to their terminology), verbal, and spatial tasks with only the spatial tasks still showing significant gender differences if the tasks were not related to mathematics or science conceptually (p. 18).

However, verbal abilities are widely recognized to be one indicator for a student's general intellectual ability level (Weinberg and Rabinowitz, 1970; Linn and Hyde, 1989). Peng and Jaffe (1979) found that verbal test scores were negatively related to entry into traditionally masculine (occupational) fields. They hypothesized that students with a lower vocabulary regardless of their sex were more attracted to science areas which required fewer verbal abilities. Not surprisingly, Peng and Jaffe could also show a positive correlation between high mathematics achievement and job entry into male-dominated fields.

Obviously, mathematical and verbal abilities are highly related to science achievement as was pointed out by Fleming and Malone (1983) who based their opinion on a meta-analysis of the relationships of individual student characteristics and performance levels in science.

However, it should not be overlooked that cross-national results on students' verbal and mathematical or numerical abilities seem to vary considerably. Another source of variation is the age or general cognitive level of development. As will be seen in subsequent chapters, data on two subtests, the Word Knowledge and Mathematics test (derived from the IEA Six-Subject Survey), provide important data when analyzing factors contributing to variation in science achievement at the different age levels.

If other possible influences are kept constant and if achievement differences in science can still be identified between the sexes, this would then seem to suggest that one or more of the individual cognitive abilities, skills, biological, genetical, and physiological determinants discussed above accounted for those gender differences.

It is also clear, that from the research review of individual characteristics associated with science achievement no major component could be isolated which would 'produce' marked gender differences. Though individual characteristics play their role influencing student achievement, it could not be concluded safely that they would specifically 'cause' the substantial differences between the sexes found in science achievement.

2.1.2 *Socio-Economic Background*

Socio-Economic Status of the Home It has long been known that the socio-economic status of the home (SES) is responsible for having a substantial influence on students' achievement in science (Husén, 1967; Comber and Keeves, 1973; Kelly, 1978; Fägerlind, 1988). Data from both IEA science studies, FISS and SISS, revealed a major influence of the family status (Kelly, 1978; Lau, 1986; Keeves and Saha, 1991; Postlethwaite and Wiley, 1991). Common criteria used for establishing the socio-economic situation of the home were father's and mother's education and occupation, the use of dictionaries, or the number of books at home (Comber and Keeves, 1973; Robitaille and Garden, 1989; IEA, 1990).

Indeed, to assess the educational and occupational status of the home renders major problems when these are intended to be compared across countries. This becomes especially true when searching for suitable indicators reflecting the economic status of the home. Too different are these indicators between developed and developing countries. While the location and length of the family's summer vacation might be a suitable status indicator for the affluence of households in the European Community, Mappa's (1982) indicators developed to measure the material conditions of homes for a study on the effectiveness of radio broadcasting programs in South Sulawesi (Indonesia) were strikingly different. Given the socio-economic situation of the people in South Sulawesi, he identified the following indicators as discriminating best between comparatively poor and relatively rich homes: quality of the floor at home (bamboo, wood, cement, tiles), type of walls (woven palm leaves, bamboo, wood, cement), type of roof (palm leaves, tiles, zinc, high quality wood), type of lighting (kerosene lamp, hurricane light, electricity).

Despite the differences to measure the socio-economic status of the home, Morgenstern (1990) was able to show the important effect of SES on science achievement across many developed and developing countries.

However, while SES seems to be effective in influencing science achievement in the industrialized countries doubts have to be cast as to whether a similar causal relationship exists in developing countries. Data released by the World Bank from a study of primary schools in Uganda suggested only a weak relationship between SES and achievement (Heynemann, 1976). Kann (1981) came to similar conclusions for Botswana, while Jimenez et al. (1988), analyzing data from a mathematics test of eighth-graders from Thailand, stated different relationships between status and achievement for students enrolled in private and public schools.

Duncan (1989), in a very comprehensive multivariate analysis of data from Botswana, using, among others, the FISS test instruments found SES to be related to achievement in science for boys rather than girls.

Sanguinetty (1983) was able to show that for seven Latin American countries (Argentina, Bolivia, Brazil, Colombia, Mexico, Paraguay, and Peru) that the importance of family factors influencing science achievement decreased as students move to higher grade levels because of less between-home variance. School factors did not seem important at any stage in explaining academic achievement as long as minimal physical facilities were made available.

Further to this, an important factor for the effectiveness of the socio-economic situation of the home in developing countries must be seen in the overall enrolment pattern (Cooksey, 1981).

As Duncan (1989) pointed out, there is evidence that for primary and lower secondary schooling SES plays a role, if not necessarily a strong one, in determining the students' achievement level in science. However, this influence appears to lessen considerably at the upper secondary school level.

While there is some indication that in the contexts of many developing countries SES seems to favour boys, in that their achievement (in science) is influenced only somewhat positively, much stronger positive effects have been found for male students in industrialized countries (Heynemann and Loxley, 1982; Keeves and Saha, 1991).

Still, no coherent picture emerged as to how strongly the status of the home affects achievement in science let alone the extent to which gender differences are related to SES. Between-student analyses of the SISS data showed a positive effect of SES of an average standardized regression weight of about .11 for seven selected countries (Postlethwaite and Wiley, 1991, pp. 125-127). However, averaged on a larger set of 20 countries, a positive but somewhat lower direct effect of .06 was found between SES and science achievement (Morgenstern, 1990, pp. 87-89).

A similar finding was reported by Keeves and Saha, who found a positive total effect of .39 for 10 countries of home background affecting the educational expectations of students (1991, Table 7.7). In an earlier study conducted in the United States, Marini and Greenberger (1978) were able to show significant gender differences for 11th-graders in their educational aspirations and expectations with parental status having the greatest effect.

In contrast, Staver and Walberg (1986) reasoned that the influence of, as they called it, 'fixed factors' such as family income, parental expectations, and family composition together would not account for more than one percent of all variance of science achievement at the lower secondary school level (10th grade) in the United States. Further to this, Walberg and Fowler (1987) stated that the socio-economic status of the school district was a much more suitable predictor for achievement in reading, writing, and mathematics than other background variables.

Finn et al. (1979) reported that lower SES significantly influenced achievement of boys but not of girls. They argued that differences of SES were distorted by the fact that girls with a lower SES who were discouraged of potential science careers because they were not available to them.

Humrich (1988 [1]) cited results from a British study in which it was found that up to the age of 16 female students performed better than male ones in mathematics and, at the same time, showed more favourable attitudes towards the subject. Yet, gender differences were even greater for girls from low and medium SES homes. This led to the assumption that the gender differences measured probably reflected some kind of 'class aspiration' which, in fact, caused the achievement level to be affected.

This hypothesis would fall in line with findings reported by Steinkamp and Maehr (1984), who in their meta-analysis of relationships between motivation and science achievement, described SES as one key factor. While girls from lower SES homes showed a more pronounced motivation towards achieving well in science, this was also true for boys with higher SES. They hypothesized that disadvantaged students, i.e. those from homes of lower SES, probably received less attention from their parents which would imply that less stereotypical attitudes would have been passed on. Hence, disadvantaged boys would not be 'taught' that academic achievement would be something to aspire because it was typically 'masculine'. Likewise, disadvantaged girls would not necessarily be educated to be 'feminine' (e.g. get married and give up any career interest) such as upper-class girls would, which, in turn, would allow girls from low status homes to more easily show a higher motivation for science.

There is general agreement that SES operates significantly on science achievement but in many instances this, more often, occurs indirectly (Humrich, 1988 [1]; Duncan, 1989; Keeves and Saha, 1991).

Peer Socialization Previous research on possible peer effects influencing science achievement of students has given some impetus to the fact that in a number of instances girls are more disadvantaged than boys due to a higher degree of concern for peer acceptance (Rosenthal and Jacobson, 1968; Kelly, 1978; Kelly and Smail, 1986; Humrich, 1988 [1]; Duncan, 1989).

Walberg (1969) showed that the rejection of traditional female role patterns was more common for girls who scored high in physics. Based on results from the United States NAEP study on science, Haertel et al. (1981) suggested that girls presumably were more interested than boys to 'fit' into role expectations set up by peers.

Apart from anxiety about peer acceptance Finn, Reis, and Dulberg (1980) using FISS data showed that adolescent girls in the United States were more concerned about being liked and respected by their teachers than their male classmates.

There are some indications that girls come under more pressure to choose between career interests ('modern girl') and traditional role expectations ('good girl') as was shown by Kelly (1978) and Humrich (1988 [1]). As this incongruence often leads to dropping advanced science and mathematics courses (Comber and Keeves, 1973; Keeves and Kotte, 1991; Keeves, 1991) this 'mistake' may be partially countered by teachers and revised course offerings as Wiley (1976), or Wolfe and Tucker (1987) suggested. Many researchers and teachers consider that by opting out of advanced science and mathematics in lower secondary schools girls forego many future career opportunities.

One possible remedy would be to have science and mathematics courses compulsory until the end of lower secondary or even upper secondary schooling. Results from the SISS study left room for the speculation that a higher exposure of students (girls) to science instruction should increase the achievement level (Postlethwaite and Wiley, 1991; Wiley and Wolfe, 1991; Keeves, 1991).

As most of the research on peer socialization had been carried out in the United States little is known about these effects in other countries although there seems to be some evidence presented by Duncan (1989), who used data from Botswana, that similarities will be found in developing countries.

Parental Expectations Parallel to role or SES-related expectations, parental expectations can have a strong effect on interest and attitudes and the resultant science achievement of students. Here, substantial

Hyp.

differences will be found in the expectations parents put on girls and boys in science or science-related matters. Moreover, the effects these expectations have vary considerably among girls, among boys, between students of different age, and, certainly, between educational or cultural systems (Marini and Greenberger, 1978; Keeves, 1991).

While Maccoby and Jacklin (1974) who reviewed research on parental behaviour were unable to detect major differentiation of treatment between sexes in the United States, a more recent study of students' achievement in mathematics in Hong Kong by Lau (1986) suggested that parental expectations and attitudes towards the achievement of their children were, inter alia, dependent of the socio-economic status of the home.

Results from a study conducted on high school students of different grades in the United States by Wigfield (1983) indicated that parents' own achievement beliefs concerning mathematics and their backgrounds in mathematics were not related to children's beliefs. However, parents' beliefs about their children were related to children's beliefs. Though parents may have contributed to girls' lower expectancies in mathematics achievement, they did not appear to influence directly either girls' or boys' intentions to take more mathematics. Children seemed to have fewer gender-differentiated views concerning mathematics than their parents.

Jayaratne (1983) showed that parents typically stereotype both their self-assessments of and subjective experiences with mathematics as well as their daughters' or sons' mathematics achievement. Mothers were more likely than fathers to hold a gender-differentiated view of their child's achievement level in mathematics. In general, parents also strongly held math stereotypes for females and males, although these stereotypes were greater on issues of performance and utility than on aptitude. Further to this, Jayaratne believed that...

☞ having a daughter would make a parent act less likely gender-stereotyped towards mathematics;
☞ parents who held an exposed positive or negative attitude towards mathematics were likely to hold such views about mathematics for their children, particularly in the case of mothers and daughters; and...
☞ parents' general stereotypes about mathematics did not relate to the gender differentiation of mathematics for their children.

Based on path analyses of an experimental study of a remedial reading program for Grade 2 pupils in Malaysia, Norisah et al. (1982) found that parents with higher SES tended to give their daughters more attention than their sons, a clear indication for gender-stereotypying of parents.

For the context of the United States, Yee et al. (1986) stressed that...

☞ parents hold gender-differentiated beliefs about their sons' and daughters' mathematics achievement, even if boys and girls show identical performance levels;
☞ parents strongly influence children's mathematics attitudes and mathematics self-concept;
☞ parents are strongly influenced by teachers' assessments of their children's abilities; and...
☞ parents have different views about their own mathematics abilities, and express different attitudes about mathematics.

Though these findings stemmed from research carried out in the United States, and though these results were, strictly speaking, valid for mathematics only, a similar influence of parental expectations on the achievement of their children is likely to be found for science. Results from Finland (Nurmi, 1988), Belgium (De Corte and Brusselmans-Dehairs, 1985), England (Kelly, 1978), or Botswana (Duncan, 1989) seem to underline this.

Following research based on FISS data, Kelly (1978) pointed out that...

> "...parents' expectations are probably important determinants of science achievement. But is too simplistic to invoke 'conditioning' or 'socialization' and assume that the problem is solved." (p. 14)

Meanwhile, different approaches have been pursued to explore the effect of parental expectations on students' achievement. However, findings seem to vary partly because of different strategies or measures chosen to reflect parental expectations. No clarity exists whether some of the proposed mechanisms exerted by parental expectations on students' achievement (e.g. Duncan, 1989 for the context of Black South African countries; Morse and Handley, 1982 for the United States) can be generalized across countries. As parental attitudes, beliefs, and expectations appear to be also a function of parents' SES and societal values, these are likely to vary considerably between countries. This would cast doubt on any simple and quick cross-national generalization.

As several conceptual and administrative obstacles excluded the collection of data on parental attitudes, beliefs, or expectations towards their children's achievement in science for FISS and SISS no direct measure will be available for this study. Yet, in interpreting the result presented in later chapters (Chapters 5 to 8) care should be taken, at least, to consider the effects of parental expectations on the manifestation of gender differences in science achievement among students.

2.1.3 *Socio-Cultural Effects*

Duncan (1989) claimed that "societal factors influencing the educational performance of girls stem largely from the socio-cultural environment and the socio-economic environment." (p. 14)

While some information has already been presented on the effect of the socio-economic status of the home on students' achievement in science in the previous section, a few interesting findings on the relationship between socio-cultural effects and students' achievement are presented here.

No one will doubt that each culture conveys distinct images and values about the role women and men are to play (Parsons and Bales, 1955; Merton, 1962) but it is evident that these images and values apparently vary between different societies.

Dupont (1981) stated that culture-specific role models generally reflect an inferiority of women to men which is immanent in most societies. Moreover, since society shapes women's values of that society, these values and perceptions will be 'traded' to the next generations in that also the attitudes, beliefs, expectations, self-images, and values of their daughters are a reflection of the society and its role models for females and males. It remains, however, unclear why a mechanism as the one suggested by Dupont should not hold culture-specific role models for boys.

While several studies highlighted the ways in which socio-cultural values were mirrored in formal educational structures, such as curriculum, textbooks, teacher behaviour and teacher expectations (see also Sections 2.1.4 and 2.1.5), or dressing regulations (e.g. Levy, 1972; Levy and Levy, 1978; Kelly, 1981; Kelly and Smail, 1986) most of these findings were based on research from industrialized countries. Quite rightly, Duncan (1989) pointed out that different mechanisms of transferring gender-role values might be found in developing countries.

In the following, some other aspects of culturally-induced gender roles as well the effects of educational policies reflecting these gender-specific values are elaborated.

Gender Role Culture According to Imsen (1987), both boys and girls develop their own gender identities and subcultures. This encompasses the construction of value systems which 'fit' the respective subculture. Imsen described girls as architects of their own situation. Imsen administered a set of self-description items to secondary school students of different age groups in the United States and found that while boys ranked 'competitiveness' highest, for girls 'orientation towards others' and 'empathy' was most important. No gender differences could be detected when ranking items measuring 'independence'. This could suggest that girls and boys do have different interests.

With respect to the very acquisition of a gender identity, Cahill (1983) showed that by the end of the preschool age, children actively seek guidance regarding the behavioural expression of their gender identity and actively attempt to confirm it in interaction with others. Cahill understands the acquisition of a gender identity as not only a process of individual development but also the process whereby the objectivity of gender within this society is inter-generationally passed on.

Matlin and Matkoski (1985) added some interesting findings on gender-role perceptions of college students in the United States. The results appeared to indicate that both female and male students judged that females would do significantly better on the verbal portion and that males would do significantly better on the mathematical portion of the SAT (Scholastic Aptitude Test), the common IQ test for high school leavers in the United States. Further analyses demonstrated that female and male subjects did not differ in the extent of their stereotyping. Gender-typed individuals (feminine females and masculine males), as identified by scores on an inventory (BSRI - Bem Sex-Role Inventory), which measured the degree to which a student was gender-typed, showed no greater tendency to stereotype than did other individuals.

Carrying the effect of gender-role identification further, Meyer (1988) looked at competitions among females themselves. In a review of the current literature he came to the conclusion that women needed to focus more on task orientation, collaborative communication, the identification of role-models, and shared input into decision-making than males. However, his results were mainly based on research carried out in the United States. It remains questionable whether similar, culturally-induced stereotypes and mechanisms of gender-role acquisition would be applicable in other countries.

This is not the place to elaborate further the psychological aspects of gender-role acquisition. For this the reader should consult the basic concepts put forward by Block (1973), Bem (1974), Eccles (1983, 1984), and Levy (1989), or the very interesting account of the research project conducted by Duncan (1989) in Botswana.

Educational Policies In an earlier cross-national study performed by UNESCO (Foshay et al., 1962) girls seemed to perform relatively well in a subtest of reading comprehension but poorly in a science subtest in comparison to boys. The careful analysis of these gender differences were interpreted in that...

> "...the aims and, more especially, the beliefs of most teachers and educational administrators lead them to expect wide differences in performance, and this is what is therefore achieved. Where, on the other hand, the grade placement system operates and especially where, within such a system, teachers do not attempt to measure innate ability and therefore do not expect their pupils' attainment to be matched to it, then the dispersion of achievement will be much less." (p. 62)

According to Lee and Gropper (1974) teachers, who are concerned about teaching without gender bias, should be aware that gender differences are a reflection of cultural differences. The authors believed that actually there were more similarities than dissimilarities between the genders as there were within each gender in terms of variation of achievement.

However, much has been said about the inequality of science teaching practices towards female and male students as well as gender-biassed curricula (Comber and Keeves, 1973; Kelly, 1978; Haertel et al., 1981; Morse and Handley, 1982; Lee and Bryk, 1986; Zerega et al., 1986; Rosier and Keeves, 1991; Keeves, 1991; Postlethwaite and Wiley, 1991) and Section 2.1.5 will review specifically the influence of teachers and teaching on the manifestation of gender differences. Even if teachers are most directly involved in the strengthening or loosening of gender gaps in the school, educational policies are also supposed to address the problem.

Based on an extensive research review, Walberg (1990; 1991) gave several promising propositions of how to improve science teaching in advanced and developing countries. Though his suggestions cannot be discussed in detail here, it should be emphasized that he saw Japan as an exemplification of successful science teaching. Walberg's conclusion can be understood if one takes into consideration the comparatively high amount of science instruction the average Japanese undergoes in his school career and out-of-school activities promoted by the society (Kanaya, 1988; NIER, 1988; Rosier and Keeves, 1991).

To decrease gender differences and to increase achievement in science Walberg recommended more effective educational practices such as direct teaching, comprehension teaching, mastery learning, and cooperative learning (pp. 63-64). But any change in teaching methods ought to be accompanied by structural and contextual changes in the science curricula of an educational system if improvements are sought.

For a more thorough analysis the reader is referred to other publications which have tackled this question from an international perspective (Keeves, 1991; Rosier and Keeves, 1991).

However, one of the major school organizational changes which has taken place in science teaching in the last decades has been the world-wide restructuring of single-sex schools into forms of coeducational schooling in most countries (Comber and Keeves, 1973; Keeves, 1991; Rosier and Keeves, 1991). As could be shown, this certainly has contributed to an increased participation of girls in science and girls advancing a career in a science-related field (Keeves, 1991; Keeves and Kotte, 1991).

Nevertheless, the research situation is unclear about which of the gender-typed forms of schooling, single-sex or coeducational, is more appropriate.

Lockheed (1975) could not find any differences in task-oriented activities between single-sex boys and single-sex girls groups, while in mixed groups boys were clearly dominant. As a result Lockheed suggested first to introduce new science concepts separately to boys and girls, while subsequent discussions might be better held in a mixed-group environment. There may, however, be doubts if this is organizationally feasible.

Hildebrand (1987) proposed that teaching of mathematics and science in single-sex classes would be advantageous for girls as they would not need to appear gender-typical, i.e. feminine, as in a mixed-class where boys expect them to behave as 'girls'.

Johnson et al. (1985) and Walberg (1991) identified more cooperative learning as a possible solution in offering gender-unbiassed instruction to both boys and girls.

In a re-analysis of the HS&B data, Lee and Bryk (1986) identified large differences between single-sex and coeducational schools but they did not arrive at an unequivocal recommendation as to which form of schooling was more appropriate for boys and girls.

Although the problems arising from mixed instruction of male and female students in science classes were acknowledged and widely discussed (for example, in Australia: Moss (1982); in England: Kelly and Smail (1986); in Germany: Heinrichs and Schulz (1989); in the Netherlands: Mottier (1986); in the United States: Humrich (1988 [1])), a general solution or advice could not be advanced. Too diverse are the educational systems and cultural settings as was shown by Comber and Keeves (1973) and Keeves and Kotte (1991).

What should be understood is that particular educational policies, be they in curricular changes in science, science teacher training reform, revised allocation of funds for science instruction, or school organizational changes, could severely contribute not only to an overall improvement of science achievement but also to a reduction of gender differences. Therefore, special attention will be given to the question of how school organization is related to achievement in science. As examples, this question can be discussed in greater detail for Australia and England (Chapter 8), where sufficient numbers of single-sex schools exist.

As a reflection of the socio-cultural situation of a society the educational system and its organization do contribute to the expression or marginalization of gender differences in school achievement. This will be made transparent in this thesis.

2.1.4 Classroom and School Environment

The previous sections reviewed research on the 'effects' of individual characteristics, family situation, and other student background factors within a certain socio-cultural and educational setting on achievement in science. This section focusses on the relationship of classroom and school environment to gender differences in science achievement.

Classroom Activities One of the key problems believed to be responsible for girls' inferior achievement in science has been the amount and type of 'exposure' to science. Exposure, here, can be defined as the contact a person has with science or science-related tasks, problems, actions, or things over the years when growing up (Kelly, 1978; Whyte, 1986). This includes playing with 'masculine' toys such as cars or model airplanes as well as the contact a science teachers exercises with a particular student by asking questions, lecturing, or correcting.

There are some indications that the use of toys promotes early childhood gender differences which may carry on to affect later attitudes to and achievement in science (Kelly, 1978; Tracy, 1987; Kroeger, 1990). However, no international longitudinal data are available to show possible cross-national relationships.

Kahle (1989) suggested that the experiences school-children have with their science teachers is crucial for their interest in science and, subsequently, for their achievement. She believes that "different expectations can contribute to teachers unconsciously calling more often on particular students, called 'target students', to answer questions." (p. 6)

Both Whyte (1986) and Tobin (1987) reported of male students showing a higher tendency to call out answers aloud, getting hold of experimental equipment more often, and demanding more of the teacher's attention than girls in science classes. Based on observational data in Australia, Tobin (1987) could show that almost exclusively target students were boys while Kahle (1987) came to a ratio of male to female target students of between 4:1 to 8:1.

This 'pushy' behaviour of male students in class is frequently observed also in the science-related field of computer instruction (e.g. Kulik, Bangert, and Williams, 1983; Adams, 1985).

Nielsen, in a study conducted with high school students in Denmark (1986), found that male dominance in the science classroom is paralleled by a lack of self-confidence of girls. As she put it, girls tended to blame themselves rather than external factors in situations of failure.

Though teachers do, of course, respond differently to students' self-expression in a variety of ways, all of these involve some kind of student-teacher interaction (Humrich, 1988 [1]). This differential interaction pattern which happens in the classroom has to be seen as part of the so-called 'hidden curriculum' (Bornstein, 1982). It is also expressed by the teacher providing different, gender-specific, learning experiences for the students (Saario, Jacklin, and Tittle, 1973; Leinhard, Seewald, and Engel, 1979; Kahle and Lakes, 1983; Duncan, 1989).

In fact, a general notion seems to exist that available equipment, both for teaching and experimenting, helps male students to achieve better in science. In many countries, special emphasis is laid on laboratory work (Kojima, 1974; Hegarty, 1987). However, results about the effectiveness of laboratory work are inconclusive (Walberg, 1991) and there are some indications that great variations occur between countries (Comber and Keeves, 1973; Keeves, 1991; Postlethwaite and Wiley, 1991). The current literature does not permit generalizations about the influence of experimental work in science laboratories nor about the effects of science textbooks across countries (Gallagher, 1987), though Comber and Keeves (1973, pp. 254) using FISS data found in eleven of the countries they analyzed that, where 10-year-old students had a science textbook to learn from, their level of achievement in science was improved.

Insufficient information is available as to which degree the availability and quality of textbooks in science classes and the conduction of experimental work contributes to gender differences in science achievement across countries.

Homework Though the IEA Classroom Environment Study (Anderson, Ryan, and Shapiro, 1989) cast some light on the classroom activities of and interactions of teachers and students in various countries, no specific results were given about possible differences between male and female students. However, one of the important findings was that across countries, teachers showed a tendency to use homework assignments to identify learning deficiencies and to use more formal tests to evaluate students.

The importance of homework, specifically for achievement in science, has been reported in the international IEA science studies (Comber and Keeves, 1973; Keeves, 1991; Postlethwaite and Wiley, 1991). Results from FISS and SISS clearly showed that girls in all participating countries reported to have done significantly more homework assigned in science than boys. Yet, Comber and Keeves caution that a simple positive correlation between the time spent on science homework per week, as was found in 15 of 19 countries for 14-year-old students, is "a sound indicator of standards of industry and effort by students at school, but that greater time given to science homework in particular does not necessarily lead to higher achievement in science." (p. 263)

In a partial least squares analysis carried out on a subset of seven countries using data from SISS, Postlethwaite and Wiley were able to show that classroom efforts were, on average, slightly more important for the achievement level of 14-year-olds in science than were experiments done in class or the amount time students spent on doing homework (1991, pp. 125-132). But the analysis also illustrated the great variability across the subset of countries so that generalizations should be made with caution.

Similar findings were made by Keeves (1991), who compared the amount of time spent on doing science in relation to science achievement in 10 selected countries. Though he identified time spent on doing science (i.e. classwork and homework) to be positively related to science achievement at all three grade levels tested (lower to upper secondary school levels), exceptions had to be noted.

In a further analysis of the SISS data, Keeves and Kotte (1991) showed that across a subset of 10 countries a number of significant gender differences appeared to exist when breaking down the effects of homework or classroom efforts on achievement for 14-year-old students. But again, patterns across countries varied considerably. However, girls seemed to have higher motivation and industry in science at that age level (1991, Table 6.9).

Extracurricular Science Activities Hardly any relevant information is available about the extent of influence of out-of-school, or extracurricular, science activities on science achievement of students. Though most of the national science curricula acknowledge an advantage of extracurricular science activities for the achievement level of students in and the attitudes towards science (Keeves, 1972; Rosier and Keeves, 1991) no reliable results were obtained so far in which way and to which extent out-of-school science activities relate to science achievement in school.

In the analysis of data from Botswana, Duncan (1989) demonstrated that it was gender-typing of school science rather than gender-typing of science in general which apparently contributed to gender differences observed in science achievement among students.

Kroeger (1990, see above) produced some evidence that the frequent use of science-related toys helps children to develop a better understanding of scientific matters. The type and frequency of toys used by students is, to some degree, also responsible for the initialization of gender differences.

That science-related extracurricular activities promote higher achievement in science is evident when comparing Japanese students with those of other countries. Even if one does not agree with Walberg (1991), who saw Japan as a model country for science education, because of the many extracurricular activities boys and girls are required or requested to join, one cannot simply reject the notion that no link exists between the enormous efforts of the Japanese society and business community to promote out-of-school science programs in order to attract students to undertake some kind of science-related career after leaving school as in an international comparison Japanese students performed extremely well (Kanaya, 1988; Keeves, 1991; Postlethwaite and Wiley, 1991).

School Factors Obviously, students like some schools better than others and certainly a number of different explanations may be applicable for the effects of this. Data from the SISS project revealed that girls, in general, had a more favourable attitude to school and school learning than boys (Keeves, 1991). If this is so, it can be questioned whether the advice from Hildebrand (1987) promotes gender differences in favour of girls, who suggested that science teachers should allow students to 'personalize' the classroom or science laboratory by decorating it with posters and magazines, so that the students develop a greater affinity for the subject matter.

According to Walberg (1991), it remains dubious whether well-equipped laboratories necessarily guarantee students' success in science. This may be supported by results from regression analyses performed by Postlethwaite and Wiley (1991), who showed that school laboratory equipment had a positive effect in eight out of nine countries for which data had been analyzed. Yet, how the availability and use of science laboratories might affect gender differences is unclear.

Though it cannot be outright rejected that school factors, such as class organization, timing of science lessons, or the type of experiments conducted, to name just a few possibilities, influence science achievement (Duncan, 1989), the gender-specific effects of school facilities on science achievement remain clouded. It is, however, obvious that much of the variation occurs not only between countries, but substantial differences will be also found within most educational systems.

2.1.5 *Teacher and Teaching Effects*

As teachers play an important role in teaching science no one will doubt their influence on their students' acquisition of knowledge. Just how teachers and the way they teach science affects the generation, constitution, or reduction of gender differences in science achievement has to be considered. Some of the important findings will be described in the following.

Teacher Characteristics Reviewing over 150 sociological and economic studies from the United States, Hanushek (1986) identified a slightly positive relationship between teacher experience as well as the amount of pre-service and in-service training received, and the achievement level of students.

Though a positive correlation may exist for certain countries or certain sub-populations within countries, teaching experience does not necessarily 'cause' higher achievement in science as was found by Comber and Keeves (1973). On the other hand, Postlethwaite and Wiley (1991) showed a very minor positive relationship between teaching experience and science achievement for 14-year-old students with teaching style even being correlated negatively with science achievement throughout all countries analyzed.

From the current state of research it remains unclear why students in many cases do not benefit in their achievement level in science if taught by a more experienced teacher.

Shukla (1974), in analyzing data from the FISS study for India, reported that teachers who had come to an in-service training program were administered the same science tests as given to the students. In comparison to the total sample of students selected from India for FISS the teachers' mean score was only marginally higher than the means for the students. However, the Indian teachers' mean score fell short of the means of students from developed countries. This underlines the importance of teacher subject matter knowledge for the achievement level of students.

Though case studies from the United States have shown that knowledgeable teachers are less likely to pass on misconceptions, are more confident in imparting information, use less time for preparation, and are able to present a wider range of examples and analogies which may help students to learn and understand a certain topic more easily (Shymanski and Kyle, 1988; Grossman, Wilson, and Shulman, 1989), only in some instances does teaching experience 'produce' higher science achievement.

If general teaching experience shows an inconsistent picture across countries, types of schools, or classrooms as was demonstrated by the analyses performed on the FISS and SISS data sets (Comber and Keeves, 1973; Keeves, 1991; Postlethwaite and Wiley, 1991), the amount and type of in-service training given to teachers could lead to a significant increase in the performance level of the students taught by these teachers (Gallagher, 1987; Walberg, 1991). However, no results were obtained with respect to gender differences in effect of teaching experience on science achievement.

Findings from a study by Jones (1989), based on the observation of dyadic teacher-student interaction in United States high school science classes, indicate that teachers of all levels of experience tend to interact more with male students than female students. Further to this, it became evident that both newly prepared and experienced teachers lack the skills needed to alter their gender-biassed behaviours and teaching strategies.

Sex of Teacher A number of studies have identified differences between male and female teachers with respect to their classroom behaviour, expectations of achievement of students, or teaching behaviour (Adams and Biddle, 1970; Good et al., 1972; Dunkin, 1985).

This opens questions about the role models of science teachers. If teachers, as some studies have shown (Kelly, 1978; Miller and Asher, 1983; Humrich, 1988 [1]) tend to show gender-stereotypical behaviours in the classroom, one would probably see this as an effect of the science teacher being male or female, thereby following his or her own gender-typical role in teaching.

In 1973, Comber and Keeves concluded that 14-year-old students in eight out of 17 countries looked at, achieved higher in those schools, where the proportion of male science teachers was higher, even after the effects of students' gender had been controlled for. A reverse trend was observed in Hungary and Italy, where the achievement of the class increased if taught by a female teacher. (p. 264)

For Kelly (1978), there is no empirical evidence that female science teachers produce better results with girls, but Ormerod (1975) attributes the relative popularity of science in single-sex girls schools in England to the possibility that the science teacher is the only male teacher in the school.

In fact, most secondary school science teachers were found to be male though some changes did occur in countering this imbalance since the early 1970s (Keeves and Kotte, 1991; Keeves, 1991). One noteworthy example from the SISS study was Hungary, where 70 percent of the science teachers teaching 14-year-old students were women. At that age group students from Hungary scored highest of all other countries (Postlethwaite and Wiley, 1991).

It seems, however, important to gain more insight about the effect, teacher's gender may have in the various countries at different grade levels on the achievement of students in science.

Teaching and Learning Styles If not much clarity can be found about effects of the sex of the teacher on students' achievement in science, several studies investigated the different methodological approaches or teaching practices in relation to the different learning styles and possible gender differences.

Postlethwaite and Wiley (1991) found a negative relationship between teaching style and science achievement for 14-year-old students in data from all but one country (Italy, Grade 9). They also provided important information about the way science was taught in the countries participating in SISS but did not relate teaching loads or teaching strategies to the achievement level of students in science.

Using a case study approach, Epler (1982) pointed out that those teachers who held the highest university degrees exhibited the most dogmatic and gender-biassed behaviours in science classrooms. Gender-stereotyping was promoted by teaching in an urban environment. Generally, the teachers observed perceived success among male students more frequently than among females.

Sadker et al. (1984) conducted a three-year research and development project which was meant to develop new knowledge about sex equity in classroom interactions, and to develop successful techniques for reducing or eliminating gender-biassed interactions with the students. The authors showed that by means of micro-teaching and collegial problem solving, teachers could be made aware of and reduce their gender-biassed classroom behaviour.

Long et al. (1986) reported significant improvements of reducing gender-biassed science instruction and classroom interaction of teachers, who underwent a specially designed training. This voluntary in-service training seminar focussed on the elimination of gender-biassed teacher-student interactions in the classroom and the distribution, precision, and quality of teacher responses to students' verbal behaviour.

Noticing gender-typical ways of teaching in Botswana, Duncan (1989) advances the view that "teachers need to be sensitized to classroom practices where boys dominate and girls tend to hang back, so that they can take steps to compensate." (p. 209)

To counter (gender-)biassed classroom interactions, school administrators, researchers, and teachers jointly developed the so-called Gender Expectations and Student Achievement (GESA) program (Grayson, 1987). GESA was designed to reduce the disparity in the treatment received by boys and girls in the classroom and to improve curriculum, materials, and the learning environment to enhance student achievement for both sexes. Of teachers participating in the program, their frequency of interactions with students, gender-stereotyping, and stereotypical interactions were observed on-site. Grayson stated that meanwhile GESA had been well-received in other areas of the United States and helped to decrease gender-biassed behaviour of teachers in the classroom.

Similarly, Montgomery (1986) offered a variety of measures aimed at reducing biassed behaviour and interactions of science teachers in vocational schools in the United States.

However, if procedures are available for making teachers aware of possible gender-typical interactions with their students during science teaching, the teaching methods applied in science gain special momentum, too.

Walberg (1991), after having reviewed numerous quantitative research syntheses and meta-syntheses from the United States, arrived at three major approaches to instruct students 'optimally' in science:

☞ learner autonomy in science;
☞ activity-based teaching; and
☞ original source papers.

Yet, no information is provided in which way these teaching strategies contribute to gender differences in students' achievement in science.

Based on findings from Finn et al. (1979), Stockard et al. (1980), and Good and Findley (1985), Bank (1985) pointed out, that it may be of greater importance to look not only at gender differences per se but also to take into account the variation among each of the sexes and the spectre of social contexts to understand the effects of different instructional methods and classroom behaviour of students on science achievement.

Though some evidence exists that the in-class behaviour of teachers and the teaching strategy applied possibly disadvantages either sex, the results are largely inconclusive and do not permit a generalization across countries. Not only is the teaching and learning style difficult to assess, as was shown by results from the IEA Classroom Environment Study (Anderson, Ryan, and Shapiro, 1989), but even more complications will arise methodologically if one wishes to characterize their effects on the promotion of gender differences in science achievement of the class or individual students.

Nevertheless, it is widely acknowledged that teachers influence students' attitudes and values towards science as well as their expectations and future career plans.

2.1.6 *Attitudinal Factors*

If, as the results from previous studies have shown, boys seem to perform better in science than girls (e.g. Comber and Keeves, 1973; Finn et al., 1979; Haertel et al., 1981; Keeves and Kotte, 1991; Postlethwaite and Wiley, 1991) at nearly all grade levels and in most subject areas, this may be less a result of underachieving girls being cognitively inferior to boys but rather an effect of boys and girls subjective perception of science being a male domain. What if girls hold negative or less favourable attitudes towards science? Do positive attitudes of students and their interest in science increase the individual achievement level in science?

Handley and Morse (1984; see also Morse and Handley, 1982) concluded that gender role attitudes were, in fact, related to both science achievement and attitudes towards science.

Eccles (1983; 1984) proposed a model for mathematics achievement and participation in mathematics for secondary school and college students in the United States. She believed that the two most important factors positively stimulating achievement and future enrolment in mathematics were the individual's expectations of success on a given task and the subjective value of this task for the individual. Though originally developed for explaining gender differences in the enrolment ratio and achievement level in advanced mathematics courses, Eccles' model could well be adapted to science.

Duncan (1989) has shown how gender-typing and science performance were related in Botswana. Boys and girls ranked the common school subjects differently with boys favouring Science, Agricultural Science, English, and Mathematics while girls indicated the highest preference for Domestic Science, English, Science, and Mathematics. Girls in Botswana thought that boys should learn practical skills at school so that they become self-reliant. Boys, however, regarded academic subjects as more important in shaping employment opportunities (p. 202-203).

Attitudes towards Science The FISS and SISS projects provided insight to the issue of the interest students expressed in studying science in various countries. Comber and Keeves reported a constantly positive correlation between the interest students expressed in science and their achievement level as measured by the total science test score for all 17 countries for the age group of the 14-year-olds (1973, pp. 258-259).

This finding was supported by partial least squares analyses carried out by Postlethwaite and Wiley (1991) on SISS data. It was found that the students' view of and interest in science showed a strong positive effect on science achievement in nearly all countries at the 14-year-old level. A relatively low interest in science as taught in the classroom was identified for Japan and the Netherlands. In eight out of 13 countries 14-year-old students expressed a higher interest in science than did students from the terminal secondary school level. However, in the remaining five countries students from the terminal secondary school level showed an increased interest in science compared with the 14-year-olds which might have been the result of a selection bias. This may be taken as a supportive argument in that the attitudes towards and interest in learning science are a function of other factors related to the students' background, method of teaching, or school environment.

Indeed, variations appear to occur across countries regarding the expression of interest in science by students of different age. Less recent studies (for example, Meyer and Penfold, 1961; Finn and Kotlewski, 1975; Ormerod, 1975; Ormerod and Duckworth, 1975) indicated that interest in science was unrelated to achievement in science up to the age of 13 but became significantly related after that.

Choppin (1974) using FISS data from England concluded that the interest of boys and girls in physics and chemistry decreased between the ages of 14 and 16, yet the gap between boys' and girls' attitudes remained nearly the same.

Investigating students in the United States in Grades 5 through 10, both cross-sectionally and longitudinally, Hoffmann (1987) found that even before any formal physics instruction had commenced at the age of 10, girls reported significantly less interest in most science areas than the boys did. With the introduction of physics these attitudinal gender differences were widened even further. But Hoffmann also pointed out that if teachers were giving attention to the interest of students this led to an increased interest in science among both boys and girls. Likewise, students interest in science could be positively affected by the teacher's ability to relate science topics discussed in class to student experiences.

Bonora (1974) explained gender differences in science achievement by showing that boys and girls performed differently on the three behavioural objectives 'knowledge', 'application', and 'higher mental processes'. Taking data from the FISS study, Bonora concluded that...

> "...sex differences which seem to increase with age come from the fact that the general environment and milieu favour boys' interest in science and engender a role whereby boys expect to know more scientific facts; but as far as higher mental processes are concerned, when one examines the performance of students receiving no specialized science education girls have a similar performance level to boys and indeed are sometimes superior to them." (p. 225)

In analyzing SISS data from Sweden, Riis et al. (1988) described the interest in and attitudes towards science as quite favourable for 10-year-old students. But a sharp decline was noticed so that the attitudes towards science of 14-year-old students turned negative. Major discrepancies were also found between the development of boys' and girls' attitudes with boys' attitudes more steadily decreasing. The Swedish report stated that one likely reason for the negative attitudes of girls towards school science might be seen in the fact that science instruction in Sweden was obviously more suited to boys than girls (pp. 395-396).

Concentrating on the analysis of the top and bottom 20 percent of 14-year-old students using data from SISS, Postlethwaite (1991) was able to show that even between these extreme groups considerable parallels appear to exist between the 23 countries examined in terms of gender differences. On a core test of 30 science items averaged across all countries boys' achievement in the top 20 percent group was more than six percent above the achievement of girls. The difference of boys' and girls' achievement in the bottom 20 percent group of students amounted to nearly four percent again in favour of the boys. In no country and in neither of the two extreme groups did the 14-year-old girls score better or equal than their male classmates. These results give way to the assumption that even among the students who seem to be more interested in science, i.e. the top 20 percent group, and those students whose achievement level and interest in science could be regarded as poor, i.e. the bottom 20 percent group, boys still outperform the girls. It may well be that both boys and girls view school science as a more male-oriented domain. This, in turn, could result in less favourable attitudes towards science of girls.

In her study of students in Botswana, Duncan (1989) came to the conclusion that "girls who see science and/or school science as a male area of activity generally have less positive attitudes to science and perform poorly." (p. 206)

Achievement and attitudes of over 80,000 students were assessed at various grades during the British Columbia Science Assessment 1982. Taylor (1982) reported that, in general, attitudes of students toward various aspects of science were quite positive. However, achievement, as judged by informed members of various panels, was rated low.

A similar contrast was identified by Postlethwaite and Wiley (1991). The between-country correlation of interest in science and achievement for 14-year-old students in 19 countries was -.46. It was reasoned that this negative relationship was due to Nigeria, Papua New Guinea, the Philippines and Zimbabwe which had higher interest in science scores than countries such as Japan and The Netherlands whereas the inverse was true for achievement.

SISS results also showed that students, in general, held very positive attitudes in terms of the beneficial aspects of science with the exception of 14-year-olds from Japan and the Netherlands. In Canada (French-speaking provinces), Hungary, and Poland the means for the 14-year-old students were higher than the mean values for students at the terminal secondary school level. For Postlethwaite and Wiley (1991) this was likely to be related to the dominance of female students in the schools at the terminal secondary level and in the samples drawn in this investigation for these countries.

It should not be overlooked, however, that attitudes towards science were not only different between countries but marked differences were observed within countries, too. Though Comber and Keeves (1973), Kelly (1978), Keeves (1991), and Postlethwaite and Wiley (1991) all found a complex relationship between attitudes towards science and achievement in science analyzing data from the two IEA science studies, hardly any comment could illustrate the importance of attitudes better than the one by Riis et al. (1988) who investigated the Swedish SISS data:

> "The attitudes among the pupils towards science and technology in society, and the prospects and hopes for the future were disheartening. These young people saw more 'cons' than 'pros' in science and technology for the happy future development of society. They did, however, agree over all ages that science and technology are important to the development of a country." (p. 396)

Only a few studies were concerned with attitudinal gender differences in science.

In a meta-analysis, Steinkamp (1982) reviewed the relationship of achievement with attitudes in science. Data on gender differences in attitude were analyzed within the context of a large number of school, individual, and methodological variables including academic discipline, age, instrument reliability, country of sample, socio-economic status, geographic region, age, and dimension of attitudes. Results indicated that gender differences in the relationship of attitudes and achievement were smaller than generally assumed, but where they did occur, with few exceptions, they tended to favour male students.

After having reviewed the existing literature, Schreiber (1984) supported the notion that female adolescents possessed significantly more negative attitudes toward middle school science than did males.

Examining the sex-related differences in attitudes of secondary school students towards computers, Collis (1985) found consistent gender differences in attitudes toward computers and mathematics, in attitudes and participation in computer study groups, and in attitudes and use of home computers.

Though various explanations may be given to understand why boys and girls sometimes hold different attitudes towards science, the impact attitudes have on achievement may be reciprocal. Bradley and Hutchings (1973) believed that girls, though holding positive attitudes towards science, quite often felt discouraged by their real or apparent lack of success. Likewise, Kelly (1978) accepts the idea that a student's achievement in science might contribute to the expression, positively or negatively, of interest in science.

However, based on evidence from theories on the acquisition, formation, and modification of attitudes recent multivariate analyses assumed a more unidirectional, i.e. causal, effect of students' attitudes and values on the individual's achievement level (Keeves, 1991; Keeves and Kotte, 1991; Keeves and Saha, 1991; Postlethwaite and Wiley, 1991).

Gender-typed Role Models Typically, a connection between the students' attitudes towards science and their gender identity is accepted (Maccoby and Jacklin, 1974; Fennema and Sherman, 1977; Fennema and Sherman, 1978; Kelly, 1978; Duncan, 1989).

Duncan (1989) saw one of the reasons for the low representation of girls in science in the gender stereotyping of female students. She concluded that girls who saw science as a male domain commonly showed less positive attitudes towards science and (therefore) performed poorly. It was interesting to note that boys despite a more marked tendency to gender-typing in science were less affected in their achievement in science by their gender-typing.

Hanna and Kuendiger (1986) conducted an assessment of gender differences in mathematics achievement and related attitude variables of 13-year-old students from 20 different countries. The pattern of achievement results indicated that girls were more successful than boys in Belgium, Thailand, Finland and Hungary but least so in France, Nigeria, Israel, and the Netherlands. With regard to attitude measures, the largest differences between countries and sexes occurred for the scale 'gender stereotyping'.

Following Duncan's conclusion (1989), 'it is time to acknowledge that the schooling process is a gendered process.' (p. 206) While gender-typed attitudes apparently have an effect on achievement and science subjects such as biology, chemistry, and physics are clearly gender-typed, the influence of this gender-typing process is not well understood (Kelly, 1978). Educators and researchers should be conscious about the likelihood of differences across countries, school-types, and, of course, between boys and girls.

If gender-typing influences students' attitudes and values towards science in that it stimulates or inhibits the students' interest and efforts, and, subsequently, their achievement level in science, then attention has to be paid to students' expectations regarding the participation in future (advanced level) science courses as well as the choice of resuming a science-related occupation after leaving school.

2.1.7 *Career Expectations and Participation Rates*

Career Expectations In many developing countries schooling is still regarded as a privilege and, unfortunately, this privilege is accorded most commonly to male students (Silver, 1985).

Results from FISS and SISS enabled Keeves and Kotte (1991) to conclude that substantial changes in many industrialized countries in the roles of women and their career expectations had occurred during the period 1970 to 1984. Girls and women are made aware of these societal changes through discussions at school and at home and through the media. Paralleled by greater encouragement of parents, peers, teachers, and often improved employment opportunities more girls intend to resume a science-related career.

Keeves and Kotte have presented several indicators of change in the roles of women in society for a subset of those 10 countries which participated in both FISS and SISS (1991, see Tables 6.1 and 6.2). So, in general, the number of women in the labour force increased in nearly all countries as did the ratio of women who became university lecturers or secondary school teachers. At the same time women reportedly were less tied to child-raising and household duties. This was underlined by the reduced fertility rate of women in 1984 when compared to 1970. Also across the 10 countries it could be observed that during the period of 14 years the modal age of mother's delivering their first baby increased, on average, by nearly two years.

However, marked differences in the types of careers pursued by male and female school leavers were noted (Finn et al., 1979; Finn et al., 1980; Silver, 1985; Keeves and Kotte, 1991).

If more women expressed an increased need for education and training during the last two decades, thus striving for their own careers in science or science-related profession then the educational and occupational opportunities offered by the educational system and labour market had changed. But, moreover, the expectations of girls and women must have undergone considerable changes at the same time, too.

Postlethwaite and Wiley (1991) reported about changes in expectations of students from secondary schools from various countries to resume a career in science or in a science-related field. Using data from the 1983/84 data collection of SISS, the authors found that with the exceptions of Hong Kong, Japan, and Korea where the same values were obtained for 14-year-olds and students from the terminal secondary school level, the perception of a career in science and of the importance of science for a future job decreased from the age of 14 onwards.

Though, at first sight, this drop, also reported in reviews by Humrich (1988 [1]; 1988 [2]) might be the result of students perceiving science as a comparatively difficult school subject, gender differences must be taken into consideration as well.

Using regression analytic techniques on data from the United States, Bridgham (1972) suggested that if science grades were no more severe than grades in other subjects, girls' enrolment would probably rise as much as 80 percent. However, grading seemed to have no impact for boys.

Given this, the decline of expectations Postlethwaite and Wiley (1991) detected could, indeed, largely be explained by gender differences in participation.

Participation in Science Courses Though, overall, more girls enrolled in science courses in 1983/84 than in 1970/71 in most countries both in relative and absolute terms (Keeves and Kotte, 1991), the ratio of male to female students participating in science appeared to constantly shift towards male students with increasing grade level.

Moss (1982) was able to show that for Australia gender differences in mathematics achievement changed drastically within the seven states and territories rather than across the whole country between 1964 to 1978, that period during which the women's liberation movement started to gain significant momentum in influencing employment opportunities and the participation of girls in education at the upper secondary and tertiary levels.

But major differences where not only found to exist between countries but also between the participation rates of students studying biology, chemistry, and physics. The biggest gender differences in enrolment were observed to occur in physics, with, on average, more than twice as many males than females taking physics courses at the terminal secondary school level across the 10 countries analyzed (Keeves and Kotte, 1991, Table 6.3). Also more male students enrolled in upper secondary chemistry courses. For biology, the enrolment patterns were nearly identical for boys and girls. Kelly (1978) speculated that this was probably due to aspects dealt with in biology which could be of interest to students in general, such as physiological changes during puberty or important facts about food and digestion ('nurturative aspects').

Though boys and girls seemed to show no significant gender differences in gender-typing occupations (Duncan, 1989), students of both sexes clearly stereotype professions (Comber and Keeves, 1973; Marini and Greenberger, 1978).

In his review of gender differences in achievement, Keeves (1988) believed that these were, at least, in part a result of societal and cultural effects. Yet, it remains largely unclear to what extent secondary school students' decisions to enrol in future science courses are influenced by other factors such as the socio-economic status of the home, student attitudes and values towards science, teaching methods, or teacher characteristics.

2.1.8 The Special Case of Cross-National Comparisons

The review of the literature identified various factors and processes which influence the manifestation of gender differences in science achievement. It could be noted that variations concerning the effects of these factors occurred between...

- ☞ male and female students;
- ☞ different grade levels and students of different age;
- ☞ classrooms and schools;
- ☞ types and gender-typed forms of schooling;
- ☞ states and regions;
- ☞ educational systems and countries.

This complex research situation would advocate analyzing the expression and reduction of gender differences in science from an international perspective. Here, IEA set milestones with large-scale assessment studies, of which FISS and SISS provided highly valuable data on science achievement from a number of structurally very different countries. Though most of the research on gender differences in science achievement has been carried out in the United States, an uncritical generalization of findings onto other educational systems would not be justified because of the different cultural, societal, and educational settings of those countries.

Despite the large number of studies which examined gender differences in science achievement, today many questions remain open. Clearly, further research is required. While several earlier studies were either restricted to a certain aspect of gender differences, a certain subset of students, schools, or districts, or restricted in other ways, methodological shortcomings often did not permit a larger generalization of those findings.

The cross-national analysis of gender differences in science achievement allows the identification of country-specific and more general effects which, in turn, can provide useful information to educators and educational planners about possible improvements in science education. It is also envisaged that by using probability samples from students, teachers, and schools of different countries more precise and generalizable inferences about gender differences in science achievement will be possible.

With the enormous pool of data from the FISS and SISS studies at hand and the availability of sophisticated statistical software as well as state-of-the art computing facilities, an attempt is made in this thesis to find answers to a number of general and specific research questions which are outlined in the following section.

2.2 The Research Questions

As could be seen from the presentation of the research situation on gender differences in science, the achievement level as well as the participation rates of male and female students is largely dependent on a set of influential factors. It was shown that next to the socio-cultural and economic context of an educational system, the students' socio-economic background and family situation were of significant importance in generating gender differences in science (Comber and Keeves, 1973; Keeves, 1991, Postlethwaite and Wiley, 1991).

Several studies described the impact of teacher-related variables on science achievement (Morse and Handley, 1982; Good et al., 1972). School organization was recognized as another major source of variation between the achievement of boys and girls (Postlethwaite, 1967; Kelly, 1978; Dupont, 1981; Sanguinetty, 1983; Walberg, 1991). Also discussed were the pros and cons of single-sex versus coeducational schooling (Ormerod, 1975; Lee and Bryk, 1986) as well as private versus public education (Staver and Walberg, 1986; Walberg and Fowler, 1987; Jimenez et al., 1988) in their effect on gender differences in science achievement.

Although achievement differences between male and female students were influenced by the level of cognitive abilities (Wittig and Petersen, 1979; Matlin and Matkoski, 1985), several studies illustrated the importance of positive attitudes and values towards science (Morse and Handley, 1982; Kelly and Smail, 1986; Humrich, 1988 [1]; Duncan, 1989). These attitudes were believed to be not only a prerequisite for achievement in science (Comber and Keeves, 1973; Kelly, 1978; Duncan, 1989; Keeves, 1991) but also very much related to the efforts students undertake in school and at home in relation to science (Steinkamp, 1982; Steinkamp and Maehr, 1984; Keeves, 1991).

Given the complexity of factors, thought to show both similarities and differences between countries and between schools within countries, the following sections list the general (Section 2.2.1) and specific research questions (Section 2.2.2) posed for analysis in this thesis.

2.2.1 General Research Questions

The general research questions formulated below are based on the data available from the FISS and SISS studies. More details about these cross-national surveys of science achievement will be given in Chapter 3. For better structuring each of the main questions is ascribed a sequential number so that after having presented the results in Chapters 5 through 8 the answers to each of the questions presented will be summarized in Chapter 9. In addition, the chapter(s) in which the results are discussed is(are) listed after each question as ➨Chapter... .

No.	General Research Question (➡*results presented in...*)
1	What differences, if any, are found between boys and girls at the 10-year-old level, 14-year-old level, and terminal secondary school level in terms of achievement in science (Biology, Chemistry, Physics) in the 10 countries? (➡Chapter 5)
2	What changes occurred in the differences between boys and girls at the three grade levels tested in terms of achievement in science between the two data collections in 1970/71 and 1983/84 in the 10 countries? (➡Chapter 5)
3	What differences, if any, are found in the attitudes towards science of boys and girls in the 10 countries? (➡Chapter 6)
4	What factors contribute to gender differences in science achievement in the 10 countries and how do they operate? (➡Chapters 6 and 7)
5	What factors contribute to gender differences in participation rates in science in the 10 countries and how do they operate? (➡Chapters 6 and 7)
6	How can the participation rate of students in science courses be increased? (➡Chapters 6 and 7)
7	Is coeducational schooling more effective than single-sex schooling for the achievement level of boys and girls in science? (➡Chapter 8)
8	What measures can be recommended to help balance gender differences in science achievement? (➡Chapter 9)

2.2.2 *Specific Research Questions*

In the following, each of the eight major research questions is outlined more precisely. Again, for each of these specific research questions consecutive identification numbers are assigned for easier reference when summarizing the results of the thesis in Chapter 9.

No.	Specific Research Question (➡*results presented in...*)
1.1	What differences are found between boys and girls in general science achievement at the 10-year-old level in the 10 countries? (➡Chapter 5)
1.2	What differences are found between boys and girls in general science achievement at the 14-year-old level in the 10 countries? (➡Chapter 5)
1.3	What differences are found between boys and girls in general science achievement at the terminal secondary school level in the 10 countries? (➡Chapter 5)

No.	Specific Research Question *(➡results presented in...)*
1.4	What differences are found between boys and girls in the achievement for the different science subjects (Biology, Chemistry, Physics) at the 10-year-old level in the 10 countries? (➡Chapter 5)
1.5	What differences are found between boys and girls in the achievement for the different science subjects (Biology, Chemistry, Physics) at the 14-year-old level in the 10 countries? (➡Chapter 5)
1.6	What differences are found between boys and girls in the achievement for the different science subjects (Biology, Chemistry, Physics) at the terminal secondary school level in the 10 countries? (➡Chapter 5)
2.1	What changes occurred in the science achievement of boys and girls between 1970/71 and 1983/84 at the 10-year-old level in the 10 countries? (➡Chapter 5)
2.2	What changes occurred in the science achievement of boys and girls between 1970/71 and 1983/84 at the 14-year-old level in the 10 countries? (➡Chapter 5)
2.3	What changes occurred in the science achievement of boys and girls between 1970/71 and 1983/84 at the terminal secondary school level in the 10 countries? (➡Chapter 5)
2.4	What changes occurred in the achievement of boys and girls for the different science subjects (Biology, Chemistry, Physics) between 1970/71 and 1983/84 at the 10-year-old level in the 10 countries? (➡Chapter 5)
2.5	What changes occurred in the achievement of boys and girls for the different science subjects (Biology, Chemistry, Physics) between 1970/71 and 1983/84 at the 14-year-old level in the 10 countries? (➡Chapter 5)
2.6	What changes occurred in the achievement of boys and girls for the different science subjects (Biology, Chemistry, Physics) between 1970/71 and 1983/84 at the terminal secondary school level in the 10 countries? (➡Chapter 5)
3.1	What differences can be identified between boys and girls in attitudes towards science at the 10-year-old level in the 10 countries? (➡Chapter 6)
3.2	What differences can be identified between boys and girls in attitudes towards science at the 14-year-old level in the 10 countries? (➡Chapter 6)
3.3	What differences can be identified between boys and girls in attitudes towards science at the terminal secondary school level in the 10 countries? (➡Chapter 6)
3.4	How are attitudes towards science related to the achievement level in science in the 10 countries? (➡Chapter 6)
4.1	What factors are associated with gender differences in science achievement for 14-year-old students at the between-student level in the 10 countries? (➡Chapter 6)
4.2	What relationships are found among the factors that influence gender differences in science achievement for 14-year-old students at the between-student level in the 10 countries? (➡Chapter 6)

No.	Specific Research Question (➨*results presented in...)*
4.3	What factors are associated with gender differences in science achievement for 14-year-old students at the between-school level in the 10 countries? (➨Chapter 7)
4.4	What relationships are found among the factors that influence gender differences in science achievement for 14-year-old students at the between-school level in the 10 countries? (➨Chapter 7)
5.1	What factors are associated with gender differences in participation rates in future science courses for 14-year-old students at the between-student level in the 10 countries? (➨Chapter 6)
5.2	What relationships are found among the factors that influence gender differences in participation rates in future science courses for 14-year-old students at the between-student level in the 10 countries? (➨Chapter 6)
5.3	What factors are associated with gender differences in participation rates in future science courses for 14-year-old students at the between-school level in the 10 countries? (➨Chapter 7)
5.4	What relationships are found among the factors that influence gender differences in participation rates in future science courses for 14-year-old students at the between-school level in the 10 countries? (➨Chapter 7)
6.1	What measures can be taken to increase the participation rates of students in upper secondary science courses in the 10 countries? (➨Chapters 6, 7, and 9)
6.2	Are different measures necessary for male and female students to increase the participation rates in upper secondary science courses? (➨Chapter 9)
6.3	In which way can recommendations regarding an increased participation in science be generalized for other educational systems? (➨Chapter 9)
7.1	What factors contribute to gender differences in science achievement in coeducational schools? (➨Chapter 8)
7.2	What factors contribute to gender differences in science achievement in single-sex schools? (➨Chapter 8)
7.3	Is coeducational or single-sex schooling more effective for the achievement of male and female students in science? (➨Chapters 8 and 9)
7.4	Can recommendations be made for the gender-typed organization of schooling in science teaching? (➨Chapter 9)
8.1	What measures should be taken to reduce gender differences in science achievement at the between-student and at the between-school level in the 10 countries? (➨Chapters 6 and 7)
8.2	Are different measures necessary to be taken for male and female students to reduce gender differences in science achievement? (➨Chapters 7 and 9)
8.3	In which way can recommendations regarding the reduction of gender differences in science achievement be generalized for other educational systems? (➨Chapter 9)
8.4	Does balancing gender differences have any (negative) side effects? (➨Chapter 9)

2.3 Summary and Conclusions

In this chapter a review of the current state of knowledge of the research into gender differences in science achievement has been presented. It is evident that many questions about which gender differences are observed, how these can be explained, and what possibilities exist for minimizing them remain unanswered.

Though many case studies and surveys based on restricted samples gave valuable indications for a possible decrease of gender differences in science achievement, most of the findings and recommendations given by various authors may not be applicable in different countries. The socio-cultural, socio-economic, curricular and school organizational settings between the educational systems as was shown recently by Rosier and Keeves (1991) are too diverse. However, little is known about the effects contributing to the manifestation, increase, or decrease of gender differences in science achievement internationally. Several large-scale international research studies in comparative education, mainly conducted by the IEA, have obscured the notion that cross-national comparisons are invalid or inapplicable. On the contrary, results from the FISS and SISS study had numerous important implications, for example, in the revision of the teacher training or for curricular changes in science, in several countries (Keeves, 1991; Postlethwaite, 1986).

Therefore, the unique possibility of analyzing two large data pools from the 1970/71 and 1983/84 science data collections will be used to investigate the topic of gender differences in science achievement in this thesis. It became clear after reviewing the current literature that a number of factors concerning the students' socio-economic situation and home, their attitudes and values towards as well as their interest in science, teacher and teaching-related variables, and school-related variables, all seem to be interrelated in influencing science achievement differences between boys and girls. These gender differences appear to vary not only across countries but are also believed to differ between grade levels and science subjects.

It is this complexity in the research situation which promotes the analysis of the various general and specific research questions outlined in the previous sections by using data from the two international science studies.

Efforts are made to describe these two projects, FISS and SISS, in more detail in the following chapter before methodological aspects will be discussed in Chapter 4.

3. International Context of Research

The major aim of this study is the analysis of gender differences in science achievement from an international perspective. Primarily, this research focusses on those grade levels in which most of 14-year-old students were enrolled at the time of data collection. Data were collected from a probability sample of schools, students, and their teachers in each of the countries.

For this dissertation, secondary analyses are conducted using the data collected during the IEA First International Science Study (FISS), which was one of the subjects embedded in the IEA Six Subject Survey (1966-1973), and also using the data from the IEA Second International Science Study (SISS) which was conducted a decade later.

Though the majority of analyses presented in this thesis are based on more recent data from SISS, it is also considered desirable to gain some impressions of the changes, if any, which took place between the data collection of FISS (1970/71) and SISS (1983/84).

The first major cross-national research project ever undertaken in comparative education, also carried out under the auspices of the IEA in the early 1960s, the First International Mathematics Study (FIMS), gave numerous indications for gender differences present among students of all age groups in nearly all countries participating (Husén, 1967, pp. 233-250). Thus, it was not surprising that in the monograph released about the results from the FISS project by Comber and Keeves (1973), special attention was also devoted to the issue of 'sex differences in science achievement' (pp. 139-153).

Despite the originality of these findings when the Comber and Keeves report was released, the results were mainly of a descriptive nature and intended to shed light on the policies whether and to which extent coeducational schooling was favourable for the achievement of the sexes. Too restricted were the technical and statistical means to conduct more detailed multivariate analyses on the FISS data sets in the early 1970s.

Inspired by the world-wide success of FISS, the educational systems that were members of IEA, decided to repeat a cross-national data collection in the field of science. This allowed, inter alia, to monitor changes in the overall achievement in science in these countries since the first data collection took place in 1970/71. Though data for the SISS project were gathered in 1983/84, it was not before 1991 that final reports were published due to the lack of international financing (Rosier and Keeves, 1991; Postlethwaite and Wiley, 1991; Keeves, 1991).

While gender differences in science achievement between educational systems were discussed briefly by Postlethwaite and Wiley (1991, pp. 73-76), more detailed multivariate analyses were carried out by Keeves and Kotte and a separate chapter was devoted to the question of gender differences in Volume 3 of the final international SISS reports (Keeves and Kotte, 1991). It became evident that further multivariate analyses were necessary in order to understand better the factors causing, increasing, or inhibiting gender differences in science achievement.

Though coeducation is very common throughout most educational systems nowadays, the question of which form of gender-typed instruction, single-sex or coeducational, is most favourable to boys and girls still persists. This question is of particular importance given the manifold contradictory research results from the related fields of instruction in computer sciences (Eastman and Krendl, 1984; Collis, 1985; Digranes, 1985; Nielsen and Roepstorff, 1985; Weinbach, 1988; Krendl et al., 1989) or mathematics (Moss, 1982; Hanna and Kuendiger, 1986; Linn and Hyde, 1989; Marsh, 1989; Walberg, 1989).

In order to discover more about gender differences in science achievement, only those data sets from systems participating both in FISS and in SISS will be used. These countries are...

- ☞ Australia,
- ☞ England,
- ☞ Finland,
- ☞ Hungary,
- ☞ Italy,
- ☞ Japan,
- ☞ The Netherlands,
- ☞ Sweden,
- ☞ Thailand, and
- ☞ The United States

(It should be noted that for the United States only the Phase 1 data sets were used in this study [see Postlethwaite and Wiley, 1991]).

The settings of the FISS and SISS projects will be outlined briefly. This is followed by an overview of the available instruments used during data collection is given. Finally, some information needs to be presented about the sampling procedures applied in the two large-scale assessment studies in science since this has implications for the multivariate analyses conducted for this study.

3.1 The IEA First International Science Study

The First International Science Study (FISS) conducted by IEA started with an initial phase, from 1966 to 1969, during which international achievement tests, attitude scales, and background questionnaires for students, teachers, and school principals as well as other material were developed, pretested, and finalized.

FISS had to be regarded as the first attempt to compare science achievement in different educational systems. All in all, 19 countries participated. Fifteen of these countries collected data from three different population levels, namely the 10-year-old, the 14-year-old, and the terminal secondary school level.

A great challenge to the researchers was the development of valid and reliable test instruments and background questionnaires (Peaker, 1975). With countries as diverse in their educational systems as Australia, Chile, The Federal Republic of Germany, Hungary, Iran, or Japan it is not difficult to imagine the enormous conceptual and contextual task of developing instruments and questionnaires, which could be administered to children at different age groups all across the world. Though interesting in themselves, these procedures cannot be described in more detail here.

The final report which summarized major results from the FISS data collection, that took place between 1970 and 1971, written by Comber and Keeves (1973), illustrated very vividly many of the original problems faced by the researchers in a variety of aspects. These, for example, were the management of thousands of students, teacher, and school data, the development of internationally valid and reliable test instruments, and the conduct of multivariate statistical analyses, which, in 'those days' had to happen without the access to powerful desktop microcomputers.

Although many of the innovative ideas of FISS could be incorporated into the successive SISS study, it should also be acknowledged that since the time of the data collection for the first science study in 1970/71, considerable changes had taken place in science education in all educational systems participating in IEA. Likewise, some technical and procedural shortcomings, which need not be described here (see Comber and Keeves, 1973; Peaker, 1975), suggested to use results from the FISS study only as an explanatory aid when interpreting findings obtained from analyses based on data from SISS. In that, the focus of the analysis of gender differences in science in this thesis is on data collected in 1983/84. However, where appropriate, results from FISS data are taken into consideration to demonstrate the changes which have taken place since 1970/71.

For more specific information about the IEA Six Subject Survey, the reader is advised to turn to the IEA monograph by Walker (1976). As indicated in Section 2.1, some interesting findings related to the issue on gender differences in science education based on data from FISS were compiled by Kelly (1978).

3.2 The IEA-Second International Science Study

IEA undertook a study of science achievement in 23 countries with the main data collection in 1983/84. Data were gathered from over 200,000 students, their teachers and school principals. Again, international comparable tests of science achievement were constructed largely based on the tests used for FISS. These were then piloted in some countries and, after some revisions, administered in all participating systems. As will be described later, two of the target populations (see Section 3.4) in each system of education took a science core test plus combinations of two out of four rotated subtests. In the third population, i.e., the last grade of secondary schooling, all students studying science took a core test and a specialist subtest of biology, chemistry, or physics. The principle of test rotation was chosen to administer a larger pool of science test items which was believed to reflect a more adequate sample of the science curriculum. However, the combining of core test and rotated test scores at the student, school, and national level still proved difficult as Keeves had rightly pointed out (1991).

In addition, data were collected on the students' home background, aspirations, study habits, and attitudes as well as information on the teachers' educational background, training, teaching practices, preparation and marking. Information on the characteristics of each school and the organisation of science instruction in the school completed the massive data bank.

Finally, raw data sets were submitted by 24 educational systems. These were: Australia, Canada (English-speaking provinces), Canada (French-speaking provinces), China (three provinces only), England, Finland, Ghana, Hong Kong, Hungary, Israel, Italy, Japan, Korea, the Netherlands, Nigeria (11 provinces only), Norway, Papua New Guinea, Philippines, Poland, Singapore, Sweden, Thailand, the United States, and Zimbabwe. Each country was, apart from funding, responsible for all data collection within the country and the preparation and publication of a national report while the development of all instruments, decisions about the definition of the target populations and many technical details were collective decisions of all countries. The heads of the participating research institutes formed an International Project Council (IPC).

The IPC Chairman and an International Coordinator, Dr. M. Rosier, were responsible for the necessary administrative, organizational, and technical arrangements which had to be made about the design of the study, instruments, sampling, data collection, and data analyses. Initially, the international coordination was situated at the Australian Council for Educational Research (ACER) at Hawthorn, Victoria. Due to financial constrains and staff requirements, the data cleaning and data analyses stages were conducted at the University of Hamburg, Federal Republic of Germany. Analyses concerning a comparison of those 10 countries which had previously taken part in FISS were conducted at the University of Stockholm, Sweden.

In early 1988, a booklet presenting some preliminary results from 17 of the 24 educational systems, based on achievement differences on a science core test administered to students at three population levels was published (IEA, 1988). It took another three years until final reports were ready for publication (Rosier and Keeves, 1991; Postlethwaite and Wiley, 1991; Keeves, 1991).

As a consequence of the sheer magnitude of the SISS project and the mammoth amount of data gathered, it will not come as a surprise that further secondary analyses would have to be undertaken separately. Hence, this study focusses on particular aspects of gender differences in science achievement.

The interested reader, who wishes to find more details about the SISS study has to be referred to the above volumes or related national reports (e.g. Keys, 1986 [England]; Riis et al., 1988 [Sweden]).

For the analyses of gender differences in this report, it will be helpful to give a short account of the test instruments and questionnaires used. Though several of these had originally been developed for the FISS study as part of the IEA Six Subject Survey (see Section 3.1), only those instruments which are of relevance for this study will be listed and described.

3.3 Instruments

The science tests administered during the FISS data collection in 1970/71 are presented in Comber and Keeves (1973, Appendix IX) while other instruments and background questionnaires may be obtained from the IEA Headquarters at The Hague, The Netherlands. It should be self-evident that neither the science tests booklets nor most of the background questionnaires could be added as appendices to this report as that would invalidate their future use (indeed, a third international study of mathematics and science is already under way).

However, a short summary of the contents of each of the instruments used for this study will be given below. More concise information concerning particular variables or scales employed will be presented as part of the methodological discourse in Chapter 4.

Student Instruments

Student Questionnaire

This instrument was designed to collect background characteristics of the students and the conditions under which they studied science.

Attitude Questionnaire

Students were requested to a number of Likert-type items regarding their attitudes and values towards science. The items were then transformed to a maximum of six separate attitudinal scales (see Appendix 4.2; Postlethwaite and Wiley, 1991, Appendix G).

Descriptive Scale Questionnaire

Students were also asked to provide data on their views about the science teaching they experienced at school. In addition, these Likert-type items were collecting information about the students' efforts to learn science both in the classroom and at home (see Appendix 4.2; Postlethwaite and Wiley, 1991, Appendix G).

Aptitude Tests

Two separate item batteries assessing the aptitude level were administered to each student. The 40-item *Word Knowledge Test* was given first. It was identical to the one used in the FISS study. Secondly, a *Mathematics Test* comprising of 20 items measuring the students' computational skills was to be answered (see Comber and Keeves, 1973, Appendix XI).

Achievement Tests

In two hours of testing time students had to respond to a *science core test* as well as one or two out of four *rotated science subtests* depending on the population level tested. All tests comprised of multiple-choice items with five answer categories each of which only one was correct. The students were instructed not to guess blindly but to proceed to another question should they encounter difficulties answering a particular item.

Provisions were made to link the science achievement test scores across population levels in order to obtain estimates on the improvement between age and grade levels. Thus, about 15 *anchor items* were common to the core tests administered to students at higher or lower population levels (see Postlethwaite and Wiley, 1991, p. 73; Keeves, 1991, Chapter 1, pp. 21-22).

In addition, by keeping at least 15, so-called *bridging items* common at each population level to the achievement tests administered during the first data collection in 1970/71, a linkage was possible between the two data points of FISS and SISS across countries (see also Keeves, 1991, Tables 2.4 and 10.2).

Teacher Instruments

Teacher Questionnaire

Each teacher of students in the sample was required to complete this questionnaire which was meant to provide, next to some personal data, information on the educational background, teaching practices in science, and emphases given in the teaching of science.

Opportunity to Learn Questionnaire

Teachers were asked to rate their students' (theoretical) opportunity to answer the individual science test items administered which would allow a more precise judgement about the achievement levels of schools in terms of science instruction by integrating the concepts of learning processes, skill acquisition, and test performance (for a more thorough discussion see Postlethwaite and Wiley, 1991, Chapter 5).

School Instruments

School Questionnaire

Information about the school, the support of and facilities for the teaching of science within the school was collected from the school's principal and head of the science department.

Some of these instruments, for example the Aptitude Tests, were optional so that some data were missing or incomplete for some countries or for some population levels. Yet, efforts were taken to select a maximum of information for cross-national comparisons for the 10 countries in question. However, where certain data had not been collected or were incomplete, this will be made clear in the respective sections of this thesis.

3.4 Target Populations and Sampling

Data for the FISS and SISS studies were collected according to specific sampling plans which followed guidelines jointly set up for all participating countries. A complete overview on the sampling procedures, filed with the *ERIC* data base system, is available in *'Administration and Sampling Report of the Second International Science Study'* (Rosier, 1989). As the sampling design has considerable impact on the multivariate analyses to be performed (see Keeves, 1991) a few explanations should be given here. The information regarding the sampling and target populations was taken from the SISS study (1983/84) but could be applied similarly to the FISS data collection in 1970/71.

3.4.1 *Target Populations*

After international agreement had been reached, three stages, or population levels, in a school system were selected for this study (Comber and Keeves, 1973, pp. 9-10; Postlethwaite and Wiley, 1991, p. 3). Below, the operational definitions of all target populations are listed:

Population 1 all students aged 10:0 to 10:11 years on the specified date of testing *or* all students in the grade in which most 10-year-old students were to be found on the specified date of testing.

Population 2 all students aged 14:0 to 14:11 years on the specified date of testing *or* all students in the grade in which most 14-year-old students were to be found on the specified date of testing.

Population 3 all students in the terminal or pre-university year of full-time secondary education.

The specified date of testing varied from country to country depending on the start of the school year and administrative constraints.

At Population 3 level a distinction had to be made between students who were specializing in the study of science (though some countries were arguing they did not have any specialization) with the intention to entering a science-related profession or occupation after further training *(science specialists)*, and students not specializing in science, who were entitled to proceed to higher education, but who did not plan to follow science-related courses *(non-specialists)*. The non-specialist students might or might not be studying science as part of general education. Hence, with respect to the sampling procedures employed Population 3 had to be subdivided.

There were two major subgroups of Population 3, namely:

Population 3N all students in the terminal year of full-time secondary education not currently studying science subjects which would permit them to continue with the study of science related courses at an institution of higher education.

Population 3S all students in the terminal year of full-time secondary education studying science at a level which would permit them to proceed with the study of science-related courses at an institution of higher education.

These students were divided into several sub-populations as were appropriate in the different countries:

Population 3E all students studying Earth Science;
Population 3B all students studying Biology
Population 3C all students studying Chemistry;
Population 3P all students studying Physics;
Population 3X all students studying a general science course.

It should be mentioned that, where necessary, slight deviations from these target definitions had to be made for certain population levels in some of the countries. This, and other more specific information can be obtained from the *'Administration and Sampling Report of the Second International Science Study'* (Rosier, 1989) or is presented in Volume 2 of the final SISS reports (Postlethwaite and Wiley, 1991, Appendix B).

As will be explained in more detail in subsequent chapters, the multivariate analyses presented in this study (Chapters 6, 7, and 8) will focus on the 14-year-old level (Population 2), while descriptive comparisons made will furnish data from all population levels (in particular, Chapter 5).

3.4.2 Sampling

All samples randomly drawn were either two- or three-stage stratified probability samples. The two-stage samples involved a sample of schools (or, in the some cases, classes) drawn at the first stage with either students or an intact class being drawn at random from within the selected schools at the second stage. In general, the samples had to be designed to yield ±2.5 percent or less of a student standard deviation for one standard error of sampling of a mean value. All sampling plans had to be approved by the sampling referee, Dr. M. J. Rosier, from ACER.

It should be pointed out that the use of intact classes in each of the 10 countries selected for this study, with the exceptions of Australia, England, and Italy increased the magnitudes of the sampling errors. To compensate for this, Japan increased the sizes of their designed samples. In turn, the use of intact classes might well have led to higher response rates than if the countries had used randomly sampled students from different classes within a school. The reason for this is that, typically, schools are more willing to participate in a research study if intact classes are selected, causing less administrative burden for the school than having to bring students together from several classes (Postlethwaite and Wiley, 1991, p. 8).

In contrast to the students, which were selected as a two-stage probability sample of all students in the defined target populations, no independent probability sampling was applied to obtain a selection of science teachers. The teachers chosen were either the class teachers (of the classes drawn) or the teachers teaching science at the defined grade level in a sampled school. For descriptive statistics about teachers all teachers for whom data were returned were used. In some cases, this included all teachers teaching science at a particular grade level. In other cases, it included only those teachers teaching the students tested. For multivariate analyses only data from one teacher teaching science were employed. In some cases the aggregation of teacher data within one school was required. Appendix 3.1 presents the sizes of the achieved samples for students, teachers, and schools for the 10 countries used in this study.

3.4.3 Weighting, Sampling Bias, Sampling Error, and Design Effect

Weighting After agreement had been reached on the target populations each participating country had to develop their sampling plans. It is common to experience some deviations from the designed samples in the actual data collection. As a consequence, corrections must be made to provide for the imbalances between the designed and achieved samples. In fact, the differences between designed and actually achieved sample sizes varied from stratum to stratum and were, of course, different from population to population and country to country. To correct for this differential loss when calculating national scores, *stratum weights* were calculated and applied.

For nearly all strata, the stratum weights ranged between .80 and 1.40. About seven percent exceeded 2.00 and about two percent fell below .40. The more extreme values were all related to small strata in countries where the number of students was large (see Postlethwaite and Wiley, 1991, p. 10).

These stratum weights were used not only for test scores but also for the production of student univariates. Teacher and school univariates were left unweighted. However, for the joint, multivariate analyses of student, teacher, and school data unweighted teacher and school variables would not be appropriate.

For more information about the achieved samples of the SISS study see Postlethwaite and Wiley (1991, Appendices C1-C7). Additional information regarding the sampling and weighting procedures applied can be found in Keeves (1991). Two different approaches were taken to arrive at an estimate about the quality of the samples taken.

Sampling Bias Target population statistics were compared against sample statistics on a particular *marker variable* in order to control for sample bias. The marker variable chosen was percent male students in the target population. The population values which were reported by nearly all countries were the percentage of male students in the population level tested. The population percentages, sample percentages and standard errors of the sample percentages for the 10 countries in question are summarized in Appendix 3.2. The standard errors of percent male students were calculated in analogy to the standard errors of sampling of mean scores by dropping a school at a time. Roughly speaking, if a difference between the population and sample statistics of more than five percent, or two standard errors, was observed this could be taken as an indication for possible sampling bias. The figures in Appendix 3.2 show that all sample estimates were close to the population values for the samples of 10- and 14-year-old students. As National Centers were, at times, having insufficient access to the population statistics on the number of males and females studying biology, chemistry, and physics separately or in some combination at the terminal secondary school level, it came not unexpectedly that intra-class correlations and, therefore, standard errors were considerably higher. Commonly, these figures were not recorded by the relevant departments in the ministries of education.

Only in the case of Australia the percentages of male students were known for the Population 3 level but for other countries, where different percentage values were given for each subject, these were *'guesstimates'*. An overall percentage value identical for each of the three subject areas had to be given for Thailand and the United States. As the standard errors recorded for the Population 3 level showed, bias might have crept into the samples of biology students in Finland (but probably not into the chemistry and physics samples), Italy and the United States. However, the United States samples were not necessarily biassed since the population statistics might have been inaccurate. For further details the reader may consult the relevant sections in the administration and sampling report (Rosier, 1989) or refer to Volume 2 of the final report of the SISS study (Postlethwaite and Wiley, 1991, Chapter 1).

Sampling Error Standard errors of sampling for mean values on the achievement tests (sampling errors) were calculated using *jackknife techniques* of repeated replication by dropping one primary sampling unit (school or class) at a time following an original proposal by Mosteller and Tukey (1968; see also Tukey, 1977) and empirical evidence presented by Rust (1984). A summary for all SISS samples can be found in Postlethwaite and Wiley (1991, Appendix C8). For the subset of the 10 countries considered in this study the sampling errors are provided in Appendix 3.2.

It is, however, important to note that no work has been done to estimate the magnitudes of sampling errors for stratified cluster samples, such as were used in this study. Although the path coefficients obtained from PLSPATH may be regarded as similar to standardized regression coefficients, no information is available about the magnitude of sampling errors when stratified cluster samples are employed.

Design Effect Complex two- or three-stage stratified probability samples of students were drawn in each of the participating countries. According to statistical theory, the complexity of such a stratification process commonly produces proportionally larger standard errors than would be expected if simple random sampling had been employed (Kish, 1965; Peaker, 1975; Ross, 1988). In general, the formulae for simple random sampling underestimate the standard errors resulting from complex cluster sampling (Ross, 1978). The *design effect*, or *Deff*, expresses the degree of this underestimation which, by definition, is:

$$\text{Deff} = \frac{V(\bar{x}_c)}{V(\bar{x}_{srs})}$$ *(Formula 3.1)*

with $V(\bar{x}_c)$ variance of the sample mean for a given complex sample design; and
with $V(\bar{x}_{srs})$ variance of the sample mean for a simple random sample of equal size.

Based on data from the FISS study attempts were made to obtain estimates of the standard errors for correlation coefficients (r) and standardized regression coefficients (β). The formulae for the estimation of standard errors for correlation coefficients, SE(r), (Formula 3.2) and standardized regression coefficients, SE(β), (Formula 3.3) for a simple random sample (srs) are presented below (taken from Guilford and Fruchter, 1978). But these formulae are inappropriate for stratified cluster samples.

$$SE(r)_{srs} = \frac{1}{\sqrt{n}} \qquad \textit{(Formula 3.2)}$$

with n — size of simple random sample

$$SE(\beta)_{srs} = \left[\frac{1-R^2_{1,2,3...m}}{(1-R^2_{1,2,3...m})(n-m)} \right]^{1/2} \qquad \textit{(Formula 3.3)}$$

with m — number of variables

with $R_{i.jkl}$ — multiple correlation coefficient associated with a regression equation which uses variable *i* as the criterion and variables *j*, *k*, and *l* as predictors

In the FISS project work was reported by Peaker (1975) in which jackknife estimates were calculated for correlation and regression coefficients. Averaged across seven countries for Population 2 data taken from the FISS study, Peaker arrived at a mean Deff value of 2.5 for correlation coefficients and 2.0 for regression coefficients (Peaker, 1975, pp. 46-48; see also Ross, 1988, p. 538).

In order to obtain crude estimates of the sampling error (i.e., the standard error of sampling) in the case of complex sample designs, such as employed in the SISS study, the sampling error for a simple random sample has to be multiplied by the value of $\sqrt{Deff}$. Typically, standard errors for correlation coefficients were found to be slightly higher than those for regression coefficients (Kish and Frankel, 1970; Ross, 1978).

For the multivariate analyses presented in this study (Chapters 6, 7, and 8), the estimation of the design effects for correlation and path coefficients, is of special interest because it needs to be determined when such coefficients, or paths, are deemed to be relevant or *'non-trivial'*.

Of the 10 countries to be analyzed at the between-student level, the sample size for Sweden was smallest with N=1461 students at the 14-year-old level (Postlethwaite and Wiley, 1991, p. 188). Given this value for N, the following example illustrates the estimation of a cut-off point for determining a correlation coefficient for student variables as being meaningless.

Given are:

N=1461 — size of sample for Sweden (Population 2, 14-year-olds)

$Deff_r$=1.4 — estimate of design effect for correlation coefficients for Sweden (Peaker, 1975, p. 46)

$\sqrt{Deff_r}$=1.2 — root of $Deff_r$

According to Formula 3.2 the standard error for correlation coefficients for a simple random sample of students would be:

$$SE(r)_{srs} = \frac{1}{\sqrt{1461}} = .026$$

As a complex two-stage cluster sample was obtained in Sweden, $SE(r)_{srs}$ is multiplied by $\sqrt{Deff_r}$ to obtain the standard error for a correlation coefficient for a complex sample design, $SE(r)_c$:

$$SE(r)_c = .026*1.2 = .031$$

Hence, three standard errors, which describe a 99 percent confidence limit, would amount to:

$$3\ SE(r)_c = .093$$

Rounding this, a value of .10 is obtained. The standard error for a correlation coefficient SE(r) is generally larger than the standard error for a standardized regression coefficient SE(β) for a simple random sample. This is also generally true for a complex or stratified cluster sample. Consequently, an estimate may be made for the cut-off level for a PLS path coefficient (p), which is similar to a standardized regression coefficient, with path coefficients $|p| \geq .10$ in between-student analyses as being non-trivial. This can be considered to be a relatively conservative cut-off point for path coefficients calculated on the basis of student-level data.

However, it must be emphasized that little is known about the magnitudes of the sampling errors for PLS path coefficients where school and teacher data are disaggregated to the student level in the analysis of the data. Following Formula 3.2, the standard error of sampling for teacher and school variables is a result of the primary sampling unit which, in most countries, were schools. Based on experience, it was deemed useful to employ a cut-off point for path coefficients for the school-level PLS models of $|p| \geq .10$. Coefficients falling below this value can be regarded as fairly 'meaningless' in that their contribution to the variance of the criterion measure is very small. Compared across the 10 countries, a cut-off point for path coefficients at the school level of $|p| \geq .10$ is equivalent to approximately two to three SE(β) given the sample sizes of schools (and, also, teachers) at the 14-year-old level. Though it is acknowledged that it might have been possible to operate with variable cut-off points of path coefficients for each level of analysis and also for each country this would have added a statistical rigidity not necessary to be applied for the type of PLS analyses to be performed. As can be seen from the results presented in Chapter 7 and with all cautious interpretations, the less conservative cut-off point of $|p| \geq .10$ not only leads to path models that provide interesting insights into the country-specific factors explaining science achievement at the school level but underscores the appropriateness of the *a priori* definition of the cut-off point of $|p| \geq .10$.

Even if this may result in the rejection of some paths which may be statistically significant at the 95 percent confidence level as being of no practical significance, the use of a more conservative limit should yield a clearer picture of the effects producing or shaping gender differences in science achievement (see also the methodological discussion about path analyses in Section 4.3.1). It may be of interest to note that given the cut-off criterion for non-trivial correlation coefficients and standardized regression coefficients of .10, a path coefficient of this minimum strength can be estimated to account for about one percent of the overall explained variance, R_i^2, in the dependent variable, as $R_i^2 = r_i * b_i$ (see Pedhazur, 1982, p. 81).

It should have become clear, that the sampling procedures applied for the data collection of the FISS and SISS studies have to be taken into consideration for the adequate calculation of standard errors. These, in turn, effect the interpretation and generalization of the multivariate analyses described in this study.

3.5 Summary and Conclusions

As this study deals specifically with the issue of gender differences in science achievement based primarily on the SISS data collection in 1983/84, it is obvious that certain methodological constraints persist as, in retrospect, no influence could be exercised on the design of the various test instruments and questionnaires or in the identification of the target populations and sampling strategies employed. Nevertheless, the data available form a rich source for the analyses of gender differences.

Fortunately, great care was taken in drawing probability samples of students, teachers, and schools in the 10 countries to be analyzed in this study, so that, in general, sampling errors were minimized, and sampling bias was slight.

Finally, a few summary statistics are given below which illustrate the sheer size of the sample of students, teachers, and schools from which data had been gathered for subsequent analyses presented in this study (taken from Comber and Keeves, 1973, p. 45, and Postlethwaite and Wiley, 1991, p. 11). Table 3.1 shows the number of students, teachers, and schools for each of the three population levels tested for the two data collections in 1970/71 and 1983/84.

As can be seen, these numbers provide a unique data base for comparisons across data points and grade levels but, furthermore, allow complex multivariate analyses within countries which are not restricted to small sample sizes.

The type of analyses performed in this thesis to assess gender differences in science achievement within the 10 countries and between the two time points are described in the following chapter, which delineates and critically discusses the methodological approaches taken.

TABLE 3.1 *Number of students, teachers, and schools for three population levels in the achieved samples for the FISS (1970/71) and SISS data collections (1983/84)*

First International Science Study (data collection 1970/71)

	Students	Teachers	Schools
Population 1	38,672	9,510	1,927
Population 2	47,732	8,216	2,055
Population 3 ▲	49,732	8,241	1,580
Total	136,136	25,967	5,562

Second International Science Study (data collection 1983/84)

	Students	Teachers	Schools
Population 1	81,855	5,065	3,096
Population 2	94,972	9,830	3,658
Population 3	85,449	7,860	2,828
Total	262,276	22,755	9,582

Note:

▲ Population 3 was actually termed 'Population IV' but was comparable to Population 3 of SISS.

4. Methodological Discussion

In Chapter 2 general and specific research questions about gender differences in science were advanced on the basis of the review of research. In Chapter 3 the international context of the comparative research in science has been described focussing on the two major science studies conducted by IEA which yielded the data pool for the research questions analyzed in this study. As could be seen, a set of largely common instruments and questionnaires had been employed in the FISS and SISS projects addressing questions to and collecting data from the students, the science teachers, and the schools. In addition, National Research Coordinators were also requested to provide key data about their educational system.

While special sampling procedures intended to minimize the sampling errors were applied when collecting the data, the subsequent analyses posed conceptual and methodological difficulties of their own, such as how to combine rotated test scores into one total score for each student or how to deal with missing values. These have already been discussed in some detail in the various international reports and volumes released about the science studies by IEA (Comber and Keeves, 1973; Rosier and Keeves, 1991; Postlethwaite and Wiley, 1991; Keeves, 1991).

To implement and analyze the research questions on gender differences in science achievement in this study, it is of importance to discuss possible statistical technologies that are adequate to cast light on the hypotheses postulated in Section 2.2. This methodological discussion should not only identify the theoretical limits and shortcomings of these statistical and analytical strategies pursued but serve as a justification for the statistical methods thought to be appropriate and feasible when dealing with the issue of gender differences in science achievement.

Primarily, this methodological discussion requires the identification of criteria for...

- ☞ the selection of variables, composite scales, and constructs;
- ☞ the discussion of problems related to the various levels of analysis; and
- ☞ the selection of statistical and mathematical procedures used for the analysis of the data.

The outline of this chapter follows this differentiation.

4.1 Variables, Composite Scales, and Constructs

In its conceptualization, the FISS and SISS projects were designed to collect information from various agents involved in science education at the different population levels. The data were obtained from students, teachers, school principals, and the National Research Coordinators of the study, the latter contributing information on the cultural, social, economic, and geographic context of a participating educational system (see Chapter 3).

A vast number of relevant variables believed to have an influence on the outcomes and processes of science education were considered by researchers from the participating countries in the development of the different background questionnaires and test instruments (see also Section 3.3). For the SISS study the total number of variables collected from all tests and questionnaires exceeded several thousand.

This enormous amount of information was obtained from probability samples in each of the 23 countries which had joined the study (see Section 3.4). Taken together, the Second IEA Science Study had collected approximately 1,500 megabytes of information.

Quite clearly, it was virtually impossible to handle this amount of data without an appropriate form of summarizing the information. Thus, after extensive cleaning and weighting procedures had been applied to the data gathered, it became imperative to scale, group, cluster, and select data before undertaking any multivariate analyses.

Clustering of Variables In a first step conceptually similar measures were clustered into constructs. This invoked the construction of composite scales, such as the attitudinal or effort scales (see Section 6.1), as well as the combination of manifest variables (MVs) into constructs or latent variables (LVs). Here, standard statistical procedures such as multiple regression, or principal components analysis were employed. As was pointed out by Postlethwaite and Wiley (1991, p. 127), the clustering of variables into constructs meant a reduction of over 90 percent in the number of relations needed to describe the data. Though this reduction significantly reduced the cognitive task of interpretation, its shortcomings should not be overlooked. The main pitfall in clustering was that if the groups were not formed in a fashion which reflected meaningful homogeneity within the clusters - both conceptually and empirically - the cluster relation would not have been meaningfully interpretable. For further details on the problem of clustering the reader is referred to the international reports about SISS prepared by Keeves (1991, Chapter 2) and Postlethwaite and Wiley (1991, Chapter 8).

It should be emphasized that despite the multitude of variables originally found in the various test instruments a large number of items had to be discarded for final inclusion in the correlation matrices prepared for the multivariate analyses which were also taken for the analyses presented in this thesis. Variables were excluded from these correlation matrices due to insufficient or incomplete data, lack of variance, lack of comparability within and across countries, problems related to the meaning or interpretation, or because they were not related to achievement and attitudinal outcomes.

Given these conditions, it must be realized that although great care had been taken to choose the most appropriate variables for a large number of multivariate analyses to be carried out after the tedious cleaning of the SISS data had been finalized, a few important shortcomings in the selection process should not be overlooked:

☞ The population- and country-specific correlation matrices did not contain information on each of the variables selected. In several cases variables had to be omitted because no or insufficient data were available across or within countries.

☞ To keep the correlation matrices to a technically and procedurally manageable size several variables considered of minor importance for the intended international multivariate analyses had to be omitted.

☞ A number of variables were deemed inadequate for inclusion as they showed no or too little variance.

☞ It was hypothesized that those variables conceptually belonging to the same construct should show some correlation with other variables believed to reflect or form the construct. Occasionally, this led to the removal of variables from the population- and country-specific correlation matrices.

☞ Although a few variables might turn out to be of further interest during the subsequent multivariate analyses, data were not obtained in any of the tests and questionnaires administered. It must be understood that even in a large-scale study like SISS individual, psychological, cultural, technical, and administrative constraints had to be taken into account.

However, most of the information related to the key variables thought to play a vital role in determining gender differences in science achievement (see Section 2.1) was furnished in the population- and country-specific correlation matrices.

Those variables and composite scales which were finally included in the analyses of gender differences are listed together with a short description about their origin (type of instrument), contents, and coding in Appendix 4.1. Additional information about the items employed in the six attitudinal and three descriptive scales are provided in Appendix 4.2.

Construct Identification for Different Models Having selected the variables and composite scales, it then became imperative to group these manifest variables (MVs), where applicable, into constructs or latent variables (LVs). To allow comparisons with the multivariate analyses reported by Postlethwaite and Wiley (1991) and Keeves (1991), it was intended to follow the strategies of forming LVs employed for partial least squares (PLS) analyses as closely as possible. For the multivariate analyses conducted for this study, a number of student-, teacher-, and school-level constructs were introduced. Table 4.1 presents an overview of these constructs. The table also lists the manifest variable(s) forming or reflecting the construct (for a more detailed discussion on 'formative', 'reflective', 'inward', or 'outward' mode see Sections 4.3.1 and 6.2.1). All in all, 11 student-related, three teacher-related, and four school-related constructs were included in the multivariate analyses. Another teacher-related construct assessing the *Opportunity to Learn Science* (OTL) was dropped after preliminary analyses showed its insufficient explanatory power within the general path model specified (see Section 7.1).

It should be emphasized that the acronyms chosen for each of the constructs listed in Table 4.1 were arbitrary. However, these abbreviations are employed throughout this report. A summary of all abbreviations and acronyms used in this thesis is provided in the *'List of Abbreviations'*.

TABLE 4.1 *Manifest and latent variables employed in the multivariate analyses of this study; by type of instrument*

Latent Variable	*Explanation,* • Manifest Variable

Student Questionnaire

HomeBack *Home background of student*

student level: measuring the socio-economic status of the student's home

school-level: describing the socio-economic status of the school's neighbourhood or catchment area

- Father's Occupation
- Mother's Occupation
- Father's Education
- Mother's Education
- Family Size
- Books in Home
- Use of Dictionary

Gender *Sex of student*

student level: identifying the student's gender

school-level: describing the gender type of the class or school group ranging from single-sex boys, coeducational to single-sex girls

- Sex of Student
- Sex of School *(variable from School Questionnaire used in HLM analyses for Australia and England only; see also Appendix 4.1)*

Aptitude *Student's aptitude*

student level: general aptitude of the student (see text for further details)

school level: general aptitude of a class or school group (see below for further details)

- Year Level or Age
- Language
- Word Knowledge
- Mathematics

SciValue *Science values of student*

student level: assessing science-related values

school level: assessing mean science-related values of a class or school group

- Career Interest
- Beneficial Aspects
- Non-Harmful Aspects

Effort *Effort of student*

student level: measuring the efforts undertaken in science at school and at home

school level: measuring the mean efforts of the class or school group undertaken by the students at school and at home

- Homework Effort
- Classroom Effort
- Hours Homework

to be continued

TABLE 4.1 *(ctd.)* *Manifest and latent variables employed in the multivariate analyses of this study; by type of instrument*

Latent Variable	*Explanation,* • Manifest Variable
SciAtt	*Science attitudes of student* **student level:** assessing science-related attitudes **school level:** assessing mean science-related attitudes of a class or school group • Interest in Science • Ease of Learning Science
Achieve	*Science achievement* **student level:** indicating the total science test score comprising of biology, chemistry, and physics items **school level:** indicating the mean total science test score of a class or school group • Biology Score • Chemistry Score • Physics Score
ExpEd	*Expected education of student* **student level:** assessing the desired educational level **school level:** assessing the mean desired educational level of a class or school group
ExpOcc	*Expected occupation of student* **student level:** assessing the desired occupational level **school level:** assessing the mean desired occupational level of a class or school group
FutEnrol	*Future enrolment in science courses* **student level:** indicating the interest in participation in future science courses **school level:** indicating the mean interest in participation in future science courses of a class or school group
Practice	*Practical and experimental work in class* **student level:** assessing the amount of time spent on practical work in science **school level:** assessing the average amount of time a class or school group spent on practical work in science • Practical Work *(HLM analyses for Australia and England only)*

to be continued

TABLE 4.1 (ctd.) *Manifest and latent variables employed in the multivariate analyses of this study; by type of instrument*

Latent Variable	*Explanation,* • Manifest Variable
Teacher Questionnaire	
TchBack	*Teacher's educational background* • Age of Teacher • Teacher Post-Secondary Education • Teacher Post-Secondary Science Education • Years of Teaching Experience • Time Spent on Inservice Training
TchGend	*Sex of teacher* • Sex of Teacher
TchEmph	*Emphasis on science content taught* • Proportion of Teaching Time Spent on Teaching Science • Emphasis on Knowledge • Emphasis on Student Needs • Emphasis on Scientific Investigation
School Questionnaire	
SchFacil	*Facilities available for science teaching at school* • No. of Science Laboratories
Region	*Region school was located in* • State *(Australia only)*; Region *(England only)*
SchType	*Type of school* • Type of School *(Australia only)*
RegType	*Combination of region and type of school* • Combination of Region and Type of School *(England only)*

As could be seen, several of the constructs were based on a single manifest variable only. Though it might have been more valid for the construct to have been represented by a set of indicators, or MVs, rather than a single MV additional data were not available. The somewhat small numbers of MVs describing a particular LV might be perceived as an explanatory weakness of the meaning of the LV but it was considered more appropriate to base the multivariate analyses on the MVs and LVs listed above. Nevertheless, great care should be taken when examining the results in order not to move into spheres of over-interpretation and speculation.

The constructs selected for subsequent multivariate analyses can be classified into three logical and chronological distinct groups, namely...

antecedents independent latent variables which are thought not be influenced by any other construct in the path model specified; *examples: HomeBack, Gender.*

mediating variables latent variables which influence and are influenced by one or more other constructs or latent variables; such latent variables are believed to play an important role in the process of modifying, positively or negatively, gender differences in science achievement; *examples: SciValue, TchEmph.*

outcomes dependent latent variables; in principle, for PLS analyses these were *Achieve and FutEnrol.*

As it is intended to throw light on the complex relationships which were hypothesized to exist among the variables influencing gender differences in science achievement, so-called *path models* will have to be developed. These path models can be perceived as the diagrammatic representation of sets of multiple regressions, and multiple regressions are used here to predict the outcome measure(s), in this case, science achievement, respectively participation in future science courses. Thus, antecedents and mediating variables need to be arranged in a certain way based on previous experiences and logical assumptions in the path model to predict and explain an outcome. As each path model is to be understood as a representation of past (antecedent) and present (mediating) effects their chronological sequence needs to be adhered to. A path model is said to be *fully recursive* if all latent variables are hierarchically arranged and where there is no feedback between them (see Pedhazur, 1982; McCullagh and Nelder, 1983). It is common in devising path models graphically to position antecedents to the far left side whereas the outcome variables are placed on the right with mediating variables in between depending on their causal relationships. An example path model is presented in Figure 4.1. For simplicity, all manifest variables related to the six constructs shown in the model have been omitted.

As can be seen from Figure 4.1, the effects hypothesized among the various LVs are indicated by an arrow where the head of the arrow symbolizes the direction of the effect. It should be stressed that by definition the effects are presumed to be of a causal nature. However, non-causal correlations may also be added to a path model which, in the example given, might exist between HomeBack and Gender. It would seem unreasonable to hypothesize that these two antecedents were causally linked. There is no reason to believe that either the birth of a boy or a girl causes the family to be of higher or lower socio-economic status (i.e., Gender→HomeBack) or that the educational and occupational situation of the family would decide about the sex of children born (i.e., HomeBack→Gender). But, it might well be that boys and girls of homes with higher socio-economic status remain longer at school than do boys and girls from less favourable home backgrounds. This, indeed, would show up in a negative correlation (following the coding of Gender introduced in Appendix 4.1) between the two latent variables HomeBack and Gender. It should also be pointed out that parents often pay more attention to sons than to daughters in terms of their schooling as was mentioned in the review of the literature (see Steinkamp and Maehr, 1984; Humrich, 1988 [1]).

FIG. 4.1 *Example path model with science achievement as the outcome measure*

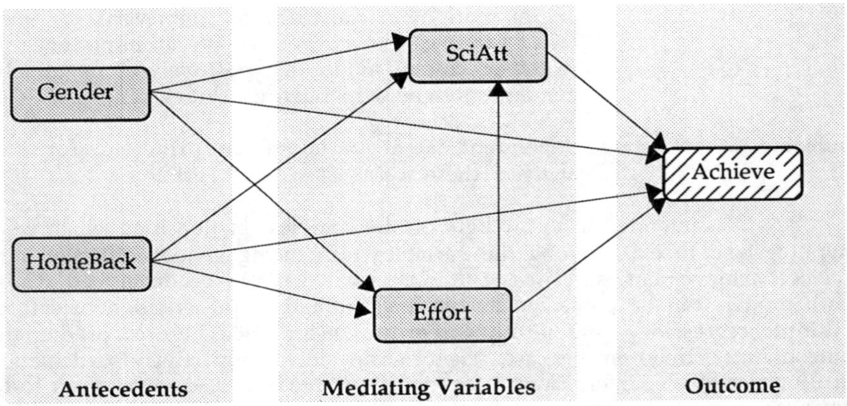

Notes:

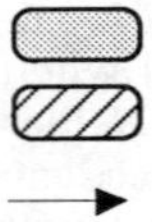

construct, latent variable (LV)

construct of outcome measure

path, with arrow showing the direction of the effect

Sometimes, the logical and chronological placement of mediating variables presents problems. While the positioning and sequencing of the constructs are discussed in more detail in the relevant sections where the full path models are presented (for example, the student-level path model for science achievement is specified in Section 6.2.1), some thought needs to be given to the construct measuring the student's general aptitude (Aptitude).

As described in Table 4.1 and Appendix 4.1, Aptitude was intended to be estimated by two manifest variables, *Word Knowledge* and *Mathematics*. These two MVs were available as summary test scores corrected for guessing from the word knowledge and mathematics subtests given to the students (see also Section 3.3). Unfortunately, the absence of data for a few countries on these two MVs made it necessary to consider alternative MVs to predict a student's aptitude level in the remaining countries (Finland, Netherlands, Sweden, Thailand) if Aptitude were to be introduced into the general student-level path model. However, preliminary analyses showed that only in the case of Sweden it was possible to use *Age* and *Language* as substitutes. This was consistent with findings reported in related publications (Riis et al., 1988; Keeves and Kotte, 1991; Keeves and Saha; 1991). Hence, Aptitude had to be omitted from the path models for Finland, Netherlands, and Thailand.

Further complications with the construct Aptitude arose from the fact that in some countries only the Word Knowledge subtest and in other countries only the Mathematics subtest had been administered (see Table 6.6). This implies a different meaning of 'general aptitude' in the various countries when making interpretations across countries. Finally, consideration had to be given to the positioning of Aptitude within the general path model. Since Aptitude was a concurrent measure to science achievement, the Word Knowledge and Mathematics subtests had been administered jointly with the rotated tests for science achievement. The idea of a 'general aptitude' indicated that the construct should be put chronologically **before** the outcome measure and all mediating variables such as SciValue, Effort, or SciAtt but **after** any antecedent as it is thought to be influenced by the latter. The general aptitude of a student is perceived as being directly dependent on the socio-economic status of the home and possibly by the sex of the student. It must be conceded that this placement is neither ideal nor fully justifiable on psychological or educational grounds. However, the Word Knowledge and Mathematics subtests, i.e. Aptitude, are seen to be crude indicators of the general cognitive ability of a secondary school student. Here, Aptitude is perceived in the light of a definition given by Reynolds and Walberg (1991). In their article on a structural model of science achievement they stated that...

> "...the student aptitude-attributes set includes (a) student ability or prior achievement, (b) motivation, and (c) developmental level (e.g. age). This dimension is usually termed aptitude, referring to the attributes of students that are related to their academic performance. In this sense aptitude includes ability, motivation, and developmental level. In a technical sense, however, aptitude is restricted to cognitive ability." (p. 97)

Thus, Aptitude should be positioned before any mediating student- or school-level variables and after any of the antecedents.

It would have certainly been more desirable to have employed a complete science achievement pre-test or standardized scholastic aptitude test which would have indicated a student's general ability level much more appropriately and with greater validity. But under the administrative and organizational constraints of the data collection in this study, that was virtually impossible. Last but not least concessions had to be made to the total time for testing the students. With respect to the interpretation of Aptitude its somehow limited validity should be borne in mind. Caution should also be used when making inferences about the strength of effects exercised by Aptitude on other constructs in the model. But it is believed that by leaving Aptitude in the general model specifications some valuable information might be gained in explaining gender differences in science achievement. Finally, it is of interest to acknowledge that the same construct used to assess the 'general aptitude' of students had been successfully employed in related multivariate analyses of data from the SISS study performed by Keeves (1991), Keeves and Kotte (1991), and Keeves and Saha (1991).

From the information presented in Table 4.1 it can be seen that no variable is included for cross-country comparisons which measures the urban or rural nature of the catchment area of a school. The variable originally built into the School Questionnaire, which was intended to identify the size of the community in which a school was located, proved inadequate for cross-national comparisons in several countries. Although the location of a school in an urban or rural area should make a difference in some countries, the direction of this difference with respect to gender differences in science achievement varies. Insufficient data for cross-country analyses suggest not to include this variable for multivariate analyses. However, as will be seen in Chapter 8, derived variables reflecting some kind of geographical differences are used in hierarchical linear modelling of the Australian and English educational systems (Region, SchType, RegType).

4.2 Levels of Analyses

It was pointed out in the previous chapter (Section 3.3) that data had been collected from students, teachers, and school principals in order to obtain a rather comprehensive body of information on key variables thought to influence differences in science achievement. Certainly, it is equally desirable to make use of data from these three sources in the investigation of gender differences in science.

Analytically, for multiple regression or other single-level of analysis techniques, combining data from two or more different levels requires either the disaggregation of school-level data to the student level or the aggregation of student-level data to the school level. The aggregation of student level data requires that for each school data set the means for each student-level variable have to be calculated and used. Vice versa, when disaggregating data from the school level, the values of teacher and school variables have to be assigned to each student of that particular school. Both techniques, aggregation and disaggregation, quite typically introduce *bias,* meaning an over- or under-estimation of the estimates of effects associated with variables which were aggregated or disaggregated.

Aggregation Bias By aggregating student data to the school level, the predictive power, or fit of the regression line, often increases, an observation commonly made (Cheung and Tsoi, 1990, pp. 245-256; Sellin, 1990 [2], pp. 267-268). Further to this, it is possible that certain variables show negative effects at one level of analysis and while a positive effect will be found on another level of analysis (Sellin, 1990 [2], p. 257).

This problem is omnipresent for all types of multivariate analyses which are confined to a single level of analysis such as Ordinary Least Squares (OLS), General Linear Modelling (GLM), Partial Least Squares (PLS), principal component, cluster, or discriminant analysis.

For this study, the question of aggregation bias is of considerable importance as the data used come from two different levels (student and school). Following the suggestion of Sellin (1990 [2]), student-level effects on gender differences in science achievement will be analyzed separately before, in another set of multivariate analyses, aggregated student-level data together with teacher and school data will be combined in between-schools path models. This would promote an approximate value of the aggregation bias since the effects of student variables will be known from the between-students path models.

Disaggregation Bias The distorting effects of the disaggregation of group level data (teacher and school variables) to the individual level (student variables), referred to as disaggregation bias, become clear when the following expressions for the correlations (r_t) and regression coefficients (b_t) for the relationships at the between-student or individual overall level are considered (taken from Pedhazur, 1982, pp. 536-538):

$$r_t=r_w\left(\sqrt{1-\eta_x^2}\,\sqrt{1-\eta_y^2}\right)+r_b\eta_x\eta_y$$ *Formula 4.1*

with

r_t correlation coefficient for the between-student overall level

r_b correlation coefficient for the between-school level

r_w correlation coefficient for the within-school level

η_x^2 variance ratio of variable x

η_y^2 variance ratio of variable y

and

$$b_t=b_w(1-\eta_x^2)+b_b\eta_x^2$$ *Formula 4.2*

with

b_t regression coefficient for the between-student overall level

b_w regression coefficient for the within-school level

b_b regression coefficient for the between-school level

where the variance ratios η_x^2 and η_y^2 are given by

$$\eta_x^2=\frac{\Sigma x_b^2}{\Sigma x_t^2} \text{ and } \eta_y^2=\frac{\Sigma y_b^2}{\Sigma y_t^2}$$ *Formula 4.3*

For Cronbach (1976) operating on overall correlation and regression coefficients, which in the case of a joint multivariate analysis of student, teacher, and school data requires the disaggregation of group-level data to the individual level, a lot of information is 'clouded'. He pointed out that by clustering both independent and dependent variables the effects on the overall correlation coefficient (r_t) as well as the effects on the overall regression coefficient (b_t) when aggregating independent variables are difficult to interpret.

It is not possible to discuss the problem of aggregation bias in more detailed aspects here. For a comprehensive examination of this 'methodological dilemma', the reader should consult the interesting considerations raised by Cheung et al. (1990), who conducted several different multivariate types of analyses using an identical data set thereby showing, among other things, the different facets and methodological implications of aggregating and disaggregating educational data.

However, it could mean a severe loss of valuable information were the multivariate analyses in this study to be restricted to one of the two analytical levels, student or school level, only.

4.3 Statistical Analyses Employed

In acknowledging the critical issue of possible aggregation and disaggregation bias, one solution could be the application of multivariate, multi-level approaches such as hierarchical linear modelling (HLM). Though two-level HLM techniques allow the simultaneous analysis of student- and school-level data, some practical and computational aspects still restrict the general, all-purpose use of this strategy. As will be seen from the discussion of HLM in Section 4.3.2, this approach is restricted to observed variables. Thus, it is so limited in the number of variables included in an analysis, that its use does not as yet meet fully the needs of exploratory or confirmatory analyses with structural equation models and latent constructs.

Given the large amount of data available for multivariate analyses, the identification of those variables found to be influencing science achievement, increasing or decreasing gender differences is of utmost importance. The use of data from different levels and from several countries clearly poses significant restrictions on the practicability multivariate approach which can be employed (see also Cronbach and Furby, 1970; Bock, 1989; Cheung et al., 1990).

In tracing relationships among the multitude of variables, separate correlation analyses of the different country data sets had to precede the multivariate analyses (see Section 4.1). With the knowledge of the correlative links between the variables, it might be possible to undertake a discriminant analysis to filter those variables which might constitute the manifestation of gender differences. Discriminant analysis is a special form of regression analysis where the dependent variable is dichotomous, such as gender (Cooley and Lohnes, 1971; Lohnes, 1988, pp. 628-633). This would certainly lead to the identification of, say, three variables X, Y, and Z which best discriminated between boys and girls in science achievement. Just how these variables might influence science achievement and how they affect, or are affected by, each other or other manifest or latent variables cannot be inferred as discriminant analyses do not take into consideration any inner path model relationships (see also Section 4.3.1)

Similarly, using classical multiple regression to identify possible gender effects on science achievement will be very time-consuming and cumbersome because of the vast amount of data and number of variables and the likely effects of multicollinearity (Pedhazur, 1982).

Also, linear structural relations analysis (LISREL) was discarded as an initial statistical procedure since LISREL in analyzing multivariate variance-covariance matrices makes a number of very stringent and restrictive model assumptions, such as distributional assumptions in the maximum likelihood estimation procedure, which could not always be satisfied by the type of data available (see Jöreskog and Sörbom, 1983).

It is, therefore, necessary to employ a multivariate technique which satisfies certain requirements. This technique should be capable of...

☞ handling many variables and composite scales simultaneously;
☞ providing an appropriate form of information reduction;
☞ being used flexible;
☞ being computationally and technically manageable.

Following earlier experiences of handling large-scale assessment data from various countries (Anderson et al., 1989; Duncan, 1989; Cheung et al., 1990; Postlethwaite and Wiley, 1991; Keeves, 1991), PLS was considered the most flexible and promising multivariate technique applicable to the setting of this study in analyzing differences in science achievement related to the issue of gender. Thus, PLS is applied as the major multivariate technique to investigate the data available in this study. This involves the testing of path models proposed for each country separately at the between-student and between-school levels. In the following section, the statistical properties as well as the methodological advantages and shortcomings of PLS are discussed.

However, in order to avoid the conceptual difficulties of aggregation bias it was considered of value to employ a multivariate two-level HLM approach in further analyzing the special cases of Australia and England. These countries were the only two countries among the set of 10 which in 1983/84 still had a sizeable number of single-sex schools in operation. The statistical approach of hierarchical linear modelling is dealt with in detail in Section 4.3.2.

Both multivariate procedures, PLS and HLM, allow the testing of complex path models for gender differences with science achievement as the outcome measure for each country. In addition to the use of these multivariate techniques, results from descriptive or inferential statistics will be presented where appropriate. But one of the aims of this study is to show that methodological possibilities exist which allow for the analysis of a large pool of educational data in a procedurally manageable and technically feasible way such that the complex question of what causes gender differences in science achievement at lower secondary schools may be examined. It should be understood that it is of primary concern to study **jointly** student-, teacher-, and school-related effects in order to throw light on the research questions.

4.3.1 Partial Least Squares Analyses

History and General Remarks Partial Least Squares (PLS) analyses are particularly useful in modelling large educational or social systems, where a large amount of data needs to be handled while at the same time the theoretical knowledge about the system is limited (Noonan and Wold, 1988, p. 710). After the introduction of LISREL as a means for analyzing complex interrelationships of manifest and latent variables by Jöreskog in the early 1970s, Wold (1982) introduced a PLS algorithm which was able to estimate path models with latent variables and multiple indicators. This technique is also referred to as 'soft modelling' because it does not make the 'hard', meaning rigid statistical assumptions about the distributions of and relationships among independent variables as are required in maximum likelihood methods (Noonan and Wold, 1988, p. 711; Jöreskog and Sörbom, 1989).

PLS became readily available only after microcomputing power was enhanced in the late 1980s. While first simple program versions were developed by Lohmoeller (1983; 1988), a more refined PLS program was released by Sellin (1989). This program called PLSPATH is used in its updated version 3.01 for all PLS analyses carried out for this study.

FIG. 4.2 *Example of a path model with science achievement as the outcome measure showing inner and outer model relationships*

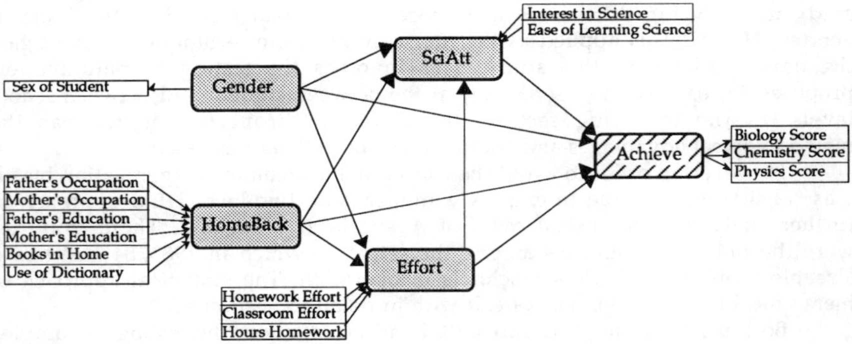

Notes:

For the labels of manifest and latent variables see Table 4.1 and Appendix 4.1.

construct, latent variable (LV)

construct of outcome measure

manifest variable (MV)

inner model path, with arrow showing the direction of the effect

outer model path, with arrow showing inward or outward estimation mode

Despite its mathematical complexity, PLS can be used efficiently to identify strong and weak relationships between latent and manifest variables. This fact should not be underestimated since this makes PLS a highly flexible and exploratory statistical tool. As a modelling approach, PLS has been used successfully in many different research fields including the sciences, sociology, and education (Noonan and Wold, 1988; Anderson et al., 1989; Sellin 1990 [2]).

PLSPATH is based on the use of ordinary two-stage least squares estimation procedures, where less restrictive assumptions apply than with maximum likelihood methods. Conceptually, PLSPATH is related to principal component analysis, canonical correlation analysis, and covariance structure analysis.

Basically, PLS and PLSPATH comprise of a set of mathematical algorithms which are used to estimate relationships between...

- ☞ manifest variables (MVs) predicting a latent construct, and
- ☞ latent variables (LVs) or constructs predicting one or more dependent manifest or latent variables (outcomes).

All relationships involving MVs and LVs can be presented in the form of a path model. For easier interpretation, a path model is commonly displayed graphically. This PLS model can be understood as the representation of all relationships identified in a given setting (for an example see Figure 4.1 above). While the relationships among the MVs related to a latent variable are specified (and subsequently mathematically estimated) in the *outer model*, the *inner model* comprises the relationships among the LVs. A full example of a path model showing both the inner and outer model is presented in Figure 4.2.

Model Specifications Variables entered into PLS are not necessarily assumed to be normally distributed and interval-scaled but at least considered to be measured on an ordinal scale. As with principal component analysis, it can be shown that even dichotomous variables can be entered into the model.

As a first step, it is necessary to define which manifest variables are supposed to form or reflect a certain latent construct. Here, two basic distinctions have to be made (Sellin, 1990 [1], p. 7):

Inward Mode

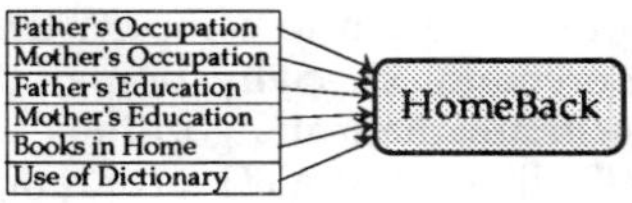

If a set of MVs 'forms' a particular LV the *inward mode* applies. As an example, when a set of socio-economic indicators (MVs) are used to measure HomeBack (LV) each variable to some extent forms the home background construct. In this instance, each MV is assigned a *standardized regression weight* (β) as the LV is estimated using a multiple regression approach. Graphically, the inward mode is indicated by arrows pointing from the MVs to the respective LV.

Outward Mode

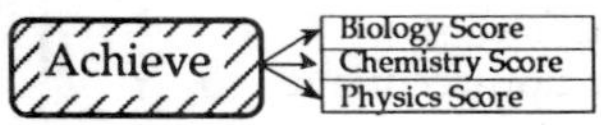

In contrast to the inward mode, the *outward mode* applies if a latent construct is said to be 'reflected' by one or more MVs. This, for instance, will be the case when a specific construct (LV), say, Achieve, is reflected by one or more independent variables (MVs). When a LV is estimated using the outward mode, the *factor loading* (a) will be calculated for each of the MVs. In graphical representations arrows are pointing from the LV to the MVs.

The major distinguishing conceptual difference between inward and outward mode is that in estimating a LV by using the inward mode the respective MVs are thought to be 'added' to form the construct. Thus in the example above, science achievement would not be sufficiently explained by using either of the three subtest scores from the biology, chemistry, or physics subtests. Each of these three scores is required to form the LV Achieve. Similarly, by using the outward mode to estimate a construct each of the MVs is believed to reflect the LV appropriately.

By assigning all MVs to their respective LV the outer model is specified. The next step with PLS is the placement of LVs within the inner model. This demands decisions about the likely causality and/or chronology of the LVs. In other words, a decision has to be made about which latent construct is 'being explained', or predicted, by which other LV or LVs. First the *exogenous* LVs are placed at the left-hand side of the model. These exogenous LVs, or antecedents, are said not to be predicted by any other LV. Following the PLS terminology, a LV which is assumed to be influenced by another LV is called *endogenous*. When specifying the inner model it is necessary to make assumptions about certain effects between and within LVs. However, it should be stressed again that the strength of PLS is its heuristic use which finally helps to improve the theoretical model. Thus, it is recommended to start by specifying a fully recursive model, then run PLSPATH and delete paths which do not contribute to explaining another LV. This procedure usually needs to be applied iteratively before a satisfactory model structure can be obtained. It is, of course, understood that the results of a PLSPATH run could also lead to rejecting the hypothesized path model completely due to model misspecifications.

Computational Procedures When running the PLSPATH program, data can either be entered as a correlation matrix (in form of a square or lower triangular matrix) or from raw data which requires the program to compute the correlation matrix. For this study, as outlined beforehand, specially prepared (square) correlation matrices were available. A set of mathematical algorithms is then applied to estimate the path model. Taking Sellin's PLSPATH program, version 3.01, routinely the following computational steps are performed sequentially in batch mode (Sellin, 1989):

☞ the model specified is checked for syntax or logical errors
☞ input data is read and the correlation matrix prepared internally
☞ the inner model LVs are identified
☞ exogenous and endogenous LVs are identified
☞ the model is tested for being recursive or non-recursive
☞ the outer model estimation algorithm is invoked
☞ block inversion and singularity checks are performed
☞ the PLS iteration procedure is initialized
☞ LVs are standardized to unit variance
☞ LV-LV correlations are computed using OLS estimation
☞ LV-MV correlations are computed
☞ inner model paths are computed
☞ a test for convergence is carried out
☞ two-stage least squares estimates are computed
☞ total inner model effects are computed
☞ inner model residual covariances are computed
☞ initial minus reproduced LV correlations are computed
☞ the output is prepared.

It would be diverging too far from the main purpose of this study to present and discuss the mathematical properties of PLS as well as the PLSPATH syntax here. For further details the reader is referred to the articles of Noonan and Wold (1988), Sellin (in Anderson et al., 1989, Appendices B and C), and Cheung and Keeves (1990). Regarding details on the application and operating of PLSPATH more comprehensive descriptions are offered by Sellin (1989; 1990 [1]).

Technical Aspects Although some of the preliminary analyses for this study were undertaken using Lohmoeller's LVPLS program (1988) all multivariate between-student and between-school analyses reported in the subsequent Chapters 6 and 7 are based on the flexible PLSPATH program (Version 3.01) developed by Sellin (1989; 1990 [1]). This program has proved very reliable and fast so that the multitude of path models using correlation matrices as input data could be analyzed efficiently.

All PLSPATH analyses were performed with the program installed on an APPLE Macintosh IIcx microcomputer. As PLSPATH was originally compiled for the MS-DOS operating system, a special software emulation mimicking the INTEL 80286 CPU and INTEL 80287 FPU had been set up on the APPLE Macintosh (SoftPC/AT™, Version 1.31, by Insignia Solutions, Inc.). With 5 Mb of RAM and a 40 Mb harddisk this configuration proved immensely successful especially because PLSPATH output could be copied and pasted immediately into the relevant word-processing documents (see Appendices 6 and 7) on the APPLE Macintosh.

Quite clearly therefore, the major technical difficulties of the analyses for this study lay with the cleaning of the raw data and the preparation of the initial correlation matrices and not with the implementation and application of PLSPATH. This permitted the investigation to concentrate on the examination of data rather than being tied down by technical or computational constraints and hardware- and software-related problems.

Interpreting PLSPATH Output It is considered useful to explain the output prepared by PLSPATH in order to allow the reader to interpret and understand the results presented in this study. Though not each and every result is deemed relevant for presentation, the most important blocks of information from the final PLSPATH outputs should be explained here.

The blocks of information presented below are, in principle, identical with those presented in the appendices related to the between-student and between-school analyses discussed in Chapters 6 and 7 (see Appendices 6 and 7). In fact, the examples shown below derive from Appendix 6.1 which contains the output for the final run of the Australian path model for the between-student level analysis with science achievement as the outcome.

LV-Correlations

	HomeBack	Aptitude	Gender	SciValue	Effort	SciAtt	Achieve
HomeBack	1.000						
Aptitude	.422	1.000					
Gender	.006	.052	1.000				
SciValue	.122	.162	-.110	1.000			
Homework	.295	.366	.151	.137	1.000		
SciAtt	.110	.132	-.128	.544	.145	1.000	
Achieve	.389	.673	-.133	.214	.302	.178	1.000

The LV correlation matrix reflects the product-moment correlations (r) of each of the inner model constructs with each other. The magnitude of a correlation coefficient provides information about the structure of the inner model and may help to set or delete inner model paths. Also, this matrix may serve as an indicator to detect constructs which can be combined to avoid multicollinearity effects or singularity of the matrix. This would be the case if the correlation of two LVs became very high.

For simplicity, only a lower triangle correlation matrix is displayed. The negative coefficient of r=-.133 implies that boys (coded '1') achieve better in the science test than girls (coded '2').

Outer Model Summary

Variable	Mode	Status	N of MVs	AvgRed	AvgCom	MinCom	MaxCom
HomeBack	Inward	Exogen	4	-	.489	.299	.764
Aptitude	Inward	Endogen	2	.136	.760	.759	.761
Gender	Unity	Exogen	1	-	1.000	1.000	1.000
SciValue	Inward	Endogen	2	.030	.735	.680	.789
Effort	Inward	Endogen	1	.176	1.000	1.000	1.000
SciAtt	Inward	Endogen	2	.176	.594	.194	.994
Achieve	Outward	Endogen	3	.343	.694	.679	.713

The *Outer Model Summary* gives an overview of the mode by which each LV is estimated as well as the number of MVs predicting a LV. Also, the status of a LV is listed (exogenous or endogenous). In addition, the *average redundancy*, a measure describing the extent to which a LV is influenced by other variables in the model, is presented. As can be seen, the average redundancy for SciValue is rather low (.030) which implies that the effects influencing SciValue are small in contrast to Achieve (.343) which is more strongly influenced by the other LVs in the path model.

The last three columns in the above table reflect the *average, minimum and maximum communality* of the MVs within a construct. The communality is a measure of that part of the variance of each variable which it holds in common with one or more other variables in the model (Spearritt, 1988, p. 648).

Outer Model Diagnostics

Variable	Mode	N of MVs	Status	Det(Block)	Min R**2	Max R**2
HomeBack	Inward	4	Exogen	.5173	.2032	.3099
Aptitude	Inward	2	Endogen	.7294	.2706	.2706
Gender	Unity	1	Exogen	1.0000	.0000	.0000
SciValue	Inward	2	Endogen	.7763	.2237	.2237
Effort	Inward	1	Endogen	1.0000	.0000	.0000
SciAtt	Inward	2	Endogen	.8644	.1356	.1356
Achieve	Outward	3	Endogen	.4375	.3730	.3920

The *Outer Model Diagnostics* table summarizes statistics for the estimation of each LV by their respective MVs. The mode differentiates between inward, outward, and unity. Unity mode is automatically invoked when only one MV is used to predict a construct (example: Sex of Student predicting Gender).

Conceptually, unity mode does not involve special considerations since unit weights are assigned. Also, for each LV the number of manifest variables is given. *Status* indicates whether a construct is exogeneous or endogenous. The three statistics on the right-hand side of the table show the minimum and maximum squared multiple correlations (R^2) between the MVs for each LV as well as the determinant of the correlation matrix. These statistics are used to detect possible multicollinearity (see also the *'Tolerance'*-statistic in the table labelled *Outer Model Results*). As Sellin (1990 [1], Appendix 2, p. 2) has pointed out, a relatively high level of multicollinearity within inward blocks may cause stability problems in the estimation of the corresponding weights for MVs in the inward mode. The determinant will be equal to unity if only one MV is used to predict the LV.

When using the outward mode multicollinearity is not a problem. It is, of course, a problem with the inward mode, but if it is not inconsistent with the theory and the logic of the composition of the LV, a change can be made to the outward mode where multicollinearity problems arise with the use of the inward mode.

```
List of Manifest Variables (MVs)
=========================================================================
MV3   Father's Occupation          MV7   Father's Education
MV8   Mother's Education           MV13  Books in Home
MV16  Sex of Student               MV20  Word Knowledge test score
MV21  Mathematics test score       MV22  Beneficial Aspects of Science
MV23  Interest in Science          MV26  Career Interest in Science
MV27  Ease of Learning Science     MV33  Hours Homework
MV37  Biology test score           MV38  Chemistry test score
MV39  Physics test score
=========================================================================
```

From the *List of Manifest Variables (MVs)* the MVs used in the path model can be inferred. As the PLSPATH input requires numeric notations, such as MV16, it is necessary to 'translate' which MV this number represents (here, Sex of Student). The numbers assigned to certain MVs vary from country to country and do also vary between-student and between-school analyses.

```
Outer Model Equations
=========================================================================
HomeBack= MV3      MV7      MV8      MV13
Aptitude= MV20     MV21
Gender  = MV16
SciValue= MV22     MV26
Effort  = MV33
SciAtt  = MV23     MV27
Achieve = MV37     MV38     MV39
=========================================================================
```

The *Outer Model Equations* show the block of MVs (or single MV) used to predict a given LV. In the table above, Aptitude, is formed by two MVs, namely MV20 (Word Knowledge test score) and MV21 (Mathematics test score). Gender is predicted by a single MV, Sex of Student (MV16).

Outer Model Results					
Variable	Weight	Loading	Communality	Redundancy	Tolerance
HomeBack	Inward	Exogen	4 MVs		
MV3	.3048	.6744	.4549	.0000	.2301
MV7	.2524	.6622	.4385	.0000	.3099
MV8	.1368	.5464	.2986	.0000	.2391
MV13	.6323	.8739	.7636	.0000	.2032
Gender	Unity	Exogen	1 MVs		
MV16	1.0000	1.0000	1.0000	.0000	.0000
Achieve	Outward	Endogen	3 MVs		
MV37	.3831	.8240	.6790	.3356	.3774
MV38	.4149	.8446	.7134	.3526	.3920
MV39	.4017	.8309	.6904	.3412	.3730
......					

The *Outer Model Results* give a more detailed idea of the goodness of the MVs predicting a LV. Both, the standardized regression coefficients or weights (for inward mode) as well as the factor loadings (for outward mode) are shown. As was pointed out beforehand, MVs were excluded from the outer model equations if their regression weight became $|\beta|<.10$ or if their factor loading was $|a|<.40$ because it could be concluded that with a 'predictive power' so small, those MVs would not contribute significantly to the LV they were meant to explain (Harman; 1976; Pedhazur, 1982; Keeves, 1991; Keeves and Kotte, 1991).

For constructs predicted by the outward mode the communalities defined as the squared loadings for each MV are also listed. The *communality* can be seen as a measure of the explained variance of a particular MV with respect to the construct it predicts (see also *Outer Model Summary* table above).

The *redundancy* index describes the squared correlation between a given MV and the set of LVs linked indirectly via inner model relationships. A high redundancy value can be interpreted as a possible misplacement of that particular MV in relation to the LV it predicts. As can be seen, the redundancy values for each of the four MVs forming HomeBack are .0000 because HomeBack operates as an antecedent and is not predicted by any other LV in the model.

The *tolerance* value indicates the squared multiple correlation of a given MV with all remaining MVs. Hence, the tolerance statistic yields information about possible multicollinearity within a block of MVs. High tolerance values should be treated with caution.

Inner Model Diagnostics and Summary								
Variable	N Pred	R-square	Det(Pred)	MinBeta	MaxBeta	MinDlta	MaxDlta	MaxTol
Aptitude	1	.178	1.0000	.422	.422	.178	.178	.000
SciValue	2	.040	.9973	-.119	.168	.014	.028	.003
Effort	3	.176	.8192	.135	.286	.018	.067	.181
SciAtt	1	.296	1.0000	.544	.544	.296	.296	.000
Achieve	3	.494	.8192	.124	.630	.013	.325	.181

From the *Inner Model Diagnostics* table the strengths and variation of the inner model paths specified can be inferred. Again, for each LV the number of predictors is listed. According to the regression equations defined for the inner model, the squared multiple correlations (R^2) for each (non-antecedent) LV are also shown.

MinBeta expresses the minimum path coefficient, or direct effect, obtained by any of the inner model predictors, while *MaxBeta* shows the value for the maximum direct effect for a given model equation. Thus, the MinBeta value for SciValue which was found to be explained by two other LVs, Gender and Aptitude, amounted to -.119, in this case showing the direct effect of Gender (see table of *Direct Inner Model Effects* below). Hence, the MaxBeta value of .168 came from the direct effect Aptitude had on SciValue.

MinDlta and *MaxDlta* provide information on the minimum, respectively maximum increase in explained variance which statistically is the relative increase in the R^2 value (often referred to as the R^2-change for a given model equation when a particular predictor is added last). The R^2-change value can be understood as a measure of predictive power for a set of LVs. In general, it can be said that a high R^2-change value indicates a high level of predictive power of the LV added to the model. In the above table the MinDlta value for Achieve was .013. As science achievement was specified to be predicted by three LVs (*N of Pred*=3), namely HomeBack, Gender, and Aptitude (see table of *Inner Model Equations* below), the smallest gain in predicting Achieve by adding one of the three LVs to the inner model regression equation would be as low as .013. However, by adding (in this example) Aptitude to the inner model equation predicting science achievement produced a MaxDlta value of .325. For a more comprehensive discussion of R^2-change see also Nie et al. (1975), Pedhazur (1982) and Norusis (1985).

The *MaxTol* value is the squared multiple correlation of a predictor LV and the remaining predictor LVs of a given inner model equation. This statistic can be taken as an indicator for the relative amount of multicollinearity (over-identification) within a given inner model equation. Typically, values of .50 and above may be indicating stability problems in the estimation of the corresponding inner model equation which, in turn, may lead to the removal of a LV from the equation (Cheung et al., 1990). Hence, the results for MaxTol in the above table indicate that the effects of multicollinearity within the LVs are minimal as the MaxTol values of .18 (for Effort and Achieve) can be regarded as uncritical.

```
Inner Model Equations
===========================================================================


Aptitude= HomeBack
SciValue= Aptitude Gender
Effort   = HomeBack Aptitude Gender
SciAtt   = SciValue
Achieve = HomeBack Aptitude Gender
===========================================================================
```

The *Inner Model Equations* provide a useful summary of how the inner model path structure is hypothesized. As an example, in the above table Effort is directly predicted by three LVs (this is actually equivalent to the number of predictors, *N of Pred*), namely HomeBack, Aptitude, and Gender. Of course, exogeneous LVs (antecedents; here HomeBack and Gender) cannot appear on the left-hand side of inner model equations as these are not predicted by any other LV within the inner model.

Direct Inner Model Effects

	HomeBack	Aptitude	Gender	SciValue	Effort	SciAtt	Achieve
HomeBack	.000	.000	.000	.000	.000	.000	.000
Aptitude	.422	.000	.000	.000	.000	.000	.000
Gender	.000	.000	.000	.000	.000	.000	.000
SciValue	.000	.168	-.119	.000	.000	.000	.000
Effort	.174	.286	.135	.000	.000	.000	.000
SciAtt	.000	.000	.000	.544	.000	.000	.000
Achieve	.124	.630	-.166	.000	.000	.000	.000

For all inner model paths the *Direct Inner Model Effects* are presented. These values correspond to the standardized path coefficients (p). Paths (deliberately) not specified are shown as '.000'. While the origin of any path is given by the LVs listed in the top row from left to right, the destination, or LV predicted, has to be taken from the LVs listed from top to bottom in the first row. Thus, the path leading from HomeBack to Aptitude (as specified according to the *Inner Model Equations* above) had a direct effect of p=.42.

Not shown are the *indirect* inner model effects (i) which can be calculated easily as the difference between the direct and total effects (see Pedhazur, 1982; Cheung et al., 1990).

Following the discussion about standard errors of sampling (see Section 3.4.3), the cut-off point for non-trivial path coefficients was set at $|p| \geq .10$. Path coefficients with $|p| < .10$ were discarded because they accounted for less than about one percent of the overall variance (Pedhazur, 1982, p. 81). It should be noted that the cut-off point of $|p| \geq .10$ is comparatively conservative in cases where very large samples of students have been employed, but is generally appropriate in those cases with smaller samples in between-student analyses. On the other hand, the cut-off point of $|p| \geq .10$ is appropriate for the larger samples while being too liberal for samples of smaller size for the between-school analyses (Sellin, 1990 [1]; Cheung et al., 1990; Keeves, 1991).

It should also be noted that where simple random samples have been employed PLSPATH provides a jackknifing routine for the rapid estimation of standard errors to replace those estimates obtained using simple random sampling formulae. However, where cluster samples have been employed, as are widely used in educational research, the jackknifing procedures which might be used are very costly of both programming effort and computing time, and their effectiveness has, as of now, not been fully explored yet. In any event, this study examines data collected using cluster samples, which involve a high degree of clustering (see Section 3.4), and as a consequence the use of simple random sampling formulae for the calculation of standard errors would be completely inappropriate.

Total Inner Model Effects							
	HomeBack	Aptitude	Gender	SciValue	Effort	SciAtt	Achieve
HomeBack	1.000	.000	.000	.000	.000	.000	.000
Aptitude	.422	1.000	.000	.000	.000	.000	.000
Gender	.000	.000	1.000	.000	.000	.000	.000
SciValue	.071	.168	-.119	1.000	.000	.000	.000
Effort	.295	.286	.135	.000	1.000	.000	.000
SciAtt	.039	.092	-.065	.544	.000	1.000	.000
Achieve	.390	.630	-.166	.000	.000	.000	1.000

In addition to the direct effects, this table shows the *Total Inner Model Effects* for each LV-LV relationship. These are the total effects (t) and are the sum of the direct effects and all indirect effects on a given LV. In contrast to the direct effects, it is, of course, possible that the values for the total effects fall below $|t|<.10$ and are associated with paths considered to be significant.

Only the key statistics necessary for interpreting PLSPATH outputs have been discussed here. It should be recalled that correlation matrices have been used as input data which allowed no jackknifing as raw data would have to be used in that instance.

However, great care had been taken to limit possible sampling errors (see Section 3.4.3). Hence, before the correlation matrices were prepared for multivariate analyses all variables had been weighted accordingly (for weighting procedures see Keeves, 1991; Postlethwaite and Wiley, 1991, pp. 9-10). For further details with respect to statistics available with PLS (e.g. jackknifing, Q^2) the reader is referred to the PLSPATH program manual (Sellin 1990 [1]) or related texts (Stone 1974; Lohmoeller 1983)

Use of PLS Finally, a few comments on how PLS was used in this study to determine the 'final' structure of a given path model should be made.

Due to the standard errors of sampling (see Section 3.4.3) and the substantial Deff values it was decided that direct paths of $|p|<.10$ should be deleted as small effects which were of minor importance for the estimation of a particular LV-LV relationship. This should be kept in mind as the key information for removing or adding paths was taken from the *Direct Inner Model Effects*. Initially, the fully recursive path model was entered into PLSPATH and successively 'trivial', or non-significant paths were removed.

However, another criterion for the stability of a given path model was the number of iterations necessary for convergence of the model. It was sought to keep the number of iterations as low as possible. In general, any path model tested finally converged in three to six iterations, a sign for the general stability of the overall model structures.

Of further interest were the regression weights for the inward mode, or alternatively the factor loadings for the outward mode, of MVs for a given LV. It was decided to exclude MVs from the outer model equations if their regression weight became $|\beta|<.10$ or if their factor loading was $|a|<.40$ because it could be assumed that those MVs did not make any significant contribution to the LV they were meant to explain (Harman; 1976; Pedhazur, 1982; Keeves, 1991; Keeves and Kotte, 1991).

In addition, it was aimed to minimize the tolerance values for MVs and LVs which, in nearly all cases, led to the removal of variables associated with a high degree of multicollinearity. Thus, in several countries it proved useful not to enter Father's and Mother's Education simultaneously to estimate the socio-economic status of the home, i.e. HomeBack, due to the high interrelation of these two MVs when the inward mode was employed.

On the other hand, one aim in arriving at the 'final' path model for a particular country and level of analysis was the maximization of the explained variance of the outcome, science achievement (or FutEnrol). This underlines the complicated process in selecting appropriate MVs to predict a given LV and to build a stable model with 'inappropriate paths' being omitted. The procedure of removing and adding MVs or, occasionally even LVs, was iterative and confirmatory, but the claim cannot be made to have arrived at **the** final path solution for a theoretically established model.

Much more, PLS requires several decisions about the various statistical parameters involved. These decisions, in many instances, may even be viewed as judgemental ones as different researchers may end up with a different model. This neither denies the seriousness of the PLS approach nor its scientific appropriateness. It merely indicates how complex the research task is to reduce the multitude of possible MV and LV relationships into a path model which is consistent with the available data. Therefore, it is of considerable importance, to understand the confirmatory nature of PLS as a statistical tool. The goodness of a path model can be expressed by various statistics in different ways. However, whether a 'final' path model is meaningful in reality depends on the theoretical foundations of the path model being tested.

Another indirect check for the statistical robustness of some of the PLSPATH results was carried out by Postlethwaite and Wiley (1991). Using Population 2 data from 14-year-old students they first undertook PLS analyses with science achievement as outcome for 19 countries with effects pooled within and between constructs. It was then decided to cross-validate the results for these analyses by implementing an alternate, two-stage procedure. First, a principal component analysis was performed for each cluster. The first factor extracted in each of these analyses was taken as a summary variable representing a particular construct. Subsequently, a multiple regression analysis was performed for the inter-construct relationships with an identical structure to that in the PLS analysis.

In a next step the within-construct and between-construct coefficients of both approaches were compared. The values were virtually identical for all comparisons, implying the stability of PLS models (Postlethwaite and Wiley, 1991, pp. 127-128). Thus, good reason existed to use PLSPATH because of its quicker and more versatile capabilities. However, the reader should be made aware that the 'procedural stability' of PLS is a statistical one implying the robustness of the procedure. It does not extend to the configuration of the path model *per se*. The postulation of a certain causal relationship, or path, within a given model can, to a large extent, be tested by examining several statistics produced by PLSPATH. Yet, this does not substitute for a careful check of the validity of paths against a theoretical understanding of the educational system.

Summary Partial Least Squares is a multivariate statistical technique which excels through low running times, relatively easy technical handling, and only very limited assumptions about the distributional properties of variables which are met by many social sciences research projects.

PLS can be used to perform complex model estimations. In that it is a helpful exploratory and confirmatory tool which helps in reducing the information associated with many manifest variables by allowing the construction of a limited number of latent variables. Key relationships among the sets of MVs can be extracted as path coefficients among the inner model LVs.

Its major shortcomings lie with the lack of a single and precise significance test and with the fact that PLS is confined to a single level of analysis only.

Its strength lies in the capacity of the technique to examine a theoretical model and to estimate the parameters of the model by using least squares procedures. Another advantage of PLS is the capability of constituting LVs from one or more manifest or observed variables.

In summary, it must be recognized that PLS is a confirmatory analytical technique. The PLS analyses are only as strong as the validity and meaningfulness of the model designed for testing. The validity of the model rests on theory and logical argument. The indices of fit of the data to the model are the amount of variance explained and the model reliability coefficients. These indices tell the extent to which reality is consistent with theory.

4.3.2 *Hierarchical Linear Modelling Techniques*

History and General Remarks The discussion of the problems caused by aggregation and disaggregation bias (Section 4.2) illustrated that within the context of social science and educational research it would be desirable to employ an analytical procedure which could treat two or more levels of analysis simultaneously. In educational settings data are often available both from student and school level. Seeking to find out about the effects of school level variables, say, the opening hours of a school library, on student achievement, perhaps the achievement in reading, which itself is influenced by a number of student-level variables would demand a two-level mode of analysis. The two levels can be viewed as arranged within a specific hierarchical structure, such as...

- ↓ students are grouped in a class
- ↓ parallel classes form a grade
- ↓ several grades are combined in a school
- ↓ schools are administered by a district school inspector
- ↓ school inspectors report to the ministry of education

This requires a hierarchical linear mode of analysis.

Raudenbush and Bryk first proposed a mathematical algorithm for a hierarchical linear model for studying school effects in 1986. Their innovative work was developed on the theoretical basis of 'slopes-as-outcomes'. The idea was to calculate separate regression coefficients based on within-group analyses, such as between-student within-class analyses, and to use these coefficients as outcome variables in a subsequent regression performed at the between-group level (Burstein, 1980; Burstein et al., 1981).

This type of analysis would involve entering all relevant student-level variables in a regression equation with the students' achievement as the criterion. This initial multiple regression analysis would have to be repeated for every class in the sample at a given grade-level in a given country. With sample sizes of several hundred schools per population per country this would yield sets of several hundred different regression coefficients which, in a second step, would serve as outcome measures in subsequent regression analysis in which teacher- and school-related variables would be entered as predictor variables.

Some of the practical difficulties of this approach are worth considering (Bryk et al., 1988, pp. 2-3):

☞ Regression coefficients are typically estimated with a sizeable proportion of measurement and sampling error. This lack of reliability unfavourably weakens the detection of between-school factors that are hypothesized to account for the regression coefficients.

☞ The sampling precision in estimating the slopes varies across schools (classrooms) depending on the sampling design within each school. This would suggest attaching weights to each slope proportional to its estimated precision.

☞ Commonly, the analytical design for multi-level data would require controlling for confounding variables at both the within-school and between-school level. The statistical model would have to be adapted to the special covariance structure among the multiple regression coefficients.

☞ Variability in the estimated slopes is accounted for by two components, true differences between schools with respect to the variables entered into the within-school regression equation and the sampling or measurement error variance. The 'slopes-as-outcomes' approach can only explain differences within schools because of the analytical segregation into two distinct sets of regression equations which are not linked statistically, thus the proportion of sampling variation cannot be estimated satisfactorily.

It would clearly be desirable for this study to employ a statistical procedure when estimating the outcomes which would take into account simultaneously the effects of variables collected at the student, or *micro level*, as well as the effects of *macro level* variables gathered from teachers or school principles.

The general class of statistical models which satisfies the conditions mentioned for the analysis of multivariate, multi-level data have become known as *hierarchical linear models*, abbreviated as *HLM*. Despite rapid developments in the theory and technical applicability of HLM techniques, it should not be overlooked that still several restrictions persist from preventing these techniques from becoming the common tool for multi-level research in the field of comparative education. As Keeves and Sellin rightly pointed out, one of the key obstacles in the widespread application of such techniques 'is not only that of formulating the analyses to be carried out but also the inversion of large and complex matrices' (Keeves and Sellin, 1988, p. 695).

Model Specifications To overcome the statistical complexity of HLM or multi-level analyses three different mathematical strategies have been proposed and successfully applied:

- a *Fisher scoring algorithm* developed by Aitkin and Longford (1986);
- a *generalized least squares algorithm* developed by Goldstein (1986);
- a general algorithm, the *EM algorithm*, which provides an iterative method of obtaining maximum likelihood estimates of variance components developed by Dempster et al. (1977).

As with PLS it does not lie within the scope of this study to discuss the mathematical properties of each of these algorithms or to evaluate them against each other. The interested reader is referred to articles and monographs of general nature on multi-level modelling (de Leeuw and Kreft, 1986; Bock, 1989; Keeves and Sellin, 1988; Cheung et al., 1990) or with respect to one of the three basic strategies mentioned above (Aitkin and Longford, 1986; Longford, 1987; Goldstein, 1986; Goldstein, 1989; Raudenbush and Bryk, 1986; Bryk and Raudenbush, 1988; Bryk et al., 1988; Bryk and Raudenbush, 1989; Raudenbush, 1988; Raudenbush, 1989).

As the research design of this study placed several constraints on the multi-level procedure to be used, a decision was made to use the multi-level program HLM-2. First, no other multi-level program was at hand at the time of analysis and secondly, the program seemed capable of handling the multitude of data collected for this study. In addition, previous experiences with the application of HLM-2 for the analyses of educational data showed some promising results (Lee and Bryk, 1986; Cheung, 1988; Cheung and Keeves, 1990). In opting for HLM-2 the decision on the mathematical algorithm employed was implicit: HLM-2 is based on the EM algorithm. However, to speed up computing a 30-40 percent decrease in running time has been achieved by implementing an *Aitkin accelerator* as part of the EM variance-covariance component (Bryk et al., 1988, p. 10).

Basically, HLM-2 operates in two separate stages. After having defined the *within-unit* (within-group, i.e., student-level) and *between-unit* (between-group, i.e., school-level) models with an identical outcome measure (here: science achievement), in a first stage regression weights, also termed *parameter slopes*, are calculated separately for each unit (school). In a second step, the parameter slopes at the micro level together with their respective Y-axis intercept, called *base*, are regressed on the macro level predictors. As pointed out beforehand, the estimation of the parameter slopes is, in general, associated with considerable measurement or sampling error because of the small N within schools. Therefore, the variation of regression slopes is decomposed into an error and parameter component. If the parameter component, which is nothing different than the fluctuation of within-unit means due to the effect of between-unit variables, can be partitioned out, the 'true' measurement or sampling error may be substantially reduced. For this reason, HLM-2 employs a special Bayesian estimation procedure (see Bryk et al., 1988, pp. 7-9).

In addition, HLM-2 allows for a differentiation into *fixed* and *random effects*. Fixed effects can involve the variation of either within-unit or between-unit constructs. To check for possible random effects a χ^2-test is performed for each within-unit construct specified to examine the null hypothesis that the residual differences between units, after taking into account the effect of all between-unit predictors, is essentially zero. Partitioning into fixed and random effects further reduces the error associated with a given HLM model (Keeves and Sellin, 1988, pp. 697-698; Bryk et. al, 1988, pp. 28-29).

It is evident that separate regression equations of micro level variables in the estimation of, say, science achievement of students, leads to different intercept terms for each unit, or school, which, in a strict mathematical sense, are not comparable if scaled predictors, such as the Word Knowledge or Mathematics subtest scores, are used. To avoid this problem, Raudenbush and Bryk (1986), proposed to centre the within-unit variables around their group mean. Centring induces some favourable conditioning of the data since the sampling variance of the intercepts and slopes reduces to zero. The generally low standard errors of sampling which arise from the sampling frames applied for the data collection for this study (see Section 3.4) would suggest that centring around the unit mean would be satisfactory.

It was already briefly mentioned in Section 4.3 that in this study the multi-level analyses will be restricted solely to the modelling of effects on gender differences in science achievement for Australia and England, those two countries where in addition to coeducational schools sizeable proportions of single-sex schools were found at the 14-year-old level at the time of the second science data collection in 1983/84. For these special multi-level analyses, described at length in Chapter 8, only one computer program was at hand. This was a PC version of a program called *HLM-2* released by the Bryk et al. research team of the University of Chicago (1988).

It should be stressed that multiple MVs or LVs can be specified as both micro and the macro variables and variables can be entered in the analysis at both levels. However, HLM-2, in contrast to PLS, does not provide for direct and indirect effects among the predictor variables at a particular level. Given the example of the path model presented in Figure 4.1, three separate model estimations would have to be performed using the three possible outcome measures:

☞ Effort = *f*(HomeBack, Gender, plus school-level variables)
☞ SciAtt = *f*(HomeBack, Gender, Effort, plus school-level variables)
☞ Achieve = *f*(HomeBack, Gender, Effort, SciAtt, plus school-level variables)

This illustrates clearly the near impossibility of using HLM as an exploratory tool in multivariate analyses. Thus, the application of HLM for this study has to be limited to only one of the research questions posed. It was decided to employ HLM for the analyses of gender differences in science achievement for school-type comparisons for Australia and England (Chapter 8) only.

Computational Procedures To run HLM-2, a special *sufficient statistics matrix* (SSM) had to be prepared which comprised two different sets of input data. Each SSM-file was created by combining a within-unit file, which contained data from student-level variables, and a between-unit file, which provided the data from school-level variables. For meaningful comparisons across all predictor variables certain standardization procedures were employed in creating the two input data sets. This involved the disaggregation of each school-level variable to the student level, multiplied by the number of cases within each school (class). These disaggregated and weighted school-level variables were then computed into z-scores ($\bar{x}$=0; sd=1). In a final step, the standardized values were aggregated back to the school level. This standardization procedure ensured that the variance components were equivalent across levels with their complementarity maintained (Cheung et al., 1990, pp. 311-312).

For the analyses of coeducational and single-sex schools in Australia and England separate SSM-files were computed for each country and each of the three gender-typical forms of schooling, namely coeducational schools, single-sex boys schools, and single-sex girls schools. The selection of variables in the within-unit and between-unit input data sets conformed to the set of variables originally chosen for the correlation matrices prepared for each country (see Section 4.1).

A general two-level path model was developed (see Section 8.1) and tested using HLM-2. Procedurally, the program performs the following steps (Bryk et al., 1988):

- ☞ the model specified is checked for syntax or logical errors
- ☞ file specifications and batch mode parameters are set
- ☞ the SSM-file is read in
- ☞ the within-unit variables are identified
- ☞ the between-unit variables are identified
- ☞ the outcome variable is identified
- ☞ variables are centred around unit means
- ☞ matrix inversion and singularity checks are performed
- ☞ the EM algorithm is invoked
- ☞ if applicable iterations are speeded up by the Aitkin accelerator
- ☞ within-unit regressions are computed
- ☞ between-unit regressions are computed
- ☞ parameter reliabilities are estimated
- ☞ for fixed effects regression coefficients, standard errors, T-statistics, and p-values are computed
- ☞ for random effects the parameter variance is estimated and the degrees of freedom together with a χ^2-test and associated p-values are computed
- ☞ the output (file) is prepared.

It would be beyond the scope of this thesis to discuss the program-specific algorithms and programming syntax of HLM-2 in greater depth. For further details the reader is referred to the relevant articles of Raudenbush and Bryk (Bryk and Raudenbush, 1988 and 1989; Bryk et al., 1988; Raudenbush and Bryk, 1986; Raudenbush, 1988 and 1989).

Technical Aspects Written in the high-level language FORTRAN, HLM-2 can be ported relatively easily to any major operating system. For greatest flexibility it was decided to use the MS-DOS version 2.0 of HLM rather than a mainframe package.

As an initial step, variance-covariance matrices (SSM-matrices) had to be prepared from raw data files. Subsequently, all HLM analyses were performed using SSM-matrices with the program installed on an APPLE Macintosh IIcx running exclusively on a 2 Mb RAM disk for maximum speed and throughput. For this, a special software emulating the INTEL 80286 CPU and INTEL 80287 FPU had been set up on the APPLE Macintosh (see remarks on Technical Aspects for PLS under Section 4.3.1). This set-up proved quite successful as the HLM-2 output could be copied and pasted easily into the relevant word-processing documents (see Appendix 8.1) on the APPLE Macintosh.

Although HLM-2 allows for an interactive use, this was considered ineffective and time-consuming. The user-unfriendly program structure suggested the operation in batch mode, meaning that special input files bearing model and output definition statements had been prepared. Even with a speedy RAM disk and a 25 Mhz clock speed of CPU and FPU run-times for the estimation of a single outcome variable with about 10 predictors took 20 minutes. As a number of runs for each outcome needed to be undertaken, HLM-2 was certainly time-consuming as an exploratory tool for multi-level analysis.

Preliminary analyses were undertaken to arrive at an appropriate cut-off point for the number of iterations to be carried out by HLM-2. It turned out that, in general, after five iterations the relative stability of the estimation could not be improved significantly (see also the following section for a discussion of model stability and significance tests).

Though HLM-2 offers a very comprehensive output only the most relevant statistics have been selected and presented in this study. It was considered inappropriate to report every change of the between- and within-unit regression equations after each iteration. Rather more informative should be the final statistics provided by HLM-2. For additional information about the output specifications of HLM-2 the reader is referred to the relevant program manuals (Bryk et al., 1988; Raudenbush, 1989).

Interpreting HLM-2 Output To interpret the relevant output statistics of the HLM-2 runs carried out for this study some information is given. As an example, a run for Australian coeducational schools estimating science achievement is presented (see also Appendix 8.1 and Sections 8.1 and 8.2). The part of the full path model (see Figure 8.3) predicted in this run is shown in Figure 4.3. Also, the regression coefficients (γ^*) calculated by HLM-2 are given.

For this part of the HLM model the following blocks of information were obtained:

```
SPECIFICATIONS FOR THIS MIXED-MODEL  RUN OF HLM OF Wed Apr 03 08:28:56 1991
---------------------------------------------------------------------------

PROBLEM TITLE: AUSTRALIA COED RUN13: ACHIEVE

THE DATA SOURCE FOR THIS RUN        = aus2ssm
OUTPUT FILE NAME IS                 = AUSMIX13
THE MAXIMUM NUMBER OF UNITS         =  188
THE MAXIMUM NUMBER OF ITERATIONS    =    5
```

FIG. 4.3 *Example of part of a two-level estimated path model with science achievement as the outcome measure for coeducational schools in Australia, 14-year-old level, 1983/84 (only significant estimated paths are shown)*

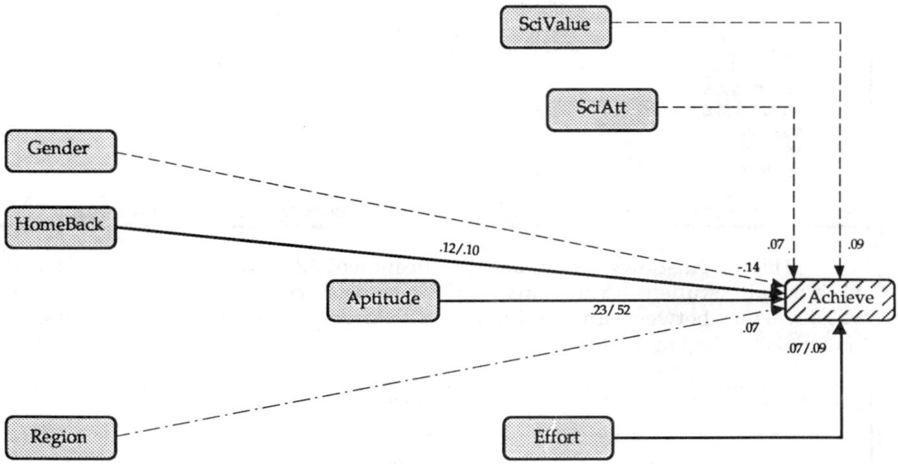

Notes:

construct (here represented by the strongest single MV in each block)

construct of outcome measure

.17/.26 — path, relevant at both student and school level, with γ^* coefficients; the first value (.17) shows the effect at the school level, the second (.26) at the student level

path, relevant on school level only

path, relevant on student level only

The initial *Specifications* list start-up information about the input data. As can be seen a *mixed-model* is to be processed by HLM-2. This indicates that both fixed (between-unit parameters) and random effects (within-unit parameters) are to be estimated by the EM algorithm (Bryk et al., 1988, p. 10). The *Problem Title* shows the data set specified, here Australian coeducational schools. The number of the run (RUN13) is arbitrary. The outcome variable to be predicted is Achieve and the input SSM-matrix used is labelled 'aus2ssm', deriving from student and school data from Population 2 (14-year-old) students of Australia's coeducational secondary schools. The name of the *Output File* was defined in the batch job (not shown). Finally, the *Maximum Number of Units* describes the number of schools for which data are available, while the *Maximum Number of Iterations* is the user-defined cut-off criterion for the number of iterations to be performed by HLM-2 (Bryk et al., 1988, p. 75-82).

```
THE HLM MODEL SPECIFIED FOR THE FIXED EFFECTS WAS:
_______________________________________________________

   WITHIN-UNIT        BETWEEN-UNIT
    VARIABLES           VARIABLES
    _________         ____________
         BASE                 BASE
                          HOMEBACK
                          REGION
                          APTITUDE
                          EFFORT
    *HOMEBACK                 BASE
    *APTITUDE                 BASE
    *GENDER                   BASE
    *SCIVALUE                 BASE
    *EFFORT                   BASE
    *SCIATT                   BASE
'*' - THIS WITHIN-UNIT VARIABLE HAS BEEN CENTERED AROUND ITS UNIT MEAN
```

This block shows the model specifications laid down in the batch file. All within-unit (student-level) constructs have been centred around their mean values. Four between-unit (school-level) variables are entered as predictors (HomeBack, Region, Aptitude, and Effort).

```
THE HLM MODEL SPECIFIED FOR THE COVARIANCE COMPONENTS WAS:
__________________________________________________________

         SIGMA SQUARED (CONSTANT ACROSS UNITS)

         TAU DIMENSIONS
                   BASE
               HOMEBACK
               APTITUDE
               GENDER
               SCIVALUE
               EFFORT
               SCIATT
```

HLM-2 allows the user to define which of the coefficients of the within-unit model are to be treated as random. Only these variables are included in the variance-covariance matrix *(tau dimensions)* used for estimating the outcome (Bryk et al., 1988, p. 20). In the example presented all student-level variables are permitted to vary randomly within each unit.

```
PARAMETER       RELIABILITY ESTIMATES:
_________   _____________________________
     BASE   TAU(1,1) / D_BAR(1,1) =  .605
 HOMEBACK   TAU(2,2) / D_BAR(2,2) =  .067
 APTITUDE   TAU(3,3) / D_BAR(3,3) =  .046
 GENDER     TAU(4,4) / D_BAR(4,4) =  .151
 SCIVALUE   TAU(5,5) / D_BAR(5,5) =  .003
 EFFORT     TAU(6,6) / D_BAR(6,6) =  .143
 SCIATT     TAU(7,7) / D_BAR(7,7) =  .102

NOTE:   THESE VALUES ARE BASED ON ONLY 186 OF 188 UNITS THAT
        HAD SUFFICIENT DATA FOR COMPUTATION
```

Based on 186 valid cases for which sufficient data could be obtained, *Reliability Estimates* are listed for each of the within-unit variables or parameters. The reliability estimates are calculated as the proportion of variance in the within-unit estimates that are parameter variance. As these are the proportions of 'true' *(tau)* to 'observed' *(d_bar)* variance, it is legitimate to treat these ratios as reliability coefficients. For the estimation of the within-unit intercepts *(Base)* a reliability of .605 is recorded. A direct comparison of the magnitude of the reliability estimates with, say, Cronbach's α is rather meaningless. Much more it seems justifiable to tolerate HLM-2 reliability estimates of <.10. It must, however, be conceded that too little experience is at hand to clearly distinguish between a 'satisfactory' and a 'poor' reliability estimate. Unfortunately, even the authors of HLM-2 do not provide more details in this respect. Therefore, emphasis is put on other statistics produced by HLM-2 to decide for the goodness of a given path model structure and of particular variables within that path model (see following tables).

THE GAMMA(*)-STANDARD ERROR-T STATISTIC TABLE:

		GAMMA(*)	STANDARD ERROR	T STATISTIC	p-VALUE
FOR BASE COEF.					
	BASE	.011765	.018802	.626	.531
	HOMEBACK	.124909	.024692	5.059	.000
	REGION	.071215	.018794	3.789	.000
	APTITUDE	.231657	.024066	9.626	.000
	EFFORT	.072580	.027091	2.679	.008
FOR HOMEBACK SLOPE					
	BASE	.101894	.013343	7.637	.000
FOR APTITUDE SLOPE					
	BASE	.519316	.013796	37.641	.000
FOR GENDER SLOPE					
	BASE	-.143525	.013666	-10.502	.000
FOR SCIVALUE SLOPE					
	BASE	.094078	.013081	7.192	.000
FOR EFFORT SLOPE					
	BASE	.092906	.016330	5.689	.000
FOR SCIATT SLOPE					
	BASE	.071366	.014633	4.877	.000

The *Gamma(*)-Standard Error-T Statistic Table* provides the most important information about the *fixed effects* and the adequacy of the model specified. The γ*-coefficients can be interpreted as path coefficients or standardized regression coefficients in ordinary least squares estimation. However, they should be seen in conjunction with their associated standard errors and p-values. HLM-2 allows for the accurate testing of the null hypothesis whether the variation of parameter intercepts and slopes occurs randomly or is influenced by a specified predictor. In the table shown above, it becomes clear that all student and school variables introduced into the model for estimating science achievement have a significant effect. Therefore, all γ*-coefficients can be viewed as 'real', meaning that the effects of the independent variables from the two levels are not arbitrary but significant and are, most likely, to be found in the reality of the coeducational schools in Australia. As probability samples are used together with weighting and standardization procedures, it is appropriate to consider p-values <.10 as a suitable cut-off point. Larger p-values are conceived as inadequate reflecting 'noise' in a given path model. In contrast to PLS, this means that even path coefficients of $p<.10$ could be considered significant which is an illustrative example for the accuracy of model estimations performed by HLM.

At the top part of the table the between-unit effects are reported for the constructs HomeBack, Region, Aptitude, and Effort. The statistics show that a sizeable direct school effect comes from the socio-economic status of the school (i.e. HomeBack) with $\gamma^*=.12$. As the related standard error is only about .02, the resultant T-statistic's value, which is the ratio of γ^* to its standard error, becomes 5.059. This suggests a very low p-value of .000 or less, and implies that the γ^*-coefficient found for the school-level effect of HomeBack on science achievement is significant with an error probability of less or equal as .000.

In contrast, the γ^*-value for the base is non-significant (p-value=.531). This means that the within-unit intercept of science achievement is unlikely to be affected by the between-unit variation, a finding which is not unexpected as most of the between-unit variation occurred because of the four school-level constructs introduced in the estimation of science achievement (see Raudenbush and Bryk, 1988, p. 28).

In the lower part of the table the results for the between-unit effects on the outcome variable are listed. Here, a significant direct influence of Gender on science achievement of $\gamma^*=-.14$ is recorded which signals that boys achieved slightly better than girls simply due to their sex.

THE CHI SQUARE TABLE:

PARAMETER	ESTIMATED PARAMETER VARIANCE	DEGREES OF FREEDOM	CHI SQUARE	P-VALUE
BASE COEF.	.03602	181	456.84	.000
HOMEBACK SLOPE	.00334	185	214.65	.067
APTITUDE SLOPE	.00288	185	208.16	.117
GENDER SLOPE	.00765	185	194.03	.310
SCIVALUE SLOPE	.00016	185	196.66	.265
EFFORT SLOPE	.01195	185	193.54	.318
SCIATT SLOPE	.00638	185	196.91	.261

NOTE: THESE VALUES ARE BASED ON ONLY 186 OF 188 UNITS THAT HAD SUFFICIENT DATA FOR COMPUTATION

This *Chi-Square Table,* also labelled *random effects* in the tables discussed in Chapter 8, is produced to show whether the model accounts for all of the variance in the within-unit parameters. In this case, again based on data derived from 186 schools, the significant *Base Coefficient* indicates that little within-unit variance in science achievement can be explained by the school-level variables specified in the model.

The *Estimated Parameter Variance* tells what percent of the total variance is accounted for by the variability in the intercepts and the slopes for each term base and the predictor variable. Here, the explained estimated parameter variance amounts to 3.6 percent, a relatively low margin. An attempt has been made to introduce further school-level variables, but without success (i.e., their p-values were non-significant).

The p-value=.310 for Gender suggests that the variation in the slopes of Gender between schools is most likely not affected by any of the school-level variables. If, on the other hand, a p-value ≤.100 had been observed, tests would have been undertaken to identify this possible *interaction effect* by introducing school-level variables stepwise into the within-unit equation. From the above table, it can be inferred that a variation of the HomeBack slopes is possibly due to an interaction effect of school-level variables because the p-value=.067 falls below the cut-off criterion set at p-value≤.100. As the above table is taken from the final run, no such school effect could be identified. Therefore, the variation in slopes is to be seen as resulting from the sampling error (see Bryk et al. 1988, pp. 28-29).

From the data of this table it is clear that the explained variance of the slopes or intercept is relatively small. This, in turn, may be taken as another indicator for the fit of the HLM model specified.

```
STATISTICS FOR CURRENT VARIANCE COMPONENTS MODEL
________________________________________________

DEVIANCE =    .867772E+04
NUMBER OF ESTIMATED PARAMETERS =   29
```

Finally, the *Deviance Index* can be taken for comparisons between different model specifications in that a likelihood ratio test may be performed. It was, however, considered appropriate to omit a full testing of the deviance indices between runs because the decision for restructuring the between- and within-unit equations were largely derived by preceding PLS results as well as the γ^*-coefficients and their associated p-values so that a major misspecification of the path model was not to be expected. Here, model misspecification may comprise errors in the chronological or logical sequencing or placement of LVs as well as the use of MVs inappropriately explaining a construct.

For additional information about the deviance measure see Bryk et al. (1988, pp. 65-69) or McCullagh and Nelder (1983), who originally devised a likelihood ratio test for model misspecifications.

In any case, the deviance indices obtained for the HLM-2 runs presented in this study are generally very close to zero so that a misfit of the path models hypothesized could largely be rejected.

Summary In using HLM-2 valuable information can be obtained about a given path model. In contrast to PLS, the conduct of HLM analyses is technically complicated and statistically more complex to handle. Therefore, analyses performed with the two-level HLM program are restricted to the detection of gender differences between coeducational and single-sex schools in Australia and England.

Though the partitioning into fixed, random, and possible interaction effects HLM-2 provides comprehensive information about the statistical significance of the paths defined in a model. As part of the set-up of HLM-2 it is not possible to detect and measure indirect effects among latent variables as would be the case with PLSPATH. The result of this inability is that in order to arrive at the final path model structure for a full path model for science achievement separate analyses have to be conducted for each of the mediating variables with a single outcome measure at a time. It can easily be anticipated that this iterative process becomes time-consuming and HLM-2, in its present form, is not recommended as an exploratory tool in multivariate, multi-level analyses.

4.4 Summary and Conclusions

In this chapter the variables, composite scales, and constructs to be included in multivariate analyses using PLS and HLM were described. It should have become obvious that by a careful selection process only those variables could be chosen from the student, teacher, and school questionnaires which allowed meaningful and promising comparisons of gender differences in science achievement.

Having arrived at a set of correlation matrices for each population of the 10 countries, it was necessary to decide on which types of multivariate analyses should be used. As was shown in detail, PLS analyses, though confined to one particular level of analysis and partly error-prone due to aggregation or disaggregation bias when combining data from the student (within-unit) and school (between-unit) levels, are used as the primary tool for the multivariate analyses of the country data sets. These analyses of gender differences in science achievement will be differentiated into two distinct steps: the analysis of student-level variables, reported in Chapter 6, and the multivariate analyses at the between-school level with selected student-level variables aggregated to the school level, which will be discussed in Chapter 7.

However, given the promising perspective of new hierarchical linear modelling techniques on the other it is deemed advantageous to investigate further the problem of gender differences in science education using two-level HLM. But, due to the technically complicated application of HLM-2 these multi-level models are restricted to a comparison of coeducational and single-sex lower secondary schools in Australia and England. All HLM analyses are presented in Chapter 8.

The discussions in this chapter should have also illustrated the complexity of the analyses to be performed in order to take into account effects on science achievement measured at student and school levels simultaneously. Though hierarchical linear modelling techniques can, at this point in time, be regarded as new and innovative, it should not be overlooked that the complexity of the multivariate multi-level procedure of the estimation of each single outcome variable makes it very time-consuming to use HLM to compare different path models for different age groups for a number of countries and, even more, types of schools. It was, therefore, decided that PLS analyses should precede subsequent HLM analyses since the PLS results provide valuable initial information for the set-up of HLM path models in Australia and England.

Before turning to the results of multivariate analyses, it is considered useful to present a number of descriptive findings on differences in science achievement for the various age groups in the 10 countries and also to discuss the relationship of students' contemporary achievement in science to their future participation in science courses.

5. Achievement and Analyses

In this chapter the findings from the First and Second International Science Studies related to the influence of gender on achievement are presented. In addition, the relationships between achievement in science and expected future enrolment of students in science courses are examined, particularly with respect to differences between boys and girls.

It was hypothesized in Chapters 1 and 2, that the effect of the women's liberation movements from the early 1970s onwards, which were active in many of the countries involved in these studies, might have provided an important stimulus to increase the expectation and interest of female students to enrol in science courses at the post-compulsory level. Moreover, such changes in expectations and aspirations might have been associated with changes in the relative levels of achievement in science of girls and boys. As a consequence, the data collected in 1970/71 (FISS) and 1983/84 (SISS), which are under survey in this chapter, provide a unique opportunity to examine the changing effects of teaching and 'learning of science in schools, particularly with respect to gender differences in both the processes and outcomes of science education.

5.1 Differences between Boys and Girls in Science Achievement

As pointed out in Chapter 3 students at the 10-year-old, 14-year-old, and terminal secondary school levels in the different countries were tested with sets of common items, the science achievement tests. Within these tests, subsets of items relative to the biology, chemistry (and earth science), and physics curricula at each of the age and grade levels sampled were identified. There were slight differences in the actual numbers of items administered to the different populations between testings in 1970/71 and 1983/84 but it was made sure that the SISS achievement tests contained at least 15 'bridge items' at each population level tested which had already been administered in the FISS study. However, these differences in the science achievement tests could be regarded as minor if the data were standardized.

From the raw data final test and subtest scores for each student were calculated. The procedures involved have been described in more detail in the published reports of these studies (Comber and Keeves, 1973, pp. 50-53; Postlethwaite and Wiley, 1991, pp. 49-81). For convenience, Appendix 5.1 has been prepared showing the science achievement test scores for each country and population level.

As this chapter focusses on the examination and explanation of the score differences between boys and girls rather than on a descriptive comparison of the total scores across countries it is considered preferable to calculate and present the standardized score differences between boys and girls rather than the raw score differences. To do this, the 'effect size' between scores of male and female students is used which, by standardization, makes allowance for the differences in the lengths of the different tests and subtests, and the differences in spread of scores between countries, between the two occasions, and between the three age or grade levels. Nevertheless, it is considered necessary to give a short explanation of what an effect size is and what its properties are.

Effect sizes When comparing effects of one or more variables on another, or between two or more different groups (here: male and female students) a unit-free measure is required to indicate the strength of the relationship. Using the simple point-biserial or product-moment correlation coefficient (r) does not seem completely appropriate as its use would ignore the possibility of different group sizes and, hence, the different contributions of the variances to the two groups to the unit or measure. Cohen (1969) has proposed the use of a unit-free measure, the effect size *d*. Its calculation is based on the score distances in units of variability (see Cohen, 1969, pp. 18-19). Thus, *d* can be expressed in terms of the correlation r. As a rule of thumb the effect size can be readily calculated as $d \approx 2r$ if $r < .50$ in cases where the two groups are equal in size (for details see Cohen, 1969, p. 20). However, where the two groups differ in size the effect size may be calculated directly from the formula (McGaw, 1985, p. 3324)

$$d = \frac{\mu_B - \mu_G}{\sigma} \qquad \textit{Formula 5.1}$$

where μ_B and μ_G are the mean scores for the boys and girls respectively, and σ is the pooled-within group standard deviation of scores.

Effect sizes can be regarded as small when $|d| \leq .20$, medium when $.20 < |d| \leq .50$, and large when $|d| > .50$ (Cohen, 1969).

Table 5.1 presents the effect sizes for gender differences in science achievement at the 10- and 14-year-old level for the data collected in 1970/71 and 1983/84 for the 10 countries. Also, the changes in the effect sizes between the two occasions have been calculated. Further to this, it is informative to average the effect sizes by occasion and level of schooling. It should be noted that the effect sizes are calculated with the score for girls subtracted from the score for boys. Thus, a positive effect size indicates the superior performance of boys with respect to girls.

The resultant mean value ($\bar{d}$) based on all effect sizes available for each row in Table 5.1 can be conceived of as a measure of average gender disparity across the 10 countries. It is, of course, noted that its generality is limited to this subset of 10 countries. Nevertheless, the mean effect size, and in particular its change across age level and occasion provides an indication of the extent to which gender differences are to be found across countries at the different levels of schooling. The change in the mean effect size between occasions was calculated as the difference of the 1983/84 mean minus the 1970/71 mean rather than averaging across the change of effect size for each of the countries for which data were obtained.

For easier interpretation of the mean effect size its standard deviation has also been reported. This allows an assessment of the variability between the countries, in other words, it permits an examination of the extent to which a country deviates from the average 'trend'.

At the 10-year-old level it can be noted that with the exceptions of Sweden and the United States the effect sizes for gender differences in science achievement dropped very slightly between 1970/71 and 1983/84. In the case of Hungary the effect size was reduced to nearly half. In 1970/71 the gender differences were largest for the Netherlands (.41) and Finland (.36) and lowest for Japan (.15) and Italy (.19). The mean effect sizes of .25 in 1970/71 and .23 in 1983/84 indicate a low but not negligible gender difference across the countries with little change taking place between the two occasions.

TABLE 5.1 *Effect sizes for gender differences in science achievement at the 10- and 14-year-old level in 1970/71 and 1983/84 and the change in effect sizes over time*

	Australia	England	Finland	Hungary	Italy	Japan	Netherlands	Sweden	Thailand	United States	average effect size mean	average effect size sd
10-Year-Old Level												
1970/71	-	.26	.36	.21	.19	.15	.41	.24	.21	.20	.25▲	.09
1983/84	.25	.21	.34	.12	.15	.12	-	.26	-	.35	.23▲	.09
change (Δ)	na	-.05	-.02	-.09	-.04	-.03	na	.02	na	.15*	-.02◆	
14-Year-Old Level												
1970/71	.40	.16	.53	.35	.38	.64	.59	.41	.04	.44	.39▲	.18
1983/84	.24	.34	.31	.17	.36	.28	.49	.25	.39	.40	.32▲	.09
change (Δ)	-.16*	.18*	-.22*	-.18*	-.02	-.36*	-.10	-.16*	.35*	-.04	-.07◆	

Notes:
positive effect sizes indicate superior performance of boys
negative effect sizes indicate superior performance of girls
Δ difference of effect sizes between 1970/71 and 1983/84
▲ mean effect size calculated as the sum of effect sizes across all countries divided by the number of countries for which data were obtained
◆ mean change of effect size between 1970/71 and 1983/84 calculated as the difference between the mean effect size of 1970/71 and the mean effect size of 1983/84
* difference significant at the 5%-level
- population not tested
na not applicable

At the 14-year-old level gender differences were noticeably larger compared with the 10-year-old level on both occasions. However, they generally decreased from 1970/71 to 1983/84, and more substantially than at the 10-year-old level. Also, the effect sizes varied considerably in 1970/71 from .04 in Thailand to .64 in Japan. Overall the picture is much less homogeneous among the countries. While the gender differences increased in Thailand (Δ.35) and England (Δ.18), they were substantially reduced in all other countries. The biggest drops occurred in Finland (Δ-.22) and Hungary (Δ-.18). The gain in England is difficult to explain. Thailand's figure could be distorted by a differently defined target population in the FISS study. However, the change in Thailand might well be related to changes occurring over the years in the decisions of students to remain at school to the 14-year-old or Year 9 level, since the participation in the study of science at this age and grade-level has increased from 19 to 32 percent between the two occasions.

When comparing the 10-year-old with the 14-year-old students, the general overall increase between age levels are surprising. These findings would seem to indicate that by the age of 14 years, students in the 10 countries would have started to be influenced by school and community expectations, that science was a subject in which boys were generally more interested than girls, and that boys were more likely than girls to enter a science-related occupation. Thus it is possible that, although the study of science is generally mandatory at this level, boys have already started to specialize in science. It is also possible that the performance of girls has already begun to lag behind that of boys as a consequence of the differences in career expectations and differences between boys and girls in interest in science. At this stage it is inappropriate to draw further conclusions with respect to possible reasons for this overall disparity between boys and girls. However, this question will be picked up again in Chapters 6 and 7. Yet it should be borne in mind that across all countries tested, science scores were greater for boys than for girls at both the 10- and 14-year-old levels, and on both occasions. Further to this, the overall gender differences were reduced over time, at both the 10- and 14-year-old levels. This latter trend could also be seen as being consistent with a general change in emphasis within the educational systems with regard to the importance of education for girls which is shared by educators and parents alike. How this societal change might operate to influence educational practice is discussed in greater detail in later chapters.

At the terminal secondary school level the gender differences in science achievement are more pronounced. This can be seen from Table 5.2 which shows the effect sizes for all students in the samples at this level and for those who were not specializing in any particular field of science (non-specialist students).

In 1970/71 the effect sizes were large for all countries except Thailand. Sampling fluctuations may have distorted the results in this case. The gender disparities were largest in the Netherlands (d=.95) and Finland (d=.94) for all students. In the non-specialist sample Hungary (d=.83) recorded the highest value. As the average effect sizes between boys and girls differed considerably in 1970/71, .67 for all terminal secondary students and .55 for the non-specialists, it can be concluded that this was probably due to the greater interest in science of male students, a trend already observed at the 14-year-old level.

For both groups, all students and the non-specialist students, the mean effect size dropped significantly between the two occasions. This decrease (Δ-.14 and Δ-.13 respectively) was, again, larger than at the previous level of schooling (Δ-.09 at the 14-year-old level). This indicates that the gender differences, which occurred across all countries, were, in fact, less marked in 1983/84. Despite this observation it must be remembered that there were still large differences for the all students group and medium size differences for non-specialist students for nearly all countries at the terminal secondary level.

Another point of note is the reduction in the variability of gender differences across countries both for all students and the non-specialist students between 1970/71 and 1983/84. This change could be the result of a general shift in policy or educational practices between the two occasions requiring that more attention is paid to girls studying science.

TABLE 5.2 *Effect sizes for gender differences in science achievement at the terminal secondary school level for all students and non-specialist students in 1970/71 and 1983/84 and the changes in effect sizes over time*

	Australia	England	Finland	Hungary	Italy	Japan	Netherlands	Sweden	Thailand	United States	average effect size mean	sd
All Students												
1970/71	.72	.82	.94	.41	.50	-	.95	.86	.29	.56	.67▲	.24
1983/84	.46	.55	.58	.44	.50	.70	-	.66	.46	.43	.53▲	.10
change (Δ)	-.26*	-.27*	-.36*	.03	.00	na	na	-.20*	.17	-.13	-.14◆	
Non-Specialists												
1970/71	.44	.62	.69	.83	.70	-	.52	.60	.14	.45	.55▲	.20
1983/84	.36	.40	-	.28	.52	.53	-	.37	.55	.32	.42▲	.10
change (Δ)	-.08	-.22	na	-.55	-.22	na	na	-.23	.41	-.13	-.13◆	

Notes:
positive effect sizes indicate superior performance of boys
negative effect sizes indicate superior performance of girls
Δ difference of effect sizes between 1970/71 and 1983/84
▲ mean effect size calculated as the sum of effect sizes across all countries divided by the number of countries for which data were obtained
◆ mean change of effect size between 1970/71 and 1983/84 calculated as the difference between the mean effect size of 1970/71 and the mean effect size of 1983/84
* difference significant at the 5%-level
\- population not tested
na not applicable

A further interesting point is that while the changes in effect sizes for all students in Hungary and Italy were small between 1970/71 and 1983/84, the changes were quite substantial for the non-specialist subgroup. While the mean change in effect sizes was well above average for Hungary (Δ-.55 compared with an average mean change of Δ-.13) it was also sizeable for Italy (Δ-.22). Why did the non-specialist groups show a marked reduction in gender differences in these two countries compared with the students in the remaining countries at the terminal secondary level? One possible explanation could be that both countries had introduced curricular changes at the upper secondary school level which required greater exposure to science by all students whether or not they intended to specialize in science at the terminal secondary stage.

Figure 5.1 records the mean effect sizes for gender differences on the science total score across the three age levels and for the two occasions.

FIG. 5.1 *Mean effect size of gender differences in science achievement from 10-year-old to terminal secondary school levels for the 1970/71 and 1983/84 samples*

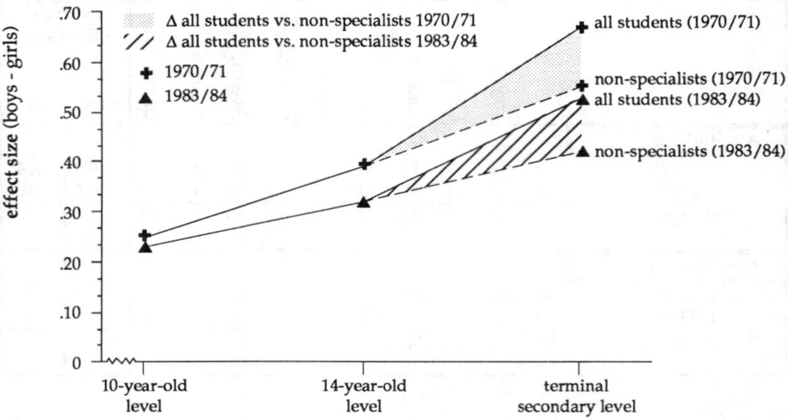

The graphs show the constant growth of gender disparities across the three levels of schooling. It can be clearly seen how the mean differences between boys and girls in science achievement decreased consistently in 1983/84 compared with 1970/71. It also highlights that the sizeable differences between all students and non-specialist students at the terminal secondary school level still remain in 1983/84.

This calls for a closer examination of the three science subscores since it might shed some light on why the gender differences between all students and non-specialist students are still to be found in 1983/84.

Moreover, as biology, chemistry, and physics all are differently 'attractive' to students and differently favoured by boys and girls it is necessary to examine gender differences for these subject subscores.

5.2 Differences in Science Achievement Subscores

In the previous section changes over time in the effect sizes associated with differences between boys and girls with respect to science total scores are recorded. However, a modified picture is to be expected when science subscores are examined, since biology, chemistry, and physics are regarded differently by boys and girls.

5.2.1 *Primary and Lower Secondary School Levels*

Table 5.3 presents the gender difference effect sizes for 1970/71 and 1983/84 in the ten countries for the students' biology, chemistry, and physics subscores.

TABLE 5.3 *Effect sizes for gender differences in achievement in science subjects at the 10- and 14-year-old levels in 1970/71 and 1983/84 and the change over time*

	Australia	England	Finland	Hungary	Italy	Japan	Netherlands	Sweden	Thailand	United States	average effect size mean	average effect size sd
10-Year-Old Level												
Biology												
1970/71	-	-.02	.11	.01	.00	.02	.16	-.04	-.01	.03	.03▲	.07
1983/84	.08	.05	-.02	.00	.00	.04	-	-.10	-	.14	.02▲	.07
change (Δ)	na	.07	-.13	-.01	.00	.02	na	-.06	na	.11	-.01◆	
Chemistry												
1970/71	-	.20	.31	.12	.09	.12	.31	.14	.09	.13	.17▲	.09
1983/84	.19	.13	.26	.03	.10	.20	-	.20	-	.31	.18▲	.09
change (Δ)	na	-.07	-.05	-.09	.01	.08	na	.06	na	.18	.01◆	
Physics												
1970/71	-	.47	.48	.37	.38	.24	.56	.50	.40	.34	.42▲	.10
1983/84	.35	.29	.59	.34	.29	.00	-	.46	-	.40	.34▲	.17
change (Δ)	na	-.18	.11	-.03	-.09	-.24	na	-.04	na	.06	-.07◆	
14-Year-Old Level												
Biology												
1970/71	.10	-.13	.15	.03	.11	.43	.11	.18	-.09	.18	.11▲	.16
1983/84	.16	.25	.16	.04	.14	.28	.17	.13	.22	.32	.19▲	.08
change (Δ)	.06	.38	.01	.01	.03	-.15	.06	-.05	.31	.14	.08◆	
Chemistry												
1970/71	.23	.00	.20	.11	.12	.34	.11	.11	-.09	.28	.14▲	.13
1983/84	.21	.22	.08	.01	.21	.33	.32	.18	.30	.33	.22▲	.11
change (Δ)	-.02	.22	-.12	-.10	.09	-.01	.21	.07	.39	.05	.08◆	
Physics												
1970/71	.58	.43	.80	.62	.58	.75	.68	.63	.23	.56	.59▲	.16
1983/84	.30	.40	.53	.31	.49	.27	.66	.41	.41	.39	.42▲	.12
change (Δ)	-.18	-.03	-.27	-.31	-.09	-.48	-.02	-.22	.18	-.17	-.16◆	

Notes:

positive effect sizes indicate superior performance of boys
negative effect sizes indicate superior performance of girls

- Δ difference of effect sizes between 1970/71 and 1983/84
- ▲ mean effect size calculated as the sum of effect sizes across all countries divided by the number of countries for which data were obtained
- ◆ mean change of effect size between 1970/71 and 1983/84 calculated as the difference between the mean effect size of 1970/71 and the mean effect size of 1983/84
- * difference significant at the 5%-level
- \- population not tested
- na not applicable

Biology At the 10-year-old level gender differences were basically non-existent in both the 1970/71 ($\bar{d}$=.03) and the 1983/84 samples ($\bar{d}$=.02). The minor difference favouring the boys in Finland in 1970/71 (*d*=.11) vanished, while the United States showed a small increase in effect size with superior performance by their male students in 1983/84 (*d*=.14).

In general, the mean effect size increased slightly for the 14-year-old students but surprisingly, as it is in contrast with the overall trend described above, the gender disparities increased from 1970/71 to 1983/84. This might be attributed to different changes in different countries. While there is hardly any change in most countries, a small increase is recorded for the United States and major increases for England and Thailand. These changes might be caused by the study of general science being made mandatory, or the introduction of a biology-oriented curriculum in lower secondary schools in the 1970s (see Keeves, 1991). In Japan where gender differences had been large in 1970/71 a significant drop took place by 1983/84. This may be accounted for by a stronger emphasis in the Japanese schools in the exposure of girls to science education.

Chemistry Across all countries in 1970/71 the gender effect on the chemistry subscore was markedly higher than for biology at the 10-year-old level. It was medium for Finland and the Netherlands (*d*=.31) and small for Italy and Thailand (*d*=.09). In all countries the effect sizes are in favour of the boys. Although in 1983/84 the mean effect size did not seem to have changed, this finding is misleading when the changes in individual countries are examined. At the end of compulsory schooling gender differences had been reduced in England, Finland, and Hungary while they had increased to a smaller extent in Japan and Sweden. Again, as with biology there was a remarkable rise in the effect size in the United States (Δ.18) probably due to a reduced attraction to and emphasis on science generally which was even more pronounced for girls. This may well be consistent with the overall observation of a relatively weak performance in science of American students at the 10-year-old level (Postlethwaite and Wiley, 1991, pp. 55, 78, 159-160).

At the 14-year-old level the spread in gender differences across countries became larger when compared with the 10-year-old level as can be seen from the increase of the sd of the mean effect size (sd=.13 for 1970/71 and sd=.11 for 1983/84 for the 14-year-old students compared with sd=.09 for both 1970/71 and 1983/84 for the 10-year-old students). Indeed, the effect sizes were noticeably different in 1970/71. Thailand reported a slightly negative effect (-.09) which implied that female students had higher achievement. In Australia, England, Finland, Japan, and the United States gender differences were medium and positive. All other countries had small effect sizes. Surprisingly, while the differences decreased slightly over time in four countries they grew considerably in others, namely in England (Δ.22), the Netherlands (Δ.21), and Thailand (Δ.39). These different developments might be attributed to changes in policies requiring female students to take chemistry courses. Another explanation could be the incorporation of more earth science items into the chemistry subtest in the second study as earth sciences is not taken as a separate subject in most countries at the 14-year-old level.

Physics Medium and large size gender differences are to be found across all countries at the 10-year-old level. In 1970/71 the gender differences were lower in Japan (d=.24) and larger in Sweden (d=.50) and the Netherlands (d=.56). On the second occasion, in general, gaps in achievement between the sexes decreased. On the one hand, they virtually disappeared in Japan with this country being the only one showing no gender differences in physics at the 10-year-old level. Japan could well have introduced some modifications in its curriculum (e.g. making more physics exposure compulsory for all students) (Keeves, 1991). On the other hand, a minor rise in the effect size could be detected in the cases of Finland and the United States.

There was a substantial increase of gender differences from the 10-year-old to the 14-year-old level with a mean effect size of $\bar{d}$=.59 in 1970/71, which dropped to $\bar{d}$=.42 in 1983/84. Again the overall trend was not reflected in each of the countries. Japan (Δ-.48) and Hungary (Δ-.31) showed noticeable reductions in gender disparities between the two occasions. However, virtually no changes occurred in England, Italy, and the Netherlands. In Thailand the effect size increased contrary to the overall changes in the other nine countries. It is interesting to note, that in 1983/84 the gender gap in physics was largest in the Netherlands (Δ-.66).

In summary, Table 5.3 illustrates the different gender effects associated with the three science subjects with achievement in biology showing basically no gender differences at the 10-year-old level while for the same age group small gender disparities are to be found for chemistry and medium ones for physics. While gender differences in chemistry across all countries appear to remain similar in size from the 10-year-old to the 14-year-old level, average gender differences in biology and physics seem to increase. Why does achievement of boys and girls with respect to chemistry differ so markedly from that in biology and physics?

Though there is no simple answer available, a possible explanation could be advanced. When introduced to students during the lower secondary school years, chemistry may be perceived as a more 'female' subject than physics. Chemistry is less mechanically oriented and therefore is more favourably chosen by girls. The same argument holds for biology. Further to this, aspects of biology are among the first science topics to be taught to students in primary school.

With the exception of physics, the overall effect sizes increased slightly or did not change significantly. In physics a large reduction in gender differences could be detected over time. This drop was even more strongly observed at the 14-year-old than at the 10-year-old level. Likely reasons for this could be that many countries either made exposure to physics compulsory at the lower secondary level (e.g. Hungary), introduced changes in the curriculum (e.g. Australia), increased numbers of female physics teachers (e.g. Finland), or experienced a general drive by parents to encourage their daughters to choose physics courses more frequently where options are available (Keeves, 1991; Rosier and Keeves, 1991).

Nevertheless, gender differences were observed for all three science subjects consistently favouring male students. This is, of course, a major finding and it must be expected that a similar result would be found at the terminal secondary school level.

5.2.2 *Upper Secondary School Level*

Table 5.4 presents the effect sizes of gender differences for the science subscores for all students at the terminal secondary school level, alongside those for students who specialized in (at least) one subject. There were no data available for the non-specialist students as these students did not take the biology, chemistry, or physics specialist tests in 1983/84.

Biology There was little difference between the all students and the science specialist student groups at the terminal secondary level with respect to the mean effect sizes both in 1970/71 and 1983/84. But, in general, the gender effects observed, with the exception of Thailand in 1970/71, favoured male students. As underlined by the large standard deviations recorded in particular for the specialist sample in 1970/71, this general trend did not hold for all countries. In Italy and Thailand, for example, the gender differences increased markedly between occasions. Alternatively, they were substantially reduced for the specialist students in Sweden in 1983/84. Biology seemed to be not at all favoured by female students in Italy and Thailand, an indicator for the more academically-oriented curricular changes in biology which took place between occasions in the upper secondary forms in these countries (Keeves, 1991). Sweden, which did not have biology as a compulsory subject in upper secondary schooling in 1970/71, modified this policy so that more girls were exposed to both science and the study of biology. This is underlined by the fact that the effect size for specialist students (.23) was even lower than that for the all students sample (.32) in 1983/84. As a general development it would seem that, except for Italy and Thailand, more female students enrolled in upper secondary education, which in many countries coincided with curricular changes, thereby exposing students to more science, and hence, to more biology instruction.

Chemistry Here, the overall picture appears a to be rather different from that for biology. On average, gender differences were reduced only for the sample of all students between the two occasions (Δ-.15). But this finding is an oversimplification as a large variation is found between the countries.

Unfortunately, complete data were not available for Japan, the Netherlands, and the United States so trends could not be reported here.

Physics This subject clearly favoured male students but changes occurred between 1970/71 and 1983/84. When looking across all countries, the mean effect sizes dropped significantly between the two data points both for all students and specialist students at the terminal secondary level. Though gender differences were high in 1970/71 (.81 for all students and .61 for the specialist ones) they declined to a medium level for the specialist sample in 1983/84 (.35). Again, these findings need to be looked at for each country.

Comparative data were not obtained for Japan and the Netherlands.

TABLE 5.4 Effect sizes for gender differences in achievement in science subjects at the terminal secondary school level for all students and specialist students in 1970/71 and 1983/84 and the change over time

	Australia	England	Finland	Hungary	Italy	Japan	Netherlands	Sweden	Thailand	United States	average effect size mean	sd
All Students												
Biology												
1970/71	.20	.17	.38	.15	.10	-	.44	.47	.03	.36	.26▲	.16
1983/84	.10	.14	.35	.26	.32	.51	-	.32	.45	nd	.31▲	.14
change (Δ)	-.10	-.03	-.03	.11	.22	na	na	-.15	.42	na	.06◆	
Chemistry												
1970/71	.65	.69	.60	.15	.29	-	.80	.49	.34	.94	.55▲	.26
1983/84	.35	.39	.45	.37	.28	.62	-	.51	.26	nd	.40▲	.12
change (Δ)	-.30	-.30	-.15	.22	-.01	na	na	.02	-.08	na	-.15◆	
Physics												
1970/71	.81	1.01	1.19	.63	.74	-	.96	1.04	.29	.62	.81▲	.28
1983/84	.68	.75	.87	.57	.72	.82	-	.85	.53	.61	.71▲	.12
change (Δ)	-.03	-.26	-.32	-.06	-.02	na	na	-.19	.24	-.01	-.10◆	
Specialist Students												
Biology												
1970/71	.19	.19	.39	.35	-.09	-	.75	.81	-.15	.23	.30▲	.33
1983/84	.04	.26	.30	.24	.35	.20	-	.23	.26	.34	.25▲	.09
change (Δ)	-.15	.07	-.09	-.11	.44	na	na	-.68	.41	.11	-.05◆	
Chemistry												
1970/71	.00	.24	.32	.58	.31	-	.58	.34	.18	.10	.29▲	.20
1983/84	.43	.30	.56	.65	.06	.26	-	.04	.32	.39	.33▲	.20
change (Δ)	.43	.06	.24	.13	-.25	na	na	-.30	.14	.29	.04◆	
Physics												
1970/71	.39	.22	.96	.69	.54	-	1.25	.75	.01	.67	.61▲	.38
1983/84	.26	.06	.62	.28	.26	.16	-	.41	.60	.47	.35▲	.19
change (Δ)	-.13	-.16	-.34	-.41	-.28	na	na	-.34	.59	-.20	-.26◆	

Notes:

positive effect sizes indicate superior performance of boys
negative effect sizes indicate superior performance of girls

- Δ difference of effect sizes between 1970/71 and 1983/84
- ▲ mean effect size calculated as the sum of effect sizes across all countries divided by the number of countries for which data were obtained
- ◆ mean change of effect size between 1970/71 and 1983/84 calculated as the difference between the mean effect size of 1970/71 and the mean effect size of 1983/84
- * difference significant at the 5%-level
- \- population not tested
- na not applicable
- nd no data available

It is difficult to explain these differences in effect sizes over time at the terminal secondary school level for biology, chemistry, and physics achievement test scores. Not only are there substantial differences in retention rates between occasions and differences in participation rates for the separate fields of science, but there are also differences in the relative enrolments of male and female students in the different fields of science (see Keeves and Kotte, 1991). Different proportions of the age group enrolling in different science courses draw upon students of different ranges of ability and aptitude (see Husén, 1967), which seem likely to account, at least in part, for relative differences in effect sizes. As a consequence it is important to examine in some detail relationships associated with gender differences in enrolment and participation in science courses at the levels when the study of science and enrolment at school cease to be mandatory.

5.2.3 *Changes in Effect Sizes Between Age Levels*

At this stage it is considered helpful to examine the three science subjects separately to analyze the development of gender differences from the 10-year-old to the terminal secondary level. Although substantial variation, especially at the terminal secondary level, occurred between countries several trends can be identified and countries can be compared with each other. By looking at the general development of gender differences with respect to biology, chemistry, and physics achievement, this could shed light on the question why such differences occur and what factors they are influenced by. On the other hand, it is also necessary to examine how each country performs when compared against general trends.

Biology Figure 5.2 depicts the changes of the mean effect sizes for the 10- and 14-year-old level as well as for the subgroups of all and specialist students at the terminal secondary level.

As the figure reveals an increase of gender differences by age level it must be recognized that at the 10-year-old level the differences appear to be largely non-existent. Even at the terminal secondary level the effect sizes are only small. Nevertheless, gender differences are to be found and rise consistently during the lower and upper secondary years of schooling.

Chemistry While gender differences do not seem to have a strong effect on achievement in biology, the development of the mean effect sizes in chemistry across the three levels of schooling shows a remarkably different picture.

As can be seen in Figure 5.3, which presents the changes of mean effect sizes in chemistry, the small effect size for the 10-year-old level both in 1970/71 and 1983/84 develops differently with increasing age. Though it is reported to have only slightly increased for the 14-year-olds in 1983/84, the huge variation between specialist and all students, found in the 1970/71 sample, has been reduced markedly in 1983/84.

FIG. 5.2 *Development of mean effect size of gender differences in the biology subscore from 10-year-old to terminal secondary school level for the 1970/71 and 1983/84 samples*

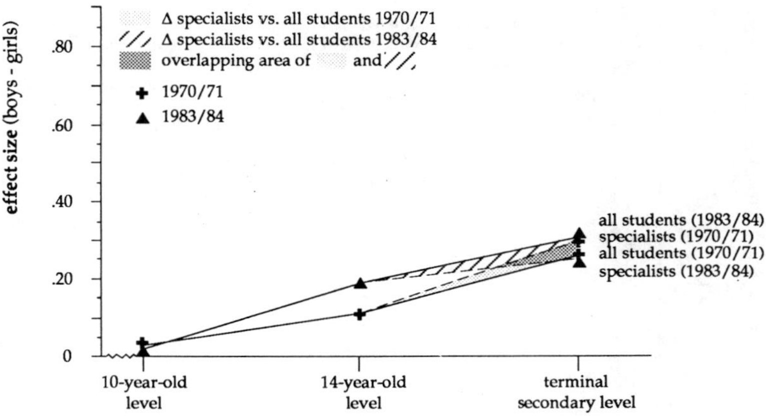

FIG. 5.3 *Development of mean effect size of gender differences in the chemistry subscore from 10-year-old to terminal secondary school level for the 1970/71 and 1983/84 samples*

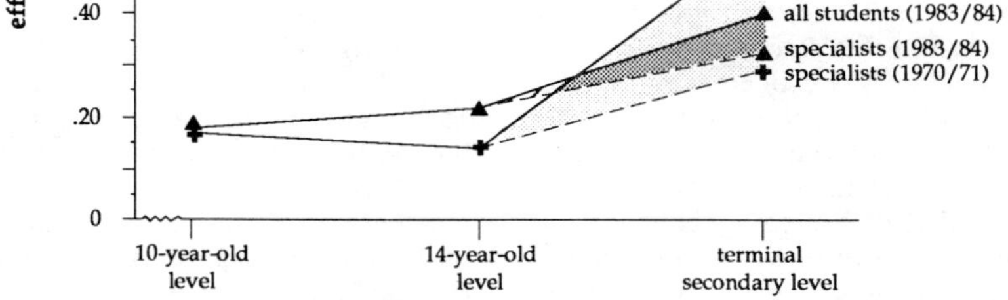

FIG. 5.4 *Development of mean effect size of gender differences in the physics subscore from 10-year-old to terminal secondary school level for the 1970/71 and 1983/84 samples*

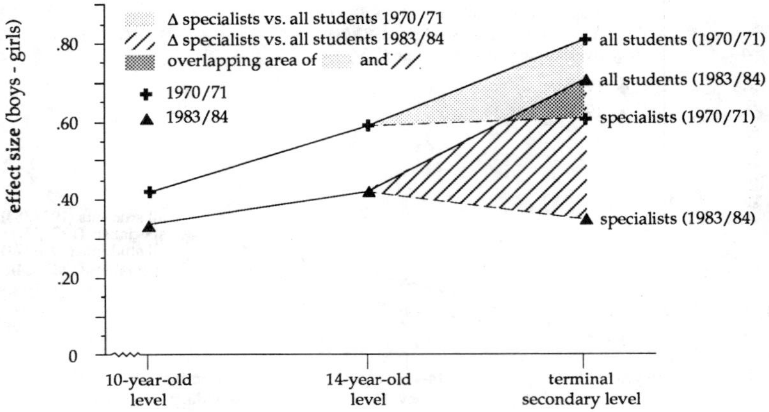

To explain this, one must remember the country-specific changes which occurred between the two data collections as described above. Compared with the mean effect sizes in biology, gender differences also seem to have increased in chemistry at the 14-year-old level in 1983/84. However, they decreased considerably during the 14-year period for the all students group which is contrary to what was found for biology. In order to understand fully the mechanisms at work which influence the growth or reduction of effect sizes with respect to achievement in chemistry each country has to be treated separately, and the performance of its students examined after taking into consideration changes in participation rates.

Physics Figure 5.4 shows the development of the mean effect sizes for the two surveys for each of the age groups in physics.

It should be noted that the marked gender differences apparent at all three levels of schooling can be seen on both occasions. Further to this, a substantial reduction in the effect size for all age groups tested is evident.

While this finding appears to be a promising indicator that gaps between boys and girls are decreasing, this impression is distorted by the fact that the variation between the specialist and all students group widened considerably in 1983/84 compared with the 1970/71 data collection. The reason for this unexpected trend may have its source in a reduced appeal of upper secondary school physics courses to boys while, at the same time, proportionally more girls were enrolled in these courses - and those girls who were specializing in physics were achieving above average, a fact supported by findings of Keeves (1991).

Physics may be still perceived by students as more difficult than biology and chemistry. Moreover, fewer girls tend to study physics. Thus, physics achievement, to a much larger extent than the other science subjects, is coupled with attitudes towards science and towards physics in particular. As a major consequence substantially higher gender differences in physics achievement are observed.

Though gender differences were found in all three science subjects these were measured using a core test and separate science subtests. This, implicitly, excludes another likely source for the manifestation of gender differences namely the gender specifity of the individual test item. As could be expected from related research, boys appear to respond differently to certain questionnaire items than girls (Erickson and Erickson, 1984; de Leeuw and Raat, 1986; Chipman, 1988; Loewen et al., 1988).

Hence, in the following attention will be given to shed some light on the aspect of item-specific differences in science achievement.

5.3 Item-Specific Differences

It has been shown that gender differences vary between countries, between levels of schooling, and between science subjects. Thus, it is safe to infer that gender differences even exist at the level of the individual items. It can be hypothesized that boys and girls score differently on different items or, in other words, the item specificity accounts for gender disparities as the content of an item may 'appeal' to boys in a different way than to girls.

However, this study does not focus on identifying possible patterns among items which effect boys' and girls' science achievement differently. While this would be a worthwhile research topic in its own right, in this study, the emphasis is placed on explaining the observed gender differences in science achievement, with science achievement as the composite of the biology, chemistry, and physics subscores.

It is considered of value to illustrate the existence of gender differences with one item as an example. The following is a chemistry item administered at the 14-year-old level core test in 1983/84.

Item No.	Content Classification	Test Location
P2M20	C33	040

Question:

Which of the following particles are gained, lost or shared during chemical changes?

A electrons furthest from the nucleus of the atom
B electrons closest to the nucleus of the atom
C electrons from the nucleus of the atom
D protons from the nucleus of the atom
E neutrons from the nucleus of the atom

The correct answer is *A*. Table 5.5 shows some item characteristics. The row *% correct* presents the percentage of students who marked the correct answer category. Likewise, the percentage of students omitting the item is given in the second row (*% omit*). The relative size of gender differences is expressed as the percentage difference between boys and girls scoring the right answer while the direction of this difference is indicated by ♂ (more male students score correctly than female students) or ♀ (more female students score correctly compared with male students). In the bottom row the ratio of the percentage gender differences to the absolute percentage of students scoring correctly is given (*relative gender difference*).

TABLE 5.5 *Selected item characteristics for a chemistry item given at the 14-year-old level, 1983/84*

Item Characteristic	Australia	England	Finland	Hungary	Italy	Japan	Netherlands	Sweden	Thailand	United States
% correct	32.1	26.6	37.0	76.0	32.3	20.3	47.0	36.1	15.4	39.4
% omit	10.8	17.8	2.0	1.8	23.9	1.5	4.2	7.7	0.5	1.7
% absolute gender difference	10.1	9.6	1.9	1.9	5.4	7.6	0.5	5.6	3.8	10.1
item easier for...	♂	♂	♂	♀	♂	♂	♂	♂	♂	♂
relative gender difference	.31	.36	.05	.03	.17	.37	.01	.16	.25	.26

Notes:
♂ boys
♀ girls

This item was a relatively difficult one for the 14-year-old students in all countries except in Hungary and the Netherlands. This is also reflected by the large percentage of omits in Australia, England, and Italy. With respect to gender differences, it was found that the percentage scores of those students choosing the correct answer varied by up to 10 percent mainly in favour of the boys. Only in Hungary did the girls do slightly better than the boys.

By calculating the ratio $\frac{\text{\% gender differences}}{\text{\% correct}}$ a crude indicator of the relative size of gender differences can be obtained because the larger the value of this ratio becomes the larger are the gender effects. In the above example, marked differences were found in Australia, England, and Japan where the ratio was >.30, as well as in Thailand and the United States which had a ratio of about .25. Contrary to this, gender differences were trivial in Finland, Hungary, and the Netherlands.

From this single example, though not being representative of all items, it can be concluded that there are considerable differences in the scores of boys and girls at the item level. These differences vary between countries. Without having presented details here it has to be stated that in each science subject there are items for which boys score a higher percentage of correct answers than their female classmates. On the other hand, there are also items for which girls achieve a higher percentage of correct answers than boys.

These item specific differences between boys and girls have to be borne in mind when talking of science, biology, chemistry, and physics. These subjects may be safely treated as separate domains of the larger field of science. Nevertheless, it must be understood that when making inferences for these subjects marked differences may occur when examining specific items within a particular field of science.

5.4 Achievement and Future Enrolment in Science Courses

The results presented in this chapter have identified gender differences on two occasions, in 1970/71 and 1983/84. It has also been shown that gender differences, when averaged across the 10 countries, changed between the two data collections. Moreover, gender differences appear to change differently in each country, sometimes even contrary to what had been observed as a general trend. Despite the fact that differences in achievement between male and female students vary across countries, levels of schooling, science subjects, and even across single items there is room for speculation as to what factors are influencing the growth or reduction of gender differences with respect to science achievement over time.

As was indicated above, societal, political, administrative, and curricular factors could change the proportion of girls enrolled in science courses in 1983/84 compared with 1970/71. Considering equal enrolment of boys and girls as being the prerequisite for equal achievement in science, Figure 5.5 shows the proportions of male to female students enrolled at the terminal secondary level in the 10 countries for biology, chemistry, physics, and non-science specialist courses in 1970/71 and 1983/84. Information is not presented for the 10- and 14-year-old level as most of the students of these age groups participated in science courses in 1983/84 if not in 1970/71. Thus, enrolment in science courses becomes a crucial factor influencing gender differences only where it is optional, namely at the terminal secondary school level.

Each of the four charts in Figure 5.5 presents the enrolment ratio for boys and girls which is calculated in such a way that the gender with the smaller N is the denominator of the ratio. As a consequence, the ratio is always larger or equal to 1.0. While a ratio of 1.0 indicates no gender differences in enrolment, any value higher than 1.0 signals a difference in enrolment which is depicted by the dots of the line graph falling into either subgroup zone (i.e. top-half zone symbolizing 'boys' and bottom-half zone symbolizing 'girls') as the larger group.

Participation in Biology Courses With the exception of Italy and Japan the enrolment ratios for boys and girls did not differ greatly from unity. This was the case on both occasions. Except for Italy, the Netherlands (in 1970/71), and the United States (in 1970/71) the enrolment ratio was slightly in favour of girls. The exceptionally high value for Italy is likely to be linked to the relatively selective nature of the biology courses at the terminal secondary school level (N=90 for 1970/71 and N=147 for 1983/84). In general, biology seems to be a subject equally attractive to boys and girls. As was seen above, the effect sizes for gender differences in biology achievement were not very pronounced either. Linking these two findings may indeed lead to the conclusion that an enrolment ratio of unity is a precondition for the absence of gender differences.

Participation in Chemistry Courses It is interesting to note that for the enrolment ratio in chemistry there are hardly any changes between occasions, except for Australia where the enrolment disproportion was reduced considerably in 1983/84. In all countries the enrolment ratio was in favour of the boys, Hungary and Thailand being the minor exceptions here. The disproportionate enrolment between the sexes was large in Australia (in 1970/71), Italy (on both occasions), Japan (no data for 1970/71), and the Netherlands (no data for 1983/84).

Chemistry courses were, in general, more commonly chosen by boys. Again this finding is in line with the reported effect size differences for chemistry achievement. As in most countries more boys tended to enrol in chemistry, on average their score was higher compared with that of girls.

Participation in Physics Courses Again, Hungary and Thailand seemed to be the exceptions as in these two countries the enrolment ratio was in favour of girls whereas in all other countries boys outnumbered girls in enrolment in physics courses at the terminal secondary level. This finding holds for both occasions. But while in Finland, Hungary, Sweden, and Thailand hardly any changes occurred between 1970/71 and 1983/84, this was not so in the other countries. In Australia and the United States the enrolment ratio decreased from 1970/71 to 1983/84 such that on the latter occasion only a minor disparity in enrolment in favour of the male students was observed. Contrary to this, there were increases in England and Italy. Japan showed an extremely disproportionate enrolment ratio favouring boys in 1983/84, but no data was available for 1970/71.

Physics, it seems, is perceived as the least attractive science subject by female students at the terminal secondary level. That this does not have to be this way is shown by the examples of Hungary and Thailand. Hungary has sought to offer a fair and equal education to both male and female students as part of its general political approach. It may, however, be speculated that in Thailand schooling is still perceived as a kind of 'luxury'. Thus boys, rather than girls, are expected to help earn money for the living of the family as soon as possible. So while boys leave school at an early age the relative proportion of girls staying on at the upper secondary school level increases. Furthermore, those girls who continue schooling until the end of the terminal secondary level may, in general, be fairly successful and therefore more likely tend to study physics.

Participation in Non-science Specialist Courses Visible changes took place between 1970/71 and 1983/84 reducing the disproportions between the sexes in enrolment at the terminal secondary school level. Hungary and Finland (for 1970/71) are the exceptions, with a high enrolment ratio for girls. In all other countries the enrolment became nearly balanced between the sexes in 1983/84. Thus, opting for a non-science course and not specializing in either science subject was slightly more attractive for girls, but the grave disproportions between the sexes which were found in 1970/71 had mainly disappeared by 1983/84.

FIG. 5.5 *Proportion of male to female students enrolled in science courses at the terminal secondary school level in 1970/71 (··○··) and 1983/84 (—●—)*

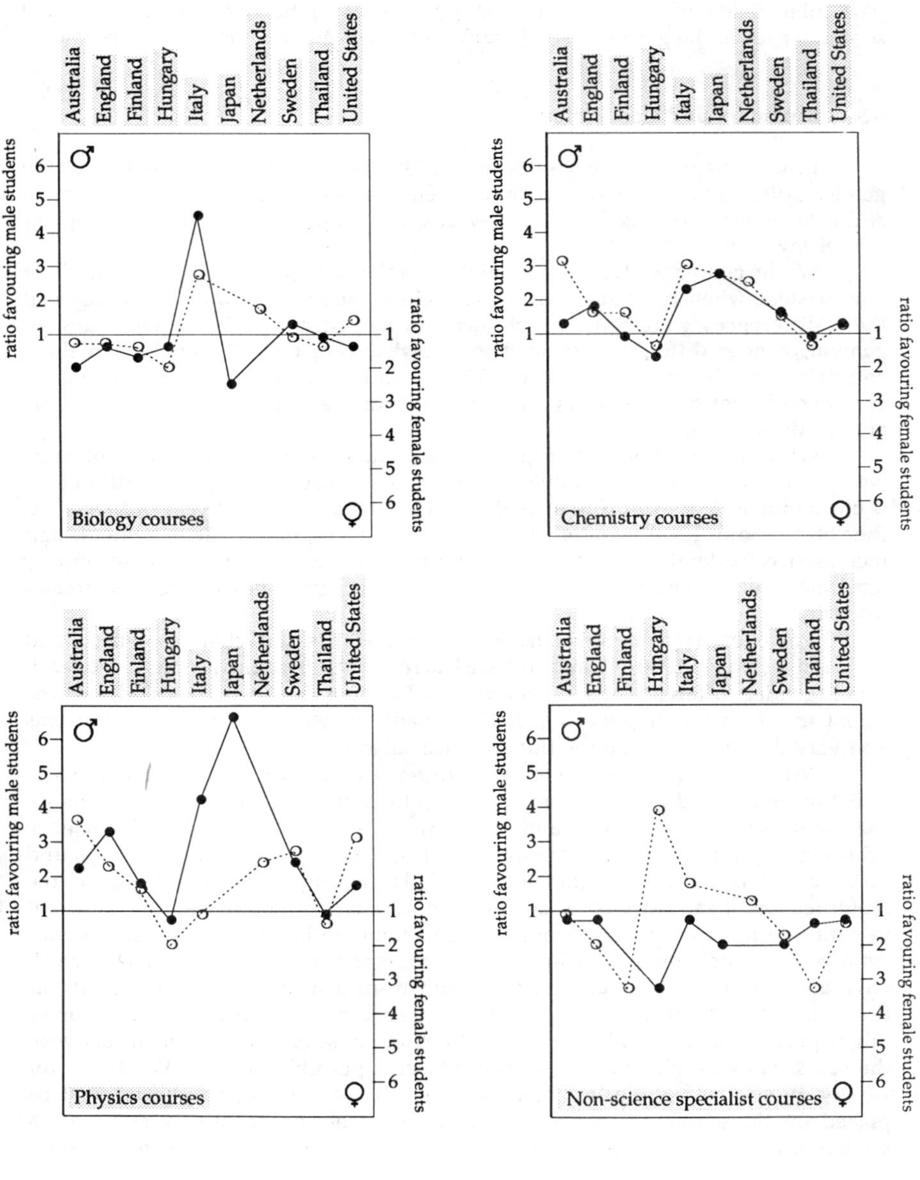

As Figure 5.5 shows, typical enrolment 'patterns' could be identified for most of the countries in the different science subjects. These 'trends' were found to be consistent with the gender differences as measured by the effect sizes and reported earlier in this chapter. This leads to the assumption that if gender differences occur, these are in fact tied to the general enrolment patterns in the particular fields of science. At least this seems to be true for the terminal secondary level, the grade level where the effects of gender are most noticeable.

5.5 Summary and Conclusions

In this chapter, evidence has been presented to show the extent to which gender differences exist and develop in science subjects with respect to enrolment and achievement starting from the 10-year-old level up to the terminal secondary school level.

While gender differences in biology achievement tended to be small or non-existent when compared with other science subjects, it was in physics that these differences were most pronounced in almost all of the 10 countries. In general, gender differences seemed to 'grow' during formal schooling in these countries in all science subjects. Thus, while science achievement varies considerably between the boys and girls, in nearly all instances female students are at a disadvantage.

With the exception of biology, which, compared with chemistry or physics, was perceived as more attractive by girls in most countries, gender differences could be identified, even if to a small degree, already at the 10-year-old level. At the 14-year-old level gender differences with respect to achievement had increased considerably in nearly all countries, while at the terminal secondary level the achievement differences between boys and girls showed an even greater disparity.

However, it is noteworthy to point towards the changes which had occurred between 1970/71 and 1983/84 across the 10 countries. As an overall trend, gender differences have been reduced during the 14-year period, but the extent to which this happened varied markedly between the countries and was also very different between the three science subjects.

With the detection of gender differences in each of the 10 countries, attention must be directed towards the identification of factors which influence the establishment, increase, and decrease of gender differences in science achievement. Without going into detail, it could not be ruled out that boys and girls scored differently at the item level. However, in concentrating on the identification of the reasons for achievement differences in science, it is sufficient to continue looking at the science subscore totals rather than individual items. As a consequence, interest is focussed on the specification of the factors which have an effect on science achievement with respect to gender. Moreover, it has been shown that the gender differences in enrolment ratio in science courses might play a vital role in influencing differences in science achievement between the sexes. Since enrolment in a science subject is perceived to be a key factor for success in science, as the findings presented above suggest, emphasis must be placed on the extraction of factors which positively or negatively stimulate a student's decision to enrol in future science courses. Thus, it is not science achievement *per se* which needs to be examined in the search for gender differences. Rather more promising would appear to be the specification of a model in which future enrolment is defined as the key outcome of science education. This implies concentrating on the group of the 14-year-olds who in the very near future in their school career have to make decisions regarding their participation in science courses at the terminal secondary level should they wish to proceed with their education. In predicting students' future enrolment in science courses it is necessary to take into account students' attitudes and values, over and above other important variables, such as home background or effort.

Chapter 6 examines the relationships between students' attitudes, expectations, and science achievement.

6. Student-Level Factors

In the previous chapter an account was given of the extent to which gender differences influenced science achievement at the time of the two data collections in 1970/71 and 1983/84. It was also shown where, when, and in which science subjects changes took place in gender differences in science achievement.

A parallel was found to exist between science achievement and the enrolment rate in science courses at the terminal secondary level where, in most instances, participation in science courses was optional. It was also shown that in subjects where there was a proportionally higher enrolment of boys, they tended to perform better than girls.

As a result, the factors influencing participation in science courses, and thus, science achievement, need to be investigated thoroughly. From related analyses (Comber and Keeves, 1973; Morgenstern, 1990; Keeves, 1991; Postlethwaite and Wiley, 1991) it can be inferred that in addition to the student's home background and aptitude, the student's attitudes, values, and motivations play an important role in determining educational and occupational expectations as well as future enrolment in science courses.

In this chapter emphasis is given to identify and evaluate the effects of student-level factors thought to influence science achievement. Further to this, the relationships among factors at work at the student level which lead to the development of a student's interest to obtain further science training, to take up a science-related occupation, or to enrol in a future science course at the terminal secondary level are also examined. Finally, the various factors influencing future enrolment in science and the relative magnitudes of their effects are specified in a set of path models and possible similarities and differences between the countries are discussed.

6.1 Results

6.1.1 Gender Differences on the Attitudinal Scales

Before discussing the effect of attitudes on future enrolment or science achievement it is necessary to investigate gender differences on the attitudinal scales. The composition, direction, and reliabilities of the attitudinal scales administered during the 1983/84 testing have already been reported in Chapter 4 (see Appendix 4.2).

Examining data from the FISS study, Comber and Keeves (1973) were able to show that attitudes had different effects depending on the level of schooling. While attitudes towards science and other school-related attitudes are developed during childhood and the years of elementary education, they are shaped considerably during secondary schooling (Ormerod and Duckworth, 1975; Schreiber, 1984; Hoffmann, 1987; Kroeger, 1990). Though this was neither new nor surprising, it was unclear from previous research to what extent gender was influential in this formative process.

TABLE 6.1 Effect sizes for gender differences on the attitudinal scales at three different levels of schooling in 1983/84

Attitudinal Scale / Level of Schooling	Australia	England	Finland	Hungary	Italy	Japan	Netherlands	Sweden	Thailand	United States	average effect size mean▲	sd
Interest in Science												
10-Year-Olds	.02	.08	.08	.11	-.10	.32	-	.08	-	.15	.09	.12
14-Year-Olds	.24	.31	.09	.15	nc	.06	.36	.39	.12	.26	.22	.12
Terminal Secondary	.12	.10	.16	.24	.14	.40	-	.24	.10	.21	.19	.10
Ease of Learning Science												
10-Year-Olds	.06	.00	.08	.11	-.04	.09	-	nc	-	.16	.07	.07
14-Year-Olds	.27	.27	.24	.08	.09	.51	.40	.46	.12	.25	.27	.15
Terminal Secondary	.12	.22	.38	.40	.14	.30	-	.00	.20	.16	.21	.13
School and School Learning												
10-Year-Olds	-.36	-.37	-.45	-.37	-.30	-.34	-	-.36	-	-.40	-.37	.04
14-Year-Olds	-.23	-.22	-.41	-.45	-.33	-.13	-.14	-.12	-.31	-.40	-.27	.12
Terminal Secondary	-.22	-.39	-.32	-.18	-.26	-.14	-	-.08	-.30	-.30	-.24	.10
Career Interest in Science												
14-Year-Olds	.25	.21	.06	.06	.17	.40	.24	.18	.14	.10	.18	.10
Terminal Secondary	.34	.32	.36	.18	.43	.51	-	.49	.12	.31	.34	.13
Beneficial Aspects of Science												
14-Year-Olds	.11	.30	-.03	-.10	.06	.25	.22	.22	.14	.05	.12	.13
Terminal Secondary	.14	.16	.18	.10	.08	.45	-	.43	.10	.20	.20	.14
Non-Harmful Aspects of Science												
14-Year-Olds	-.20	-.03	-.06	-.25	-.13	-.10	-.06	.06	-.12	-.18	-.11	.09
Terminal Secondary	-.02	-.06	.12	.10	.02	.02	-	.34	-.04	.07	.06	.12

Notes:
positive effect sizes indicate that boys hold a more favourable attitude
negative effect sizes indicate that girls hold a more favourable attitude
▲ mean effect size calculated as the sum of effect sizes across all countries divided by the number of countries for which data were obtained
\- population level not tested
nc data not collected or scale unsatisfactory

The 1983/84 data were used to calculate the effect sizes for gender differences on the six attitudinal scales measured across the three age groups tested. There were three distinctly different attitudinal scales *(Interest in Science, Ease of Learning Science, School and School Learning)* as well as three scales aimed at assessing students' values with respect to science *(Career Interest in Science, Beneficial Aspects of Science, Non-Harmful Aspects of Science)*. Each scale was administered to at least two of the three population levels. In addition, mean effect sizes and standard deviations introduced in Section 5.1 have been calculated as a basis for discussion. Table 6.1 presents these results.

TABLE 6.2 *Means and standard deviations for selected attitudinal scales administered at the 14-year-old level, 1983/84*

Attitudinal Scale No. of Items (score range/midpoint)	Statistic	Australia	England	Finland	Hungary	Italy	Japan	Netherlands	Sweden	Thailand	United States	grand mean
Interest in Science												
4 items	mean	8.53	8.99	8.50	10.22	n c	7.65	7.87	8.54	9.71	8.49	8.72
(4-12 / 8.00)	sd	2.38	2.27	2.89	1.64	n c	2.30	2.16	2.21	1.60	2.55	2.22
Ease of Learning Science												
4 items	mean	8.00	7.98	8.51	7.69	8.07	7.29	7.72	7.79	7.04	8.07	7.82
(4-12 / 8.00)	sd	2.16	2.07	1.95	2.19	2.13	2.24	2.22	2.10	2.00	2.43	2.15
School and School Learning												
8 items	mean	16.14	16.52	14.26	16.07	16.71	16.70	15.00	14.83	19.89	16.37	16.25
(8-24 / 16.00)	sd	3.95	3.94	3.76	3.34	3.87	3.60	3.92	3.93	2.74	3.88	3.69
Career Interest in Science												
7 items	mean	15.09	15.93	15.05	17.47	17.21	13.82	14.52	15.80	18.40	16.11	15.94
(7-21 / 14.00)	sd	3.15	3.00	2.74	2.16	2.45	3.13	3.15	2.47	2.03	3.08	2.74
Beneficial Aspects of Science												
7 items	mean	16.60	16.79	16.62	18.47	17.67	14.45	15.24	15.97	18.18	16.98	16.70
(7-21 / 14.00)	sd	2.81	2.77	2.57	1.83	2.57	2.72	2.64	2.39	2.05	2.73	2.51
Non-Harmful Aspects of Science												
6 items	mean	12.84	13.08	13.46	13.70	12.76	12.10	13.11	11.22	13.03	13.67	12.90
(6-18 / 12.00)	sd	2.58	2.65	2.53	2.50	3.02	2.75	2.75	2.53	2.65	2.57	2.65

Notes:
nc data not collected or scale unsatisfactory

TABLE 6.3 *Product-moment correlations for means of selected attitudinal scales and science achievement (core test and two rotated test forms) across 10 countries for the 14-year-old level, 1983/84*

Attitudinal Scale	Science Achievement	Interest in Science	Ease of Learning Science	School and School Learning	Career Interest in Science	Beneficial Aspects of Science
Interest in Science	.06					
Ease of Learning Science	-.25	-.18				
School and School Learning	-.19	.44	-.71*			
Career Interest in Science	-.28	.92**	-.24	.62		
Beneficial Aspects of Science	-.18	.95**	.05	.44	.92**	
Non-Harmful Aspects of Science	.31	.40	.28	.11	.28	.52

Notes:
* significant at .05 level (2-tailed)
** significant at .01 level (2-tailed)

To provide a better understanding of the results in Table 6.1, it was considered of value to present the means and standard deviations for each of the 10 countries for each of the six attitudinal scales. Also reported are the grand means, its standard deviations, the number of items, the score range, and midpoint of each scale. Table 6.2 shows these results for the 14-year-old students.

Further to this, Table 6.3 gives the correlations between the six attitudinal scales as well as their correlations with the country mean science achievement test-scores (core test plus two rotated test forms). Again, Table 6.3 is based on the data obtained at the 14-year-old level. The product-moment correlations are calculated using country grand mean values for each of the scales.

It is of interest to discuss each scale separately before providing a summary for all attitudinal measures.

Interest in Science The items of this attitude scale measured the interest students had in science. A higher scale score indicated that students expressed a more positive interest in science.

Except for the 10-year-old students in Italy, being a boy rather than a girl led to the expression of a greater interest in science taught in schools and classrooms across all countries and at each level of schooling. In the case of Italy, the girls in the sample tested were more interested in science at the 10-year-old level than were the boys. No ready explanation for this can be given although it might be speculated that this finding is partly resulting from sampling bias.

It can be clearly seen that across all countries a sharp increase in gender differences takes place between the 10- and the 14-year-old level (Δ.13) which more than doubles the effect size. In contrast, Japan showed a sizeable drop within these two age levels (Δ-.28). It should also be noted, although the data are not presented here, that there is a marked decline in the average level of interest between these two age levels (see Keeves, 1991; Kida, 1991). It is possible that the Japanese students at the 14-year-old level were extremely reluctant to express their attitudes and interests on the attitudinal items to which they were asked to respond. Hence, a vast majority of Japanese students chose the *uncertain*-category within the attitudinal scales. The reason for this cannot be readily explained.

Another interesting finding is the slight overall decrease in gender differences at the terminal secondary school level compared with the 14-year-old level (Δ-.03). It would seem that two different developments might have occurred among the group of countries. While in the majority of countries the effect size dropped, it increased in Finland (Δ.07) and Hungary (Δ.09). This may be associated with the relatively high proportions of girls proceeding to the terminal secondary school level in these two countries. It should also be stressed that Hungary showed by far the highest level of interest in science ($\overline{x}$=10.22). This could also account for relatively small gender differences.

The overall decrease in the effect size at the terminal secondary level might have, in general, been caused by a declining interest in science among boys. With the introduction of biology, chemistry, and physics boys might more rapidly develop an interest in these subjects than girls. But throughout formal schooling these subjects might also have lost some of their initial attractiveness as students had to work harder and do fewer experiments. At the terminal secondary level the content in science becomes much more theoretically-oriented than in elementary or lower secondary school (Rosier and Keeves, 1991). This, in turn, could have a more pronounced effect on boys limiting their interest and enthusiasm.

Another possible explanation is the differentiation of interests across the various science subjects. Interest in science may be too crude a measure to use at the terminal secondary level. To overcome this problem, three further scales were employed to measure interest in biology, chemistry, and physics separately. The results are not reported here as these scales were only administered at the terminal secondary level and cannot be compared with those at the 14-year-old level (for details see Keeves, 1991).

As seen in Table 6.3, interest in science did not correlate with science achievement (r=.06) across the 10 countries. This is contrary to what was expected.

Ease of Learning Science In the items for this attitude scale the students were asked whether they perceived learning science as easy or difficult. A higher scale score indicated that students perceived learning science as easy.

On the whole, the effect sizes for this scale, also administered at all three levels of schooling, were similar to those found with the Interest in Science scale. But a few exceptions should be noted.

Finland (Δ.14) and Hungary (Δ.32) showed large gains in effect sizes between the 14-year-old and terminal secondary level which meant that boys perceived science learning to be considerably easier than their female fellow students did at the higher level of schooling. While at the terminal secondary level the magnitudes of the effect sizes were small for most other countries, they were medium for Finland (.38) and Hungary (.40). This result is perhaps not surprising as both countries had high proportions of female students enrolled in upper secondary forms and the study of science was mandatory for all students. Thus, ease of science learning might indeed have been perceived differently between boys and girls despite the fact that the difference in achievement scores between the sexes were small in Finland and Hungary (see Postlethwaite and Wiley, 1991, pp. 75-77).

It should be added that at the 14-year-old level, on average, learning science was perceived as easiest by the Finnish students ($\bar{x}$=8.51) and as most difficult by the students from Thailand ($\bar{x}$=7.04). Although this finding was in line with the overall poor achievement of Thai students (see Postlethwaite and Wiley, 1991, pp. 75-77) there was no significant correlation between ease of learning science and science achievement. Rather, the correlation calculated across countries in the sample tended to be negative (r=-.25).

Overall, learning science was rated easier by boys throughout all countries and populations tested (with, again, Italy's 10-year-olds deviating from this trend) but it must be remembered that science was not generally considered to be an easy-to-learn subject by students at the 14-year-old level.

School and School Learning The items of this scale were meant to find out to which degree students liked attending school. A higher scale score indicated that students held positive attitudes towards school and school learning.

Here, without exception, female students very clearly showed a much higher affiliation towards school and learning. Across all countries and at each of the levels tested girls enjoyed being at school more strongly than their male classmates. However, the effect sizes dropped from the 10-year-old level (-.37) to the 14-year-old level (-.27), with a further marginal overall decrease at the terminal secondary level (-.24).

Compared with other countries, girls in Finland and the United States enjoyed school more than boys. It should be kept in mind, though, that these figures do not say anything about the absolute perceived attitude towards school and school learning. Rather, the measure of effect size describes the relative difference between the genders with respect towards attitudes. Indeed, attitudes to school and school learning varied widely between countries and between students within countries at the 14-year-old level as can be seen from the figures in Table 6.2 where it is shown that attitudes in Finland are extremely low and in the United States relatively high.

While Finland ($\bar{x}$=14.26), Sweden ($\bar{x}$=14.83), and the Netherlands ($\bar{x}$=15.00) expressed a negative attitude towards school, the scale's midpoint was 16.00, Thai students liked school a great deal ($\bar{x}$=19.89). Further to this, Thailand showed the smallest variation in students' attitudes towards school and school learning. These results must be seen in relation to the economic situation in the countries with Thailand being the only developing country of the 10 under survey. Here, schooling was still understood to be a privilege as not every 14-year-old had the chance to attend school. Students obviously perceived school and school learning as a great opportunity for their future careers which, in turn, might have been the result of the social desirability. This seemed especially true for a career interest in science (see below).

On the contrary, students from the two Nordic countries and Netherlands did not express favourable attitudes towards school. Their opinion of school and school learning would appear to have been influenced by economic needs to a much lesser extent. In these three countries, the GNP, per capita income, and social security were well above average. Students, in general, enjoyed all amenities. Economically, they could 'afford' to dislike school. It could also be suggested, that the freedom allowed to the students within their school system was high. These students probably enjoyed a considerable amount of personal and legal freedom in school in comparison with other countries. In turn, this might have given rise to a more critical view about their school environment and school learning.

This seemed to be especially true for Sweden, which was the country, in which the gender differences towards school and school learning were reduced most during formal schooling (Δ.28). This is possibly an indicator of the liberality and equality of the Swedish school system.

Again, school and school learning was not significantly correlated with science achievement across all countries (r=.-19).

Career Interest in Science Students were asked to respond to a set of items showing their interest to pursue a career in science after leaving school. A higher scale score indicated that students expressed an increased interest to resume a career in science. The items of this scale were only administered at the 14-year-old and terminal secondary level as it was contended that at the age of 10 students' career interests would not be firmly developed.

Except for Japan ($\bar{x}$=13.82), students of all countries showed an interest in a future scientific career (grand mean=15.94). The Japanese results deserve attention as this was completely contrary to what could be expected of students who, on the international tests, performed very well in science (see Walberg, 1989; Postlethwaite and Wiley, 1991; Walberg, 1991). It, therefore, could be speculated that the strong pressure put on students at the 14-year-old level to engage in science develops a personal antipathy in some students. The pressure would appear to be simply too much for some students and many would rather opt for a different career. But as other results show (see Keeves and Kotte, 1991), it might have been highly suitable for others because many students who leave upper secondary school take up a science-related career. It should be noted, too, that the rejection to engage in a science career was particularly strong for female Japanese students as seen by the large effect sizes.

In general, the effect sizes were medium to large and a sharp increase between the two levels of schooling could be observed (Δ.16). Across all countries, the boys showed a much higher interest in undertaking a science-related career. At the terminal secondary level, this trend was found to be somewhat less strongly expressed in Hungary (.18) and Thailand (.12). In Hungary, as shown in Chapter 5, fairly equal proportions of male and female students were enrolled in science in 1983/84, and equal opportunities would appear to exist for a science-related career. In Thailand, the proportion of students entering science classes in upper secondary school was comparatively small and the proportions of boys and girls was balanced. Hence, those girls who did proceed to science at this grade level probably showed a realistic career interest in science.

Career interest was negatively correlated with science achievement across countries though only to a moderate degree (r=-.28).

Beneficial Aspects of Science This scale was measuring students' values about the beneficial aspects of science. A higher scale score indicated that students held more positive values about the beneficial aspects of science.

When asked about their views on the usefulness of science, male students at the 14-year-old and terminal secondary level expressed a stronger acceptance of scientific benefits than the girls. This gap widened towards the end of schooling.

However, there were a few exceptions. Gender disparities were hardly found in Italy. In Finland and Hungary, at the 14-year-old level slightly more girls recognized the benefits of science than did boys. Moreover, gender differences nearly doubled towards the end of schooling with high values in Japan (.45) and Sweden (.43); in the United States the effect size also increased but was less strong (.20). The Swedish case deserves some consideration as, in general, attitudinal differences between the sexes seemed to decline during schooling. Beneficial and, as will be seen, harmful aspects were an exception. At this stage, the reason can only be speculative. It might be argued that attitudes show a different pattern of individual affective development than values. Career Interest, Beneficial and Harmful Aspects were all measures of students' values and, as will be seen later in this chapter, are combined into one construct of *Science Values* while interest in science, the ease of learning science, as well as the enjoyment of school are more specific attitudes with respect to what occurs in schools and classrooms.

Students in all 10 countries, on average, acknowledged strongly the benefits of science (grand mean=16.70). Very surprisingly, the Japanese 14-year-olds, in particular the girls, did not share the widespread positive view on the beneficial aspects of science. In stark contrast, Hungarian ($\bar{x}$=18.47) and Thai students ($\bar{x}$=18.18) overwhelmingly 'accepted' the benefits of science.

Nevertheless, there seemed to be no correlation between beneficial aspects and achievement in science. This is another indicator that attitudes and values may be distinctly different from achievement.

Non-Harmful Aspects of Science The items of this scale have to be seen in some kind of contrast to the items of the previous scale. Though the items were originally worded negatively (see Appendix 4.2), they were recoded in the process of data-cleaning. Thus, a higher scale score indicated that students held less negative values about the harmful aspects of science.

This scale was shown in previous analyses to be measuring a different psychological trait than did the Beneficial Aspects of Science scale (Noonan, 1982; Keys and Matsubara, 1986). Hence, students' responses to the set of items of this scale were, not unexpectedly, different from those for the beneficial aspects of science.

At the 14-year-old level, girls were more likely to acknowledge non-harmful aspects, the only exception here being Sweden (+.06). These gender differences disappeared largely at the terminal secondary level with Finland and Hungary reporting only small positive effect sizes. However, Swedish students rated the harmful aspects of science remarkably differently. Quite markedly, girls judged science to cause harm (+.34). This, indeed, is a very interesting observation. It may be that because girls in Sweden had a strong exposure to science, they became more critical throughout their formal education, thus rejecting science as being harmful more readily than boys did.

This critical stance was underlined by the widespread notion that science was harmful which Swedish youth expressed at the 14-year-old level ($\bar{x}$=11.22). Sweden turned out to be the only country taking this critical view. The pollution of the Baltic Sea which constantly hits the Swedish coastline, the general anti-nuclear policy imposed in the aftermath of the Three-Mile Island accident (testing occurred before the Chernobyl disaster), and an overall ecological awareness certainly add to Swedish students being very sensitive and sceptical about the non-harmfulness of science.

But in general, students at the 14-year-old level may have had greater difficulties responding to the items of this scale other than neutrally. This, in part, might have stemmed from the complexity of the attitudinal items as being more difficult to understand - harder than the ones regarding the beneficial aspects. It should be noted that all items of this scale were worded negatively and it may be doubted whether all 14-year-olds were able to judge whether *'Scientific inventions have made the world too complex'* as one of the items was phrased (item no. P2ATT17, see Appendix 4.2).

Thus, it does not come as a surprise that, across all countries, students' attitudes towards the non-harmfulness of science correlated slightly positive with achievement which is in contrast to the other attitudinal scales. This could be understood in terms of the relative cognitive complexity of the items. It would be the better-achieving students who were able to answer the items. The less knowledgeable opted for the *uncertain*-category.

Relationships among the Attitudinal Scales As the data in Table 6.1 revealed, consistent and sizeable gender differences were found across all six scales presented. The relatively low standard deviations for the mean effect sizes calculated support the finding that, in general, no major differences were found between the countries. This would imply a rather congruent or parallel development and manifestation of attitudes and values towards science across the countries. Further to this, it is probable that gender differences in values (career interests, beneficial aspects) as distinct from attitudes (interest in science, ease of learning science) were less susceptible to teaching and school-related factors. Consequently, it can be expected that the student's primary socialization affected by the socio-economic status might well account for most of the gender differences exhibited with respect to values towards science.

Except for the School and School Learning scale, where students were asked how they liked school and learning in general, it was the boys who voiced a stronger acceptance of science. Boys were more interested in science, found learning science-related tasks easier, showed a more pronounced interest in starting a career in science, and rated science to be more beneficial than their female classmates. On the contrary, girls favoured school and learning. Once again, it should be noted that the latter scale was measuring a student's general attitude towards school - not particularly towards science instruction.

It would therefore appear justified to conclude that, on average, boys had a stronger affinity towards science. But if this was so, and as at the same time considerable gender differences were found in science achievement favouring the boys (as reported in Chapter 5), it becomes necessary to look at the linkage between science attitudes and values and science achievement.

Using countries' grand mean values from the 14-year-old level, product-moment correlation coefficients have been calculated for the effect sizes of gender differences on attitudes and a country's mean science achievement score (core test plus two rotated test forms). Table 6.4 exhibits these correlations.

TABLE 6.4 Product-moment correlations for effect sizes of gender differences between selected attitudinal scales and science achievement (core test and two rotated test forms) across 10 countries for the 14-year-old level , 1983/84

Attitudinal Scale	Science Achievement	Interest in Science	Ease of Learning Science	School and School Learning	Career Interest in Science	Beneficial Aspects of Science
Interest in Science	-.42					
Ease of Learning Science	.16	.31				
School and School Learning	-.14	.45	.82**			
Career Interest in Science	.08	.02	.70*	.82**		
Beneficial Aspects of Science	-.30	.43	.68*	.89**	.75*	
Non-Harmful Aspects of Science	-.45	.45	.60	.63	.22	.65*

Notes:
* significant at .05 level (2-tailed) ** significant at .01 level (2-tailed)

As was mentioned, the countries mean effect sizes for gender differences were correlated with each other and also with a country's mean science achievement score (calculated on the basis of the core test plus two rotated test forms). Two examples should illustrate how to 'read' these correlations.

Interest in Science
↕
Science Achievement

Across all 10 countries effect sizes for gender differences on interest in science were positive (i.e., boys had a higher interest in science) at the 14-year-old level (see Table 6.1). Thus, the negative coefficient of r=-.42 indicates that in those countries where gender differences on interest in science were smaller, these countries tended to achieve higher. However, this correlation was not found to be significant.

School and School Learning
↕
Career Interest in Science

The results in Table 6.1 showed that for all 10 countries girls had a more favourable attitude towards school and school learning while throughout all countries boys expressed a more pronounced interest than girls to resume a career in science. Thus, the correlation of r=.82, found to be significant at the one percent level, indicates that when gender differences were large on either of the two scales gender differences tended to be also large on the other, or vice versa, that countries with small gender differences on either scale showed also small gender differences on the other. As girls held more favourable attitudes to school and school learning, it could also be concluded that, in general across all 10 countries, their career interest was lower than that of boys. In contrast, as boys expressed a higher interest in a science-related career, boys' attitude towards school and school learning was, in general, lower than that of girls.

As can be seen, there seemed to be a significant, positive interrelationship of gender differences between ease of learning science, school and school learning, career interest, and beneficial aspects. There was also a slightly weaker relationship between gender differences on these four scales with those of the Non-Harmful Aspects of Science scale.

Obviously, across countries boys and girls each showed a relative similarity in answering items on these scales 'gender identically' (see also Maccoby and Jacklin, 1974; Fennema and Sherman, 1978; Cahill, 1983; Duncan, 1989). In other words, it looks as if boys' attitudes were distinctively different from those of girls. With respect to the positive signs of the relationships, it can be noted that the more pronounced gender differences were on ease of science, career interest, and beneficial (non-harmful) aspects the less negative they were on the general attitude towards school (the only attitude which, throughout all countries, was favoured by female students).

On the contrary, gender differences on interest in science did not correlate significantly with gender differences on any other attitudinal scale. This was surprising because the overall means showed significant positive correlations of interest in science with career interest (r=.92) and beneficial aspects of science (r=.95). Furthermore, gender differences on interest in science showed a moderate correlation with those on the School and School Learning scale (r=.45), a signal that the Interest in Science scale was probably a too classroom-specific measure to exhibit a similar gender identity in the answering patterns as were found with the Ease of Learning Science, School and School Learning, Beneficial and Non-Harmful Aspects of Science scales. This is supported by the fact that the scale only comprised four items which were specifically related to the classroom and the teaching of science experienced by the students.

Another interesting finding was the significant positive relationship between career interest and beneficial aspects. Both, gender differences (r=.75) and the overall means (r=.92) correlated significantly across countries on these two scales. This, in turn, could be attributed to the fact that both scales had a quite high grand mean while, at the same time, their average gender differences were small. It seemed that girls and boys alike shared a career interest in science and had a positive view on scientific benefits. The gender identity was less different than that observed with the Ease of Science and School and School Learning scale and hence, a gender specific answering pattern less common.

However, the results given in Tables 6.3 and 6.4 should be interpreted cautiously. As the correlations were from grand mean values at the country level, some aggregation effects might have influenced the results. On the other hand, those correlation coefficients even to be found significant at this highly aggregated level do indeed appear to reflect relationships among attitudinal scales and gender differences on attitudes across all 10 countries. This would not only be a further indication of the validity and meaningfulness of the attitudinal measures but supports the hypothesis that there exist distinct 'gender typical' attitudinal answering patterns for boys and girls in response to the underlying or latent attitudinal differences.

Relationships with Science Achievement Finally, consideration should be given to the correlations of attitudinal mean values and gender differences with science achievement.

In Table 6.4, when all the effect sizes across countries are positive then a positive correlation between effect sizes and achievement indicated that boys scored better than girls and negative signs indicated that girls scored better than boys. As Table 6.3 compared pooled-within country means on the attitudinal scales and science achievement of all students, the respective positive coefficients should be read as the stronger an attitude was voiced the higher the science achievement score. Conversely, a negative sign signalled that positively expressed attitudes (i.e. choosing the statement *'agree'*) paralleled a low science score respectively students with a high science test score often disagreed with the attitudinal statements given.

A first observation is, that at the between-country level none of the attitudinal scales showed a significant correlation with science achievement. Rather, interest in science (r=-.42) and non-harmful aspects of science (r=-.45) tended to correlate negatively with achievement. However, it must be borne in mind that these are results found at the between-country level. It may well be that different interrelationships would exist at the between-school and between-students level. Here, the reader is referred to the results presented in Chapters 7 and 8.

Nevertheless, gender differences on the Interest in Science and Non-Harmful Aspects of Science scales tended to correlate negatively with science achievement. But these relationships were not found to be of significant order. Apart from this, none of the attitudinal gender differences corresponded noticeably to science achievement differences across countries.

The overall picture was not much different with respect to the correlations of attitudinal means with science achievement. Again, neither of the correlations was significant even on the 5 percent level. Despite this, some findings need to be discussed.

Interest in science differences did not seem to correspond to science achievement differences at all. This result is, at first glance, surprising because the negative coefficient of r=-.42 indicates that although students expressed an interest in scientific matters their achievement in science was comparatively poor. This obvious mismatch may be explained in different ways. It might well be that the form of the attitudinal items which only allowed students to answer *'agree'*, *'uncertain'*, or *'disagree'* was too crude to pick up subtle attitudinal differences. Another explanation could be that the instruction provided was, in many cases, insufficient to turn a student's present interest into scientific knowledge and, finally, higher achievement.

Of all attitudinal means only those on the Non-Harmful Aspects of Science scale correlated positively, if somewhat low, with science achievement (r=+.31). While interest in science did not correlate at all, the other four scales showed low negative correlations. The positive correlation of non-harmful aspects with science achievement seemed to underline as was mentioned beforehand, that the items of this scale were difficult to comprehend and, most likely, they could only be sufficiently understood and answered by better achieving students. This does not necessarily imply that the recognition of non-harmful effects of science was causally 'linked' to science achievement.

While educational and societal factors obviously influenced attitudes towards science differently for boys and girls, and while the observed gender differences in many countries tended to be largest at the 14-year-old level there appeared to be only a very minor relationship between attitudes towards science and science achievement when using grand means from countries. However, Keeves (1991) has shown that the strength of this relationship generally increased when looking at the between-student within-school level of each country.

The gender specific patterns in attitudes and values towards science were shown by the significant intercorrelations of most attitudinal scales administered at the 14-year-old level. However, it appears justified to conclude that in several countries there were distinct gender differences in science attitudes not related to science achievement.

Yet, the above findings suggest that consideration must be given to how gender might influence science achievement. This, in turn, can be investigated using multivariate analytical procedures.

It should also be borne in mind that, for subsequent multivariate analyses, there was strong evidence of the need to split the five science-oriented attitudinal scales into two separate constructs, which, in the following, are differentiated as *Science Attitudes* and *Science Values*. The former are more proximal to the science classroom, and are more influenced by the science teacher than the latter which are more distant and more likely to be affected by home and societal factors.

Science Attitudes comprised...	*Interest in Science* and *Ease of Learning Science.*
Science Values comprised...	*Career Interest in Science* *Beneficial Aspects of Science* and *Non-Harmful Aspects of Science.*

The School and School Learning scale was considered inappropriate for combining it with the science-related scales as it measured attitudes towards school rather than science. It should also be borne in mind that girls in all ten countries and at each population level tended to have a considerably higher attitude towards school and school learning though their attitude towards science and science learning was different. It was, therefore, decided to drop the School and School Learning scale from subsequent analyses.

6.1.2 Gender Differences on Effort Scales

From previous research on gender differences in science achievement it was inferred that the student's motivation, or effort, could play an important role in influencing achievement. Hence, two separate sets of items were employed to assess the student's homework and classroom efforts with respect to science and were administered along with the attitudinal items at the 14-year-old and terminal secondary levels.

Table 6.5 shows the effect sizes for gender differences on these two scales. Unfortunately, several countries did not collect these optional data, so the picture across countries was rather incomplete and the average effect size means and standard deviations less readily interpretable.

Despite these shortcomings, in those countries where data were obtained student's motivation in school and at home obviously hardly varied between the sexes. Across all countries, on average, both boys and girls alike expressed their strong efforts undertaken in school and at home with respect to science (for further details see Keeves, 1991, or Postlethwaite and Wiley, 1991).

Nevertheless, it should be noted that with the exception of the Classroom Effort scale in Japan and Sweden at the 14-year-old level, it was the girls rather than the boys who indicated greater efforts both at home and in the classroom. In Japan the boys' motivations in the classroom were higher than those of their female classmates if only to a slight extent.

TABLE 6.5 *Effect sizes for gender differences on homework and classroom effort at three different levels of schooling in 1983/84*

Effort Scale / Level of Schooling	Australia	England	Finland	Hungary	Italy	Japan	Netherlands	Sweden	Thailand	United States	average effect size mean▲	sd
Homework Effort												
14-Year-Olds	nd	nd	nd	-.14	-.08	nd	-.04	-.10	-.10	-.08	-.09	.03
Terminal Secondary	nd	nd	nd	-.08	-.10	nd	-	nd	-.10	-.11	-.10	.01
Classroom Effort												
14-Year-Olds	nd	nd	nd	-.12	-.06	.07	.00	.01	-.10	-.07	-.04	.07
Terminal Secondary	nd	nd	nd	-.10	-.09	.03	-	nd	-.12	-.11	-.08	.06

Notes:
positive effect sizes indicate that boys hold a more favourable attitude
negative effect sizes indicate that girls hold a more favourable attitude
▲ mean effect size calculated as the sum of effect sizes across all countries divided by the number of countries for which data were obtained
\- population level not tested
nd data not collected

Home- and classroom-related efforts could well contribute significantly to students' achievement in science. Therefore, it was considered important to incorporate a construct labelled 'Effort' comprising the scales of Homework and Classroom Effort as component variables. As a third manifest variable, the number of hours spent weekly on homework (*Hours Homework*), was also included because its strong predictive power had been found from many previous studies and it was seen to be related to the efforts and motivations in learning science and to achieve well at school.

6.2 Towards the Design of a Multivariate Model

6.2.1 A Path Model for Science Achievement

From the above results it was concluded that gender had a strong impact on students' attitudes towards science. In fact, science-related attitudes and values showed gender-typical patterns. Nevertheless, at the between-country level no significant correlations of either science attitudes or science values with science achievement emerged. But since gender differences, as was shown in Chapter 5, were observed with respect to science achievement it was important to examine how these were caused, using a path model for science achievement.

Following preliminary path analyses of the SISS data carried out by Keeves (1991), Keeves and Kotte (1991), and Postlethwaite and Wiley (1991), a path model for gender differences in science with science achievement as outcome measures was developed. In setting up this general model the findings of the preliminary PLS analyses, where appropriate, were incorporated. This became especially important when deciding about the question whether separate models should be employed for boys and girls as suggested by Duncan (1989) who examined science achievement and attitude data from Botswana (see also discussion in Chapter 2). As Keeves and Kotte (1991) have shown, using simultaneous LISREL analyses of the SISS data, similar processes seemed to operate for boys and girls influencing achievement in science as well as in the effects attitudes have on achievement. It was concluded to conduct further PLS analyses with Gender as a separate construct rather than operating on simultaneous LISREL modelling techniques.

As was pointed out, the set of manifest variables (MVs) available for this study was due to various reasons limited to those described at length in Section 4.1 (see also Appendices 4.1 and 4.2). Implications for the design of the basic path model for gender differences in science achievement also came from the review of research (see Chapter 2).

Figure 6.1 shows the fully recursive path model for the between-student within-country level of analysis. As can be seen, all theoretically possible paths are specified. This reflected the initial setup when starting to examine the relationships among the constructs predicting science achievement using PLS. Stepwise, as described in Section 4.3.1, insignificant paths were removed during the analyses. For simplicity, manifest variables are omitted in Figure 6.1.

From preliminary analyses and the data available it seemed advisable to select a set of 20 MVs to be combined into a fully recursive path model involving six constructs, or LVs, assumed to explain science achievement. These six constructs, highlighted by a shaded background pattern for easier recognition in the following tables, were:

HomeBack	*Home background of student*
Aptitude	*Student's aptitude*
Gender	*Sex of student*
SciValue	*Science values of student*
Effort	*Effort of student*
SciAtt	*Science attitudes of student*

The outcome measure,

Achieve	*Science achievement,*

was a composite of the biology, chemistry, and physics test scores. The formation of the constructs by MVs has already been illustrated in Section 4.2. Preliminary analyses not reported here suggested to operate on a combined score for science achievement rather than to run separate analyses for the three science subjects (Keeves, 1991; Keeves and Kotte, 1991; see also the discussion in the previous chapter).

In seeking to predict science achievement, two exogeneous LVs, namely HomeBack and Gender are to be found on the far left side of the model. These constructs were to be exogeneous because a student's sex and socio-economic status could be seen as being inalterable and thus, could not be explained causally by any other variable for which data had been obtained.

FIG. 6.1 *Between-students within-country path model (inner model) having science achievement as an outcome measure for the 14-year-old level, 1983/84*

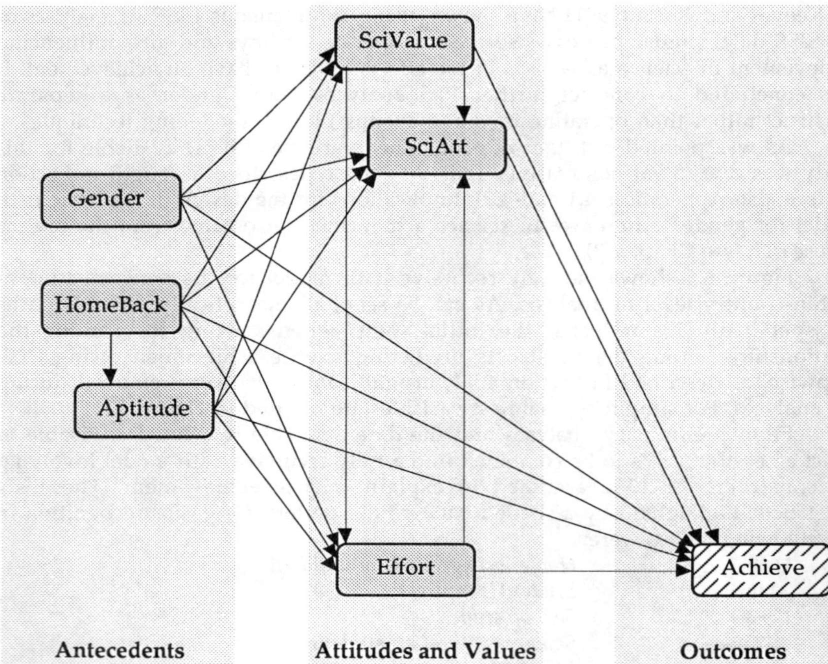

Notes:

construct, latent variable (LV)

construct of outcome measure

path, with arrow showing the direction of the effect

Aptitude was supposed to be directly dependent upon HomeBack, since it could be assumed that no practically relevant reason could be advanced to link it causally to student's gender. Nevertheless, there was justification, slightly different from what Keeves (1991) had argued, to place the student's aptitude to the far left side of the model (Duncan, 1989; Linn and Hyde, 1989; Reynolds and Walberg, 1991). The model proposed in Figure 6.1 suggests the three constructs HomeBack, Aptitude, and Gender jointly influence the manifestation of motivations (i.e. efforts), attitudes, and values towards science. This would be in line with findings by Humrich (1988 [1]), Duncan (1989) and Keeves (1991). It is, of course, understood that teacher, peer group, and school effects also play an important role influencing science achievement. These will be dealt with in Chapter 7 when discussing the effects of student-, teacher- and school-related factors on science achievement at the between-school level.

To some extent, this process is directly connected with the efforts a student undertakes in studying science, in school and at home. As there exists some justification from cognitive and socialization theories to explain attitudes by values, and not vice versa, because of a causal relationship towards student's attitudes (e.g. see Krathwohl, Bloom, and Masia, 1964, pp. 154-164), the LV SciAtt is influenced by SciValue and Effort, thus being placed slightly to the right of the latter two to reflect the chronology and causal influences inherent in the path model.

Preliminary exploratory PLS and multiple regression analyses, not reported here, have suggested rather to predict students' attitudes by students' values than vice versa. This supported the hypothesis that values and beliefs, once developed, are more durable psychological traits than attitudes. However, it ought to be mentioned that a reverse placement of these two LVs would have yielded only marginally different results with respect to the development of an overall path model for gender differences on science achievement.

Further preparatory analyses were carried out to test the possibility of combining SciValue and SciAtt into a single construct. However, it could be seen that in nearly each country data set investigated, the combining of these LVs into a single attitudinal construct resulted in less variance explained ($R^2<.10$). Hence, good empirical and theoretical reasons (see Duncan, 1989; Keeves and Kotte, 1991) existed for keeping values and attitudes (chronologically and causally) separated.

Thus, each of the resultant six LVs were believed to influence science achievement, which, as the outcome measure, had been placed to the far right of the model.

In order to obtain an indication of the relative effect of the manifest variables predicting each construct in question factor loadings were calculated for each manifest variable for each country. These data are presented in Table 6.6. Though the conceptual difference of using inward or outward mode estimation would have resulted in showing either weights or loadings for the manifest variables, it has been decided to confine Table 6.6 to the presentation of loadings only as the relative differences between manifest variables for a certain LV are fairly similar for weights or loadings.

Also, grand means and standard deviations across the 10 countries are given for each MV. The between-country mean of the weights of a manifest variable can be understood as the relative strength of a particular variable 'forming' the respective construct. The resultant between-country standard deviation is used to obtain an estimate of how much variation a certain MV showed across the 10 countries. A low standard deviation would indicate the suitability of the manifest variable for cross-country comparisons as that respective MV then appears to be of importance throughout all or at least most of the countries being analyzed.

The rather low standard deviations shown in Table 6.6 do indeed allow for the conclusion that a between-country comparison of the model as presented in Figure 6.1 seems justified. Moreover, these data, once more, underline the appropriateness of administering the variables, scales, and achievement tests as was done in 1970/71 and 1983/84 in the different educational systems.

TABLE 6.6 Factor loadings for manifest variables for between-students within-country PLS analyses for the 14-year-olds, 1983/84

Latent Variable / Manifest Variable	Australia	England	Finland	Hungary	Italy	Japan	Netherlands	Sweden	Thailand	United States	average mean▲	average sd
❑ **HomeBack**												
↑ Father's Occupation	.67	.73	.72	.86	.72	nc	.64	.56	.55	.69	.68	.09
↑ Mother's Occupation	d	.67	.68	.80	d	nc	d	d	d	.52	.67	.11
↑ Father's Education	.66	nc	d	.88	.64	nc	.51	d	d	.74	.69	.14
↑ Mother's Education	.55	nc	.71	.88	.61	nc	d	d	.62	.67	.67	.12
↑ Books in Home	.87	.84	.85	nc	.87	1.00	d	.76	.82	.78	.85	.07
↑ Use of Dictionary	d	nc	d	nc	d	nc	.78	.69	d	d	.74	.06
❑ **Aptitude**												
↑ Year Level or Age	d	d	d	d	d	d	d	.54	d	d	.54	
↑ Language	d	d	d	d	nc	nc	d	.86	d	d	.86	
↑ Word Knowledge	.87	.86	nc	nc	1.00	nc	nc	nc	nc	1.00	.93	.08
↑ Mathematics	.87	.88	nc	1.00	nc	1.00	nc	nc	nc	nc	.94	.07
❑ **Gender**												
↕ Sex of Student	1.00	1.00	1.00	1.00	1.00	1.00	1.00	1.00	1.00	1.00	1.00	
❑ **SciValue**												
↑ Career Interest	.89	.79	.62	.68	.80	1.00	.87	.88	.82	.87	.82	.11
↑ Beneficial Aspects	.82	.91	.88	.79	.87	d	.82	.76	.83	.87	.84	.05
↑ Non-Harmful Aspects	d	d	.70	.63	d	d	d	d	d	d	.67	.05
❑ **Effort**												
↑ Homework Effort	nc	nc	nc	.85	.51	nc	d	.86	.65	.70	.71	.15
↑ Classroom Effort	nc	nc	nc	.74	.82	.94	.83	d	.89	.81	.84	.07
↑ Hours Homework	1.00	1.00	d	.67	.75	.64	.76	.79	.50	.74	.76	.16
❑ **SciAtt**												
↑ Interest in Science	.99	.97	.88	1.00	nc	1.00	.94	.95	1.00	.93	.96	.04
↑ Ease of Learning Science	.44	.59	.79	d	1.00	nc	.67	.71	d	.71	.70	.17
❑ **Achieve**												
↓ Biology Score	.82	.83	.76	.80	.77	.82	.77	.79	.71	.83	.79	.04
↓ Chemistry Score	.84	.82	.77	.81	.82	.85	.84	.82	.79	.85	.82	.03
↓ Physics Score	.83	.83	.79	.85	.83	.85	.87	.84	.81	.82	.83	.02

Notes:

❑ construct, variate, latent variable
↑ manifest variable estimated by inward mode
↓ manifest variable estimated by outward mode
↕ single manifest variable; value becomes unity
▲ mean loadings calculated as the sum of loadings across all countries divided by the number of countries for which data were obtained
d deleted prior to analysis
nc data not collected

With the exception of the LV Achieve, which was estimated using the *outward mode*, all other constructs were estimated in *inward mode*. As the LV Gender was represented by just one variable, namely *Sex of Student*, the regression weight of the MV is equal to the factor loading both taking a value of 1.00. The three achievement test scores for biology, chemistry, and physics are supposed to 'reflect' the latent dimension of science achievement, hence the factor analytic approach to estimate the LV Achieve by its science subtest scores. For a more detailed discussion on PLSPATH estimation modes see Section 4.3.1.

During preliminary multivariate analyses problems were encountered with a few manifest variables only. *Family Size* (not shown in Table 6.6) had to be deleted prior to the analysis in all countries due to its poor correlation with the other MVs in the HomeBack construct. *Year or Age* and *Language* proved not to be suitable MVs with the exception of Sweden, where no other data were available for Aptitude. Attitudinal data from the *Non-Harmful Aspects* scale appeared to match with the two remaining MVs in the SciAtt construct only for Finland and Hungary. In the remaining eight countries this variable was deleted prior to analyses.

In the following, the loadings of each MV within a construct will be discussed with respect to the latent variable to which it contributes.

❑ **HomeBack** Apart from Family Size, which had to be deleted prior to analysis, the mean loadings of the remaining six variables varied between a=.67 (Mother's Occupation, Mother's Education) and a=.85 (Books in Home). As can be seen from Table 6.6, the number of predictors differed considerably between countries ranging from one in Japan to five in the United States. For interpreting subsequent PLS analyses it should be noted, however, that slight misfits might have to be expected for Japan using only one MV other than the number of books in home to predict the status of the student's home. Family Size correlated slightly negatively with HomeBack but failed to contribute to the formation of the construct within the path model as was found in preliminary PLS analyses. Another noteworthy point is that Mother's Occupation was not found to be a suitable predictor (i.e. $|\beta|<.10$; see Section 4.3.1) in five countries. This observation coincided with findings from previous cross-national studies (see Comber and Keeves, 1973; Thorndike, 1973; Fennema and Sherman, 1978; Kelly, 1978; Sanguinetty, 1983). Apart from this, the data in Table 6.6 show that all other MVs were well suited to predict and form a student's home background or socio-economic status, a fact which is supported by the small overall standard deviations for each of the MVs.

❑ **Aptitude** Unfortunately, not very much information had been collected cross-nationally. In most country, only one or two MVs could be assigned to predict aptitude. Moreover, the data available were scarcely comparable: can the age of a student predict aptitude as well as does mother tongue or word knowledge? But, as will be seen from later results, the doubt cast upon the suitability of the MVs and the construct Aptitude within the path model largely disappears as both the predictive power of Aptitude, i.e. the resultant path coefficient, and the variance explained by it turned out to be high and fairly stable across countries. Nevertheless, caution should be excised when interpreting 'Aptitude' across countries.

Both, the Word Knowledge (a=.93) and Mathematics scores (a=.94) loaded highly on Aptitude. It was hypothesized that these LVs constituted verbal and numerical abilities (see Comber and Keeves, 1973). As was discussed in Section 4.1, these abilities were regarded as a necessary precondition for a student's 'general aptitude level' (Thorndike, 1973; Reynolds and Walberg, 1991). Thus, the high loadings could serve as an indication that both measures were equally suitable predictors of the LV. This provides further justification that even for those countries where only one of these MVs was available Aptitude would still be adequately predicted.

Only for Sweden, Year Level and Language had to serve as substitutes for the missing Word Knowledge and Mathematics test scores. It is, of course, acknowledged that this brought up certain conceptual difficulties in the interpretation of Aptitude across countries. For the sake of at least some comparability, the use of Year Level and Language in Sweden was tolerated. However, great care should be taken in the interpretation of the LV. For all other countries, the measures of age and language were taken out in the course of the analysis. It should be added, however, that throughout all countries Age correlated, if only somewhat low, negatively with Aptitude. This could be explained in that students who had to repeat a grade, most likely, tended to show a lower general aptitude. These grade-repeating students, in general, were older than their classmates, hence the resultant slightly negative correlation between Age and Aptitude.

❑ **Gender** Throughout all countries, the sex of the student was taken to indicate Gender. Hence, this LV comprised just a single manifest variable. It should be recalled that boys were coded '1' and girls '2'.

❑ **SciValue** As previously discussed, the LV consisted of the three scales Career Interest in Science, Beneficial Aspects, and Non-Harmful Aspects of Science. With the exception of Finland and Hungary, the Non-Harmful Aspects showed constantly low loadings on SciValue which led to the deletion of this MV in the remaining eight countries. This general 'misbehaviour' of the Non-Harmful Aspects was unexpected from the attitudinal results given above (Section 6.1.1). The correlation of Career Interest with Beneficial Aspects was the only significant one of the three MVs at the between-country level which was consistent with the high loadings of both MVs across all countries. This was also reflected by the high value of the mean loadings, a=.82 for Career Interest and a=.84 for Beneficial Aspects, and low standard deviations.

❑ **Effort** In general, the loadings of the three MVs were satisfactorily high. It should be noted, though, that for Finland this LV had to be excluded due to missing data and in the cases of Australia and England only Hours Homework could be used to represent this construct.

FIG. 6.2 *Fully recursive within-country path model (outer and inner model) for science achievement at the 14-year-old level, 1983/84*

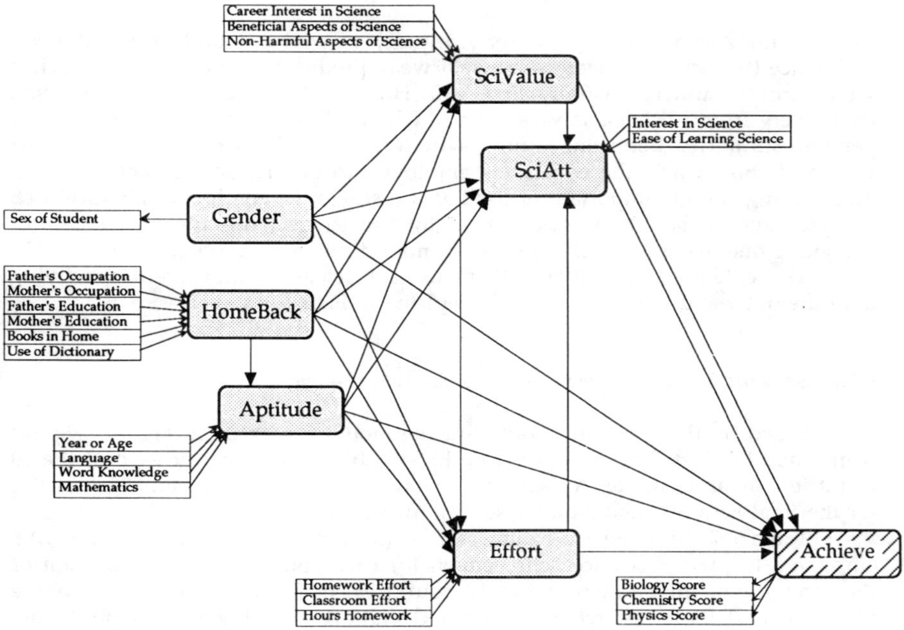

Notes:

construct, latent variable (LV)

construct of outcome measure

manifest variable

inner model path, with arrow showing the direction of the effect

outer model path, with arrow showing inward or outward estimation mode

❑ **SciAtt** The data given in Table 6.6 indicate that SciAtt was primarily composed by the Interest in Science scale ($\bar{a}$=.96) as its loadings were considerably higher than those of the other MV, Ease of Learning Science ($\bar{a}$=.70). This was true for all countries except Italy where the data for Interest in Science were incomplete.

❑ **Achieve** As can be seen by the mean loadings for the three MVs predicting science achievement, little variation occurred between countries. All three predictors appeared to be about equally strong with a high average loading of $\bar{a}$=.81.

However, while the loadings were marginally stronger for physics and chemistry than for biology, the high level of similarity within and across countries of the factor loadings for the three subtest scores strongly supports the use of a science total score in most analyses reported in this volume.

While Age and Language (for Aptitude), as well as Non-Harmful Aspects of Science (for SciValue) were somewhat weak predictors the other MVs worked well across countries. Family Size (for HomeBack) needed to be removed completely from further analyses. On the whole, Table 6.6 showed the apparent general similarity between countries in terms of adequacy of the manifest variables chosen and the constructs employed to predict science achievement. This finding is quite important in itself as it justified the conduct of identical PLS analyses across the 10 countries. It might also suggest the appropriateness of designing one overall model to explain gender differences in science.

The extended recursive path model for science achievement specifying both the outer and the inner model is depicted in Figure 6.2.

6.2.2 *Student Variable Effects on Science Achievement*

Based on the conceptualization of the model presented in Figure 6.2, and using the MVs just described, separate PLS analyses were run for each of the 10 countries. As an example, Appendix 6.1 shows the PLSPATH summary statistics for the final between-student analyses for Australia.

Table 6.7 gives an overview of the direct and total effects of the LVs employed to predict science achievement for each country. As an indication of the strength of effects across the 10 countries, direct and total effects were averaged and their respective standard deviations calculated to illustrate the variations for each LV across the countries. It is, of course, understood that in a strict mathematical sense, the calculation of means and standard deviations of various path coefficients from different samples is not possible.

Also, in Table 6.7 the percent of variance explained by the model is presented for each country. All figures are based on between-students within-country data of the 14-year-olds obtained from the 1983/84 data collection.

Of the six LVs forming the inner model, Aptitude showed, not unexpectedly, the strongest predictive power. On average, its direct path coefficient on science achievement was more than twice as high ($\bar{p}$=.47) as any of the other predictors. In Sweden, Aptitude did not turn out to be the strongest predictor which could almost certainly be attributed to the substitute variables, Year Level and Language, taken to predict Aptitude. Variation on both of these MVs and their correlation with other MVs and LVs in the model was obviously too low to account for a strong effect of Aptitude. This shortcoming was discussed above and was, therefore, not a surprise. However, the slightly erratic case of Sweden does not defy the general notion of Aptitude being most powerful in predicting achievement in science. It is, however, interesting to note that the direct and total effects were about the same. This may be taken as an indicator that student's aptitude exerts a very direct influence on science achievement, not mediated by any other construct employed in the path model.

TABLE 6.7 *Direct and total effects derived from PLS analyses on science achievement as outcome measure for between-students within-country data for the 14-year-olds, 1983/84*

Relationship to Science Achievement	Australia	England	Finland	Hungary	Italy	Japan	Netherlands	Sweden	Thailand	United States	average mean▲	average sd
Direct Effects												
HomeBack	.12	.12	.27	.20	.12	.11	.13	.23	.14	.17	.16	.05
Aptitude	.63	.58	nd	.48	.43	.51	nd	.13	nd	.49	.47	.16
Gender	-.17	-.13	-.17	-.14	-.18	-.11	-.21	-.11	-.24	-.18	-.16	.04
SciValue	...	...	.17	...	...	...	.18	.22	...	...	.06	.09
Effort	...	...	nd	...	.12	.11	.16	...	.22	...	.07	.09
SciAtt	...	.12	.14	...	.11	.15	.12	.17	...	.16	.10	.07
Total Effects												
HomeBack	.39	.41	.30	.34	.32	.29	.22	.35	.14	.40	.31	.09
Aptitude	.63	.60	nd	.48	.46	.56	nd	.13	nd	.51	.48	.17
Gender	-.17	-.14	-.19	-.14	-.16	-.16	-.27	-.19	-.21	-.20	-.18	.04
SciValue	...	...	...	...	...	...	.27	.29	...	...	.11	.11
Effort	...	...	nd	...	.13	.15	.18	...	.22	...	.09	.09
SciAtt	...	.12	.14	...	.11	.15	.12	.17	...	.16	.10	.07
% Variance Accounted for Science Achievement	47	43	17	32	29	37	18	21	12	38	29◆	12■

Notes:

▲ mean effects calculated as the sum of effects across all countries divided by the number of countries for which data were obtained (effects with |β|<.10 have been included)

◆ mean variance accounted for science achievement across 10 countries

■ standard deviation of mean variance

... the effect was discarded with |β|<.10 (see also Appendix 6.1)

nd data not collected, or deleted prior to analysis, or all manifest variables not significant

HomeBack came through as a medium-sized LV in almost all countries with a total effect nearly double ($\bar{t}$=.31) the size of the direct effect ($\bar{p}$=.16). Clearly, the student's home, apart from its direct effect on achievement becomes important by having an impact on the other constructs, such as Aptitude, SciValue and Effort, which, in turn, effect science achievement as will be seen later on. While the direct effects of the status of the student's home was most influential in Finland and Sweden as compared with the other LVs, it played only a marginal role in Thailand. Likewise, the direct effect on achievement was small in Japan. The Finnish result might be explained by the absence of the Aptitude construct so that HomeBack took most of the explanatory power. The reason for the Swedish figure, as mentioned beforehand, has to be seen in the selection of less suitable MVs representing Aptitude. As this caused Aptitude to have less impact on science achievement, HomeBack obviously picked up more variance. Thailand's minor influence of home background might be explained by its relatively small variation among Thai students and the comparatively low status reported for the Thai sample (see Postlethwaite and Wiley, 1991, pp. 24-31).

Gender, negatively associated with achievement, proved to be of similar strength across all countries with an average direct regression weight of $\bar{p}$=-.16. The negative sign indicated that boys performed better. This was the case throughout all 10 countries as indicated by the very low standard deviation. Due to the lack of adequate data for student's aptitude in Thailand, Gender took over the role of being the most potent predictor. Gender differences had a largely direct impact on science achievement as can be seen from the very similar mean regression coefficients for the direct ($\bar{p}$=-.16) and total effects ($\bar{t}$=-.18). This was not quite the case for Japan and Sweden, where the total effects were considerably larger than the direct ones, suggesting that Gender obviously worked through effort, values, and attitudes to a sizeable extent as will also be shown in the discussions of the Japanese and Swedish path models with future enrolment in science courses as the outcome (see Sections 6.2.3 and 7.3).

As discussed earlier, the LV SciValue showed a rather heterogeneous picture across countries. This, as was expected, was partially due to the MVs incorporated in the construct. SciValue had sizeable direct effects on achievement only in the Netherlands (p=.22), Sweden (p=.18), and Finland (p=.17). While the total effects were higher than the direct effects in the Netherlands and Sweden, the total effect had to be discarded for Finland with $|\beta|<.10$. This, indeed, is interesting as it may be concluded that in Finland the influence of science values on achievement seemed to be countered by the attitudes towards science. However, it could also be related to the absence of any data for the LVs Aptitude and Effort leaving a slightly distorted path model. Generally speaking, it seemed that science values, and hence, gender differences in science values, became of marked importance only in the Netherlands and Sweden where SciValue had a considerable total effect. It should be borne in mind, though, that this might reflect a somewhat 'artificial' picture due to missing or inadequate manifest variables.

Only in Italy, Japan, the Netherlands, and Thailand sizeable direct effects were found with the Effort construct ($\bar{p}$=.07). Even this overall mean may have been overestimated as the Netherlands (p=.16) and Thailand (p=.22) were lacking another key construct (Aptitude) which probably led to the variance on Effort serving to be a substitute for the absent construct Aptitude. Thus, only for Italy (p=.12) and Japan (p=.11) students' motivations played a seemingly, if not very large, direct role influencing achievement in science.

In contrast to values and motivations, SciAtt had an effect on achievement in most countries ($\bar{p}$=.10) with little variation across countries. Interestingly enough, in England, Sweden, and the United States science attitudes, apart from antecedent variables, were the only important factors predicting science achievement. Here, values and motivations did not seem to be of importance.

When looking at the total effects, also presented in Table 6.7, it can be seen that the most powerful predictors of science achievement were, despite the shortcomings in some countries, Aptitude (t=.47) and HomeBack (t=.31). But it must be kept in mind, and what was surprising, was that, in general, the next strongest LV was Gender (t=-.18). Its regression weight, when seen in relation to the hardly less pronounced direct effect of $\overline{p}$=-.16, must be interpreted such that at the age of about 14 the boys, regardless of their status of home, family background, cognitive abilities, and even their personal values, attitudes and efforts in class and at home, achieved better in science. Yet it would be misleading to assume that this reflects a 'gender superiority' of the boys or a 'gender inferiority' of the girls. Until other effects, for instance school- or teaching-related factors, as well as the measurement and specification errors implicit in the model have been carefully sorted out and determined, the advantage that boys seemingly have when looking at student-related factors to estimate science achievement remains largely unexplained (see Section 7.2.2).

Thus, overall the three background constructs HomeBack, Aptitude, and Gender played the most significant part in predicting students' achievement in science. The three mean total effects taken together accounted for more than twice as much explained inner model variance as did the intermediate LVs SciValue, Effort, and SciAtt. But it could be argued that, where the students' individual background was less influential (e.g. in Finland, the Netherlands, or Sweden), countries which could be perceived as comparatively liberal with respect to education and as 'transparent' societies offering a high level of social security, students' values and/or attitudes gained a larger impact on achievement in science.

This leads to the cautious conclusion, that the model proposed in Figure 6.2 and modified according to the results of the PLS runs served its purpose in explaining the effects of latent variables at the between-student within-country level for 14-year-olds. Moreover, it should be noted, that the relative effects of the constructs considered in the above model were very similar to the effects obtained by Keeves and Kotte (1991) when analyzing student data at the between-students within-school level. The apparent similarity of Keeves and Kotte's results is taken as a supportive argument for the adequacy of comparing student data across countries further to comparing them between schools.

But when considering the appropriateness of a theoretical path model with science achievement as the outcome measure, one should also pay attention to the overall explained variance of the model. The data recorded in the bottom row of Table 6.7 indicates that student factors did vary considerably in their total effect on science achievement (range of R^2: 12% in Thailand to 47% in Australia). This allows ample room for speculation which other main factors or variables might have influenced achievement other than student-related ones. As can be seen, student-related factors had a strong impact in Australia (R^2=47), England (R^2=43), the United States (R^2=38), and Japan (R^2=37), while being much less powerful in Thailand (R^2=12), Finland (R^2=17), the Netherlands (R^2=18), and Sweden (R^2=21).

It may be argued that the relative effect of student-related factors decreases in countries with a rather open societal structure and where formal education is regarded as being fairly 'liberal' (Finland, the Netherlands, and Sweden; see Riis, 1988; Keeves, 1991). Likewise, student-related effects appear also to be of less importance in countries where the background and socio-economic status of the homes are more homogeneous. A situation like this, as found in Finland, the Netherlands, Sweden, and Thailand, apparently diminishes the overall variance accounted for by student-related variables in predicting science achievement.

However, the mean variance accounted for in the model outlined in Figure 6.2 of 29 percent underlines the general importance of student-related variables in fostering students' achievement in science.

Despite minor shortcomings due to insufficient or missing data it is appropriate to present a path model for science achievement across the 10 countries. For this purpose, the direct and total path coefficients have been 'averaged' (see comment above) across all 10 countries as far as data were obtained. Table 6.8 shows these standardized regression weights, or path coefficients, both for the direct and total inner model effects. Again, paths with $|\beta|<.10$ have been discarded assuming they reflect rather trivial effects.

In addition, Figure 6.3 pictures the modified path model showing the direct effects while Figure 6.4 displays the resultant total effects on science achievement.

TABLE 6.8 *Path coefficients (standardized regression coefficients) for direct and total inner model effects derived from PLS analyses with science achievement as outcome measure for between-students data averaged across 10 countries at the 14-year-old level, 1983/84*

Direct Inner Model Effects averaged	HomeBack	Aptitude	Gender	SciValue	Effort	SciAtt
Aptitude	.34					
Gender						
SciValue	...	.12	...			
Effort	.16	.20	...	.17		
SciAtt	...	...	-.10	.37	.18	
Achieve	.16	.47	-.16	...	...	.10

Total Inner Model Effects averaged	HomeBack	Aptitude	Gender	SciValue	Effort	SciAtt
Aptitude	.34					
Gender						
SciValue	.12	.12	...			
Effort	.24	.22	...	.17		
SciAtt	.11	...	-.12	.40	.18	
Achieve	.31	.48	-.18	.11	...	.10

Notes:
all figures are based on data from at least seven countries
... the effect was discarded as trivial with $|\beta|<.10$
nd data not collected, or deleted prior to analysis, or all manifest variables not significant

As can be seen, several paths of the inner model suggested in Figure 6.2 turned out to be trivial or irrelevant. However, marked overall direct effects were encountered from HomeBack to Aptitude ($\bar{p}$=.34) which, in turn, strongly influenced the students' motivation ($\bar{p}=-.20$) and achievement ($\bar{p}=.47$). Furthermore, there was a strong path from SciValue to SciAtt ($\bar{p}=.37$). The latter path is not only an indication of the appropriateness of the chronology which specified that values influence attitudes, but underlines the implicit link between these constructs as hypothesized in the previous section. However, preliminary analyses supported the keeping of these LVs separate.

Cross-nationally, gender appeared to influence directly the manifestation of science attitudes ($\bar{p}=-.10$) and achievement ($\bar{p}=-.16$) in that boys showed both more favourable attitudes towards science and better results on the science tests administered than girls. It should be emphasized that these were **direct** effects without taking into account any mediating or inhibiting influences of other variables.

Though gender had a direct influence in students' attitudes towards science it appeared not to affect the manifestation of science values. This may be taken as further supportive evidence that the differentiation between values and attitudes was theoretically justified and not of a purely speculative nature.

Moreover, it can be seen that the gender differences were much the same when comparing direct with total effects. This underscores the immediate impact gender differences have on science achievement.

Across all 10 countries, science achievement was largely dependent upon the students' home environment and parental support (as measured by the MVs Books in Home or Use of Dictionary) which considerably shaped students' aptitude ($\bar{p}=.34$) and individual motivation ($\bar{p}=.24$), though aptitude itself exerted the strongest influence on achievement ($\bar{p}=.48$) while students' efforts largely contributed directly to attitudes towards science ($\bar{p}=.18$). Although the direct effect of the home background on science achievement appeared to be of medium size ($\bar{p}=.16$) its total effect was rather more substantial ($\bar{p}=.31$).

As Aptitude was composed of two manifest variables, namely the scores on the Word Knowledge and Mathematics subtests, it should be considered to be a measure of a more general cognitive ability and not as one assessing science-related pre-dispositions. Given this, it is important to acknowledge that student's science achievement is obviously closely dependent on his or her general performance in school. This suggests that science may not be as specific a subject as many educators would like to think it is, at least not in terms of science performance. On the contrary, science seemed to be related not only to mathematical and numerical abilities, which as research has shown, was indeed often the case (see discussion in Chapter 2), but, likewise, to verbal capability. This finding is supported by the fact that both the Word Knowledge and Mathematics scores were equally suitable predictors for the LV Aptitude.

Science-related values and attitudes, on the other hand, exercised a less pronounced effect on science achievement. Their total effects were even lower ($\bar{t}=.11$ for SciValue; $\bar{t}=.10$ for SciAtt) than that of gender ($\bar{t}=-.18$).

No significant gender differences were hypothesized with respect to the student's home and aptitude level, and only to a minor extent with student efforts. This suggests that, in addition to factors deriving from the different socialization of boys and girls, other factors, such as student's attitudes and values towards science as well as expectations regarding a possible science-related career, need to be taken into account to explain the gender differences found in science achievement at the 14-year-old level.

FIG. 6.3 *Path model for direct inner model effects with science achievement as outcome measure averaged across 10 countries at the 14-year-old level, 1983/84*

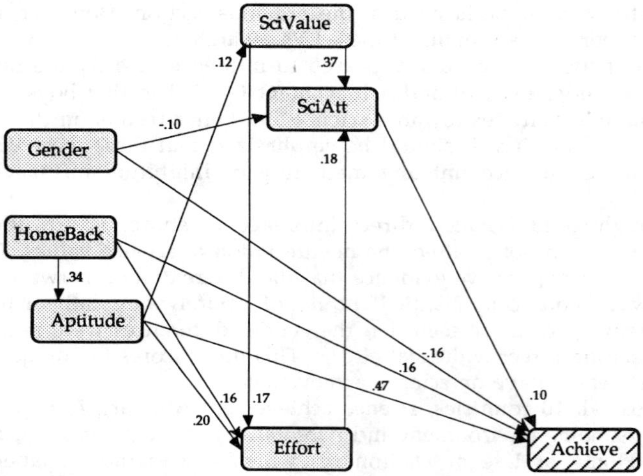

Notes:

paths with |β|<.10 have been omitted; for further details see legend of Figure 6.1

FIG. 6.4 *Path model for total inner model effects with science achievement as outcome measure averaged across 10 countries at the 14-year-old level, 1983/84*

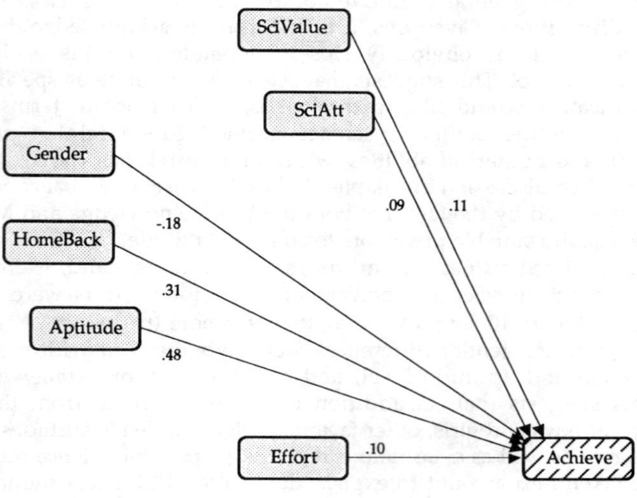

Notes:

for further details see legend of Figure 6.1

Apparently, school-, teaching-, and classroom-related factors, not yet considered, might be expected to play an important role in determining science achievement. This was certainly possible for Italy and Sweden, where the variance in science achievement accounted for by the path model above turned out to be low, despite the fact that sufficient data were available for each of the constructs employed.

While main effects at the school level are reported in detail in the next chapter, the path model given in Figure 6.2 is extended in the following section to incorporate students' expectations with respect to science.

6.2.3 Student Variable Effects on Participation in Science

In Chapter 5 it was shown that enrolment in science courses would appear to be related to the success in science, measured in terms of science achievement. Thus, emphasis must be given to identifying factors stimulating, positively or negatively, a student's decision to enrol in future science courses. It is, therefore, desirable, to specify a model, in which future enrolment in science is taken as the outcome measure.

This is achieved by extending the model presented in Figure 6.2. Students' educational and occupational expectations are integrated into the model while employing the students' future participation in science courses as the new outcome variable.

As was stated beforehand, Keeves (1991, Figure 7.1) had hypothesized that students' expectations depended, among other factors, directly on students' science achievement. This would imply placing students' expectations on the right hand side of the path model shown in Figure 6.1 to indicate the dependent relationships predicting students' educational and occupational expectations. Following the terminology of LVs given in Section 6.2.1, the new constructs employed in the path model are:

ExpEd	*Expected education of student*
ExpOcc	*Expected occupation of student*

The new outcome measure is labelled:

FutEnrol	*Future enrolment in science courses*

As reported in Table 4.1, each of these newly introduced constructs was represented by just a single MV.

Figure 6.5 presents the extended fully recursive path model of student-related factors to be analyzed at the 14-year-old level.

It is assumed that the students' educational expectations determine the occupational ones. Both, in turn, influence the students' plans for future enrolment in science courses.

In the following, it is considered useful to discuss briefly the path model for each of the 10 countries which results from the final PLS runs before attempting to summarize the results across countries. Thus, for each country the diagrammatic path model showing all direct effects is presented. It should be repeated that all paths found to have a standardized regression weight of $|\beta|<.10$ have been discarded as being non-significant. A construct was omitted completely from the PLS analyses where no manifest variable was available for a certain LV.

FIG. 6.5 *Extended, fully recursive student-level path model with future enrolment in science courses as outcome measure; 14-year-old level, 1983/84*

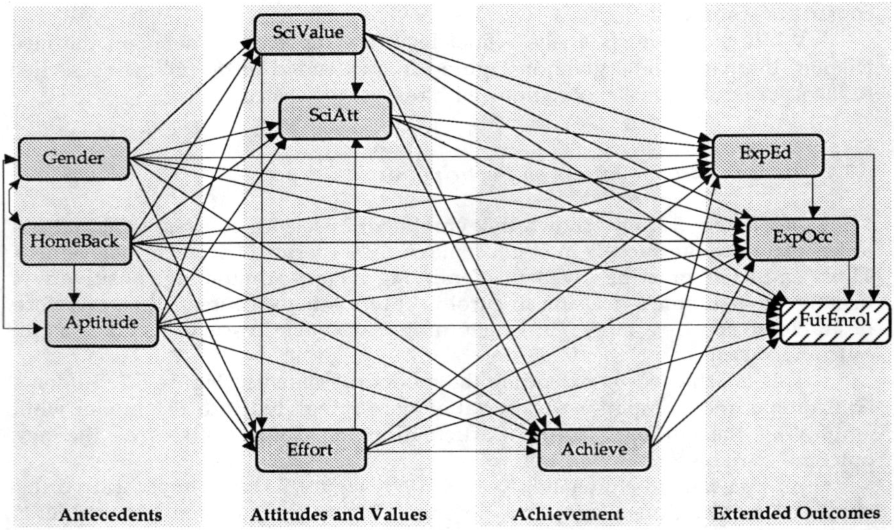

Notes:

construct, latent variable (LV)

construct of outcome measure

path, with arrow showing the direction of the effect

correlation

Unfortunately, Japan provided only data for expected education (ExpEd) and did not administer the questions regarding expected occupation (ExpOcc) and future enrolment in science courses (FutEnrol). Hence, in contrast to all other countries the new outcome measure for Japan had to become ExpEd rather than FutEnrol. In addition to this shortcoming, no information was obtained for ExpEd and ExpOcc in England, while Italy did not collect data for ExpOcc.

As an example, the PLSPATH summary statistics for Australia are shown in Appendix 6.2.

TABLE 6.9 *Direct and total effects derived from PLS analyses on future enrolment as outcome measure for between-students within-country data for the 14-year-olds, 1983/84*

Relationship to Future Enrolment	Australia	England	Finland	Hungary	Italy	Japan	Netherlands	Sweden	Thailand	United States	average mean▲	average sd
Direct Effects												
HomeBack	...	.14	...	...	...	.21	...	...	...	...	.02	.05
Aptitude	.12	...	nd	.12	...	.15	nd	...	nd	...	.04	.06
Gender	...	...	...	...	-.11	-.10	...	...	...	...	-.01	.04
SciValue	.14	.25	...	.16	.11	...	.27	...	...	.28	.13	.12
Effort	...	.10		...	...	.21	...	...	...	...	.01	.03
SciAtt	.10	.18	.24	...	...	...	.25	...	.20	.15	.12	.10
Achieve	...	...	.13	...	...	.17	...	...	...	...	.01	.04
ExpEd	.12	nd	.28	.20	.43	nd	.13	.64	.35	.33	.31	.17
ExpOcc	...	nd	...	.11	nd	nd	...	...	...	...	.02	.04
Total Effects												
HomeBack	.11	.21	.16	.18	.21	.35	.13	.14	...	.25	.16	.06
Aptitude	.18	...	nd	.20	.11	.30	nd	...	nd	.14	.12	.07
Gender	...	...	...	...	-.10	-.14	-.10	...	...	...	-.03	.05
SciValue	.21	.35	.18	.17	.12	...	.43	.15	...	.37	.23	.12
Effort	...	.10		...	...	.23	...	...	.10	...	.06	.03
SciAtt	.10	.18	.26	...	...	...	.25	.10	.20	.16	.14	.10
Achieve	...	...	.22	...	...	.17	...	...	...	...	.03	.07
ExpEd	.12	nd	.28	.24	.43	▼	.13	.64	.35	.33	.31	.17
ExpOcc	...	nd	...	.11	nd	▼	...	...	...	...	.02	.04
% Variance Accounted for Future Enrolment	8	16	17	11	21	21▼	20	41	16	24	19◆	9■

Notes:

▲ mean effects calculated as the sum of effects across all countries divided by the number of countries for which data were obtained (effects with |β|<.10 have been included)

◆ mean variance accounted for FutEnrol across nine countries (Japan has been excluded)

■ standard deviation of mean variance

▼ the outcome variable for Japan was ExpEd; there were no data collected for ExpOcc or FutEnrol

... the effect was discarded as trivial with |β|<.10 (see also Appendix 6.2)

nd data not collected, or deleted prior to analysis, or all manifest variables not significant

FIG. 6.6 *Path model for Australia with future enrolment in science courses as outcome measure; 14-year-old level, 1983/84*

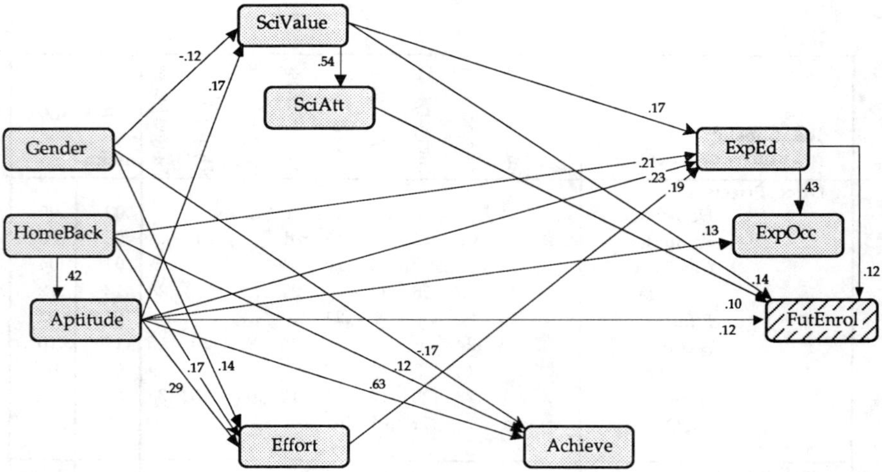

FIG. 6.7 *Path model for England with future enrolment in science courses as outcome measure; 14-year-old level, 1983/84*

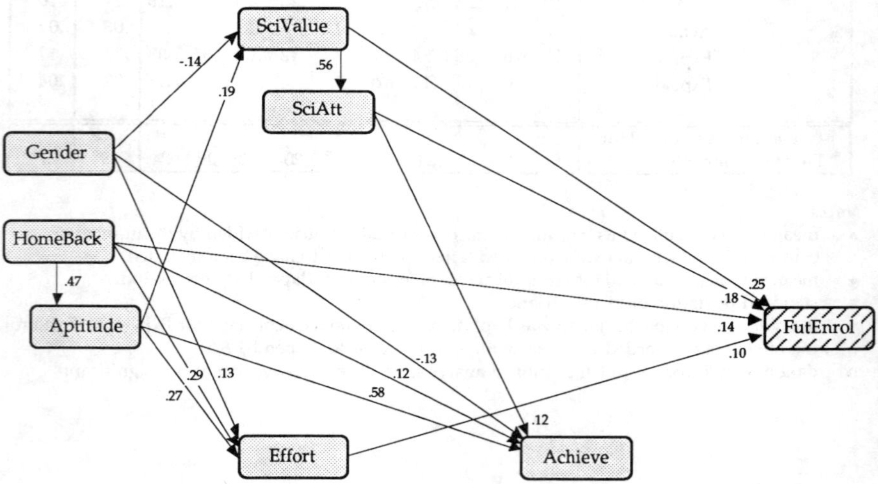

Notes for Figures 6.6 and 6.7:
paths with |β|<.10 have been omitted; for further details see legend for Figure 6.5

FIG. 6.8 *Path model for Finland with future enrolment in science courses as outcome measure; 14-year-old level, 1983/84*

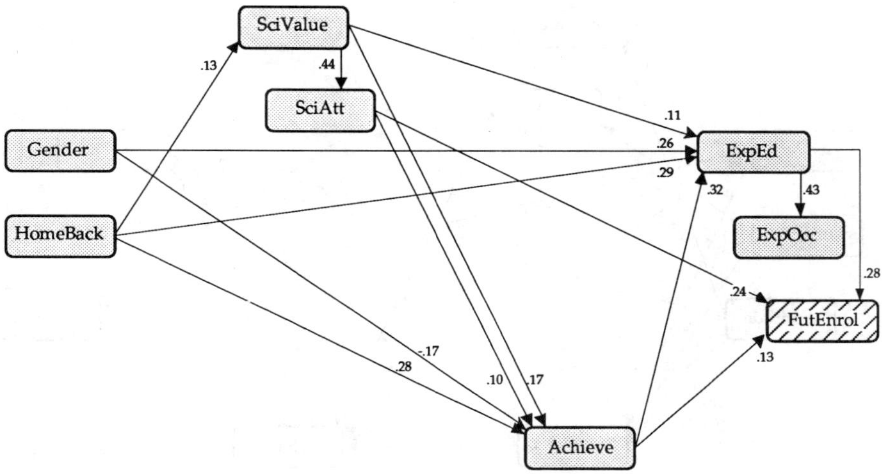

FIG. 6.9 *Path model for Hungary with future enrolment in science courses as outcome measure; 14-year-old level, 1983/84*

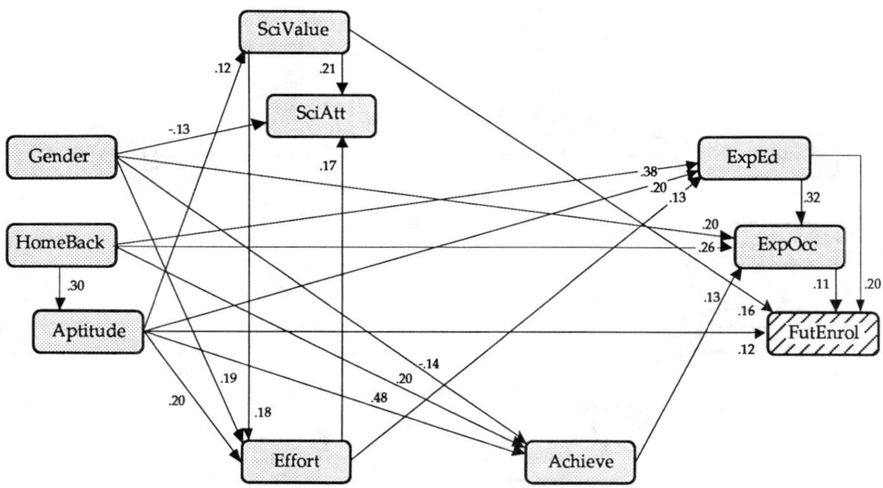

Notes for Figures 6.8 and 6.9:
paths with $|\beta|<.10$ have been omitted; for further details see legend for Figure 6.5

FIG. 6.10 *Path model for Italy with future enrolment in science courses as outcome measure; 14-year-old level, 1983/84*

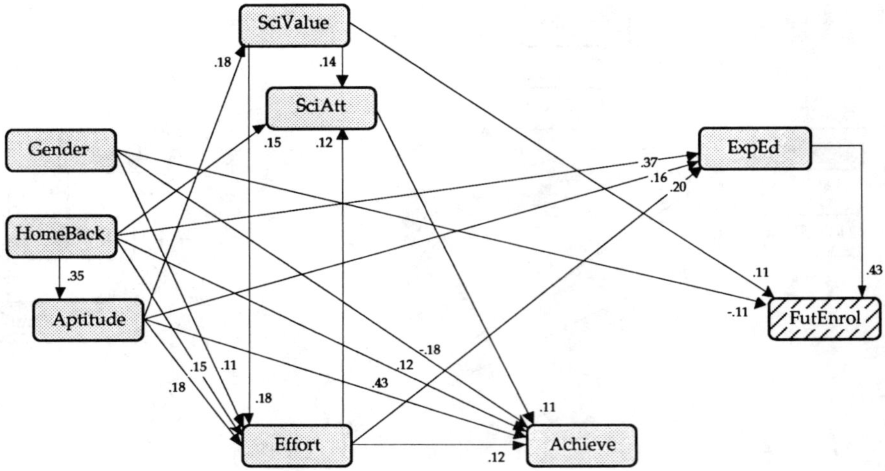

FIG. 6.11 *Path model for Japan with expected education as outcome measure; 14-year-old level, 1983/84*

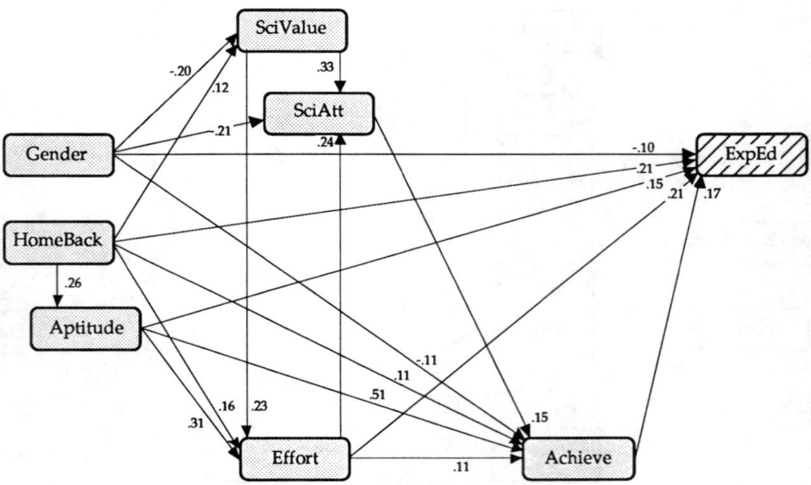

Notes for Figures 6.10 and 6.11:
paths with |β|<.10 have been omitted; for further details see legend for Figure 6.5

FIG. 6.12 *Path model for the Netherlands with future enrolment in science courses as outcome measure; 14-year-old level, 1983/84*

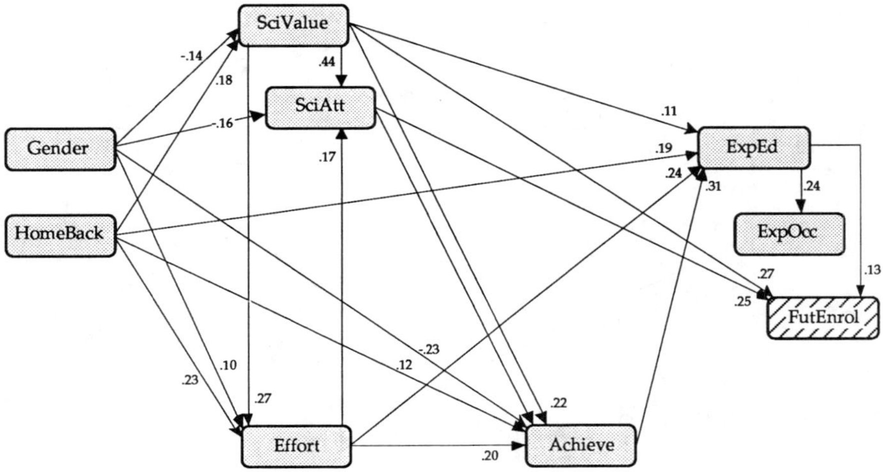

FIG. 6.13 *Path model for Sweden with future enrolment in science courses as outcome measure; 14-year-old level, 1983/84*

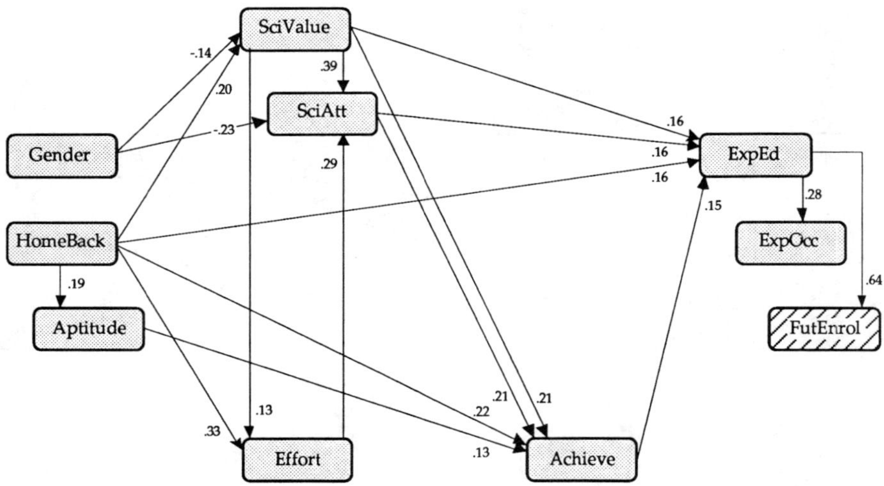

Notes for Figures 6.12 and 6.13:
paths with $|\beta|<.10$ have been omitted; for further details see legend for Figure 6.5

FIG. 6.14 *Path model for Thailand with future enrolment in science courses as outcome measure; 14-year-old level, 1983/84*

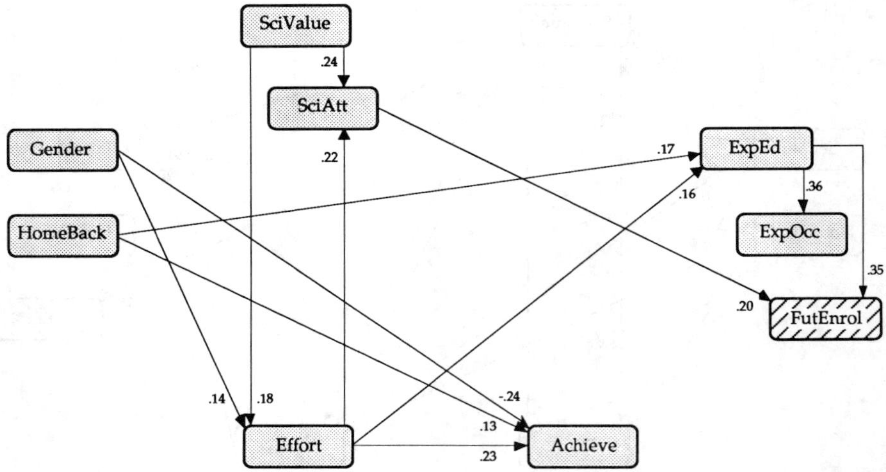

FIG. 6.15 *Path model for the United States with future enrolment in science courses as outcome measure; 14-year-old level, 1983/84*

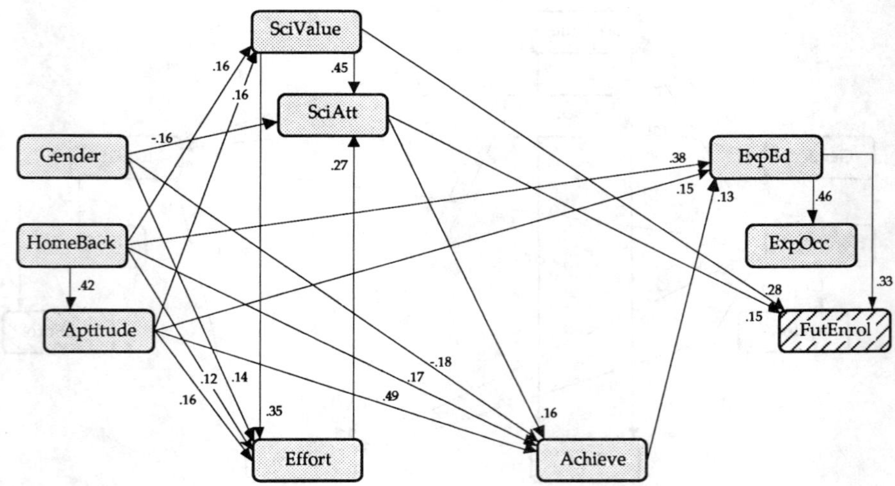

Notes for Figures 6.14 and 6.15:
paths with |β|<.10 have been omitted; for further details see legend for Figure 6.5

Although it had been shown that a number of paths from the fully recursive inner model predicting science achievement (Figure 6.1) were non-significant with their regression weights $|\beta|<.10$ to obtain the averaged 'final' model (Figure 6.3), these paths, which marginally explained science achievement, were introduced again in the fully recursive model estimating future enrolment in science courses. This was done to cater for any possible deviations within the model should the introduction of students' expectations affect the stability of the 'final' model for science achievement. As will be discussed later, this safeguard proved unnecessary: no additional direct effect occurred within the left-hand part of the model predicting achievement. Quite the opposite, a few effects found to have small impacts within the model for science achievement were deleted from the averaged 'final' model predicting future enrolment.

In a similar way to Table 6.7, which summarized the direct and total effects for each country on the outcome measure (science achievement), Table 6.9 gives an overview of the effects of the independent LVs within the model predicting future enrolment of students in science courses. Also, the percent of variance accounted for in FutEnrol in each of the countries has been calculated.

The data presented in Table 6.9 allows the conclusion that, despite the fact that the initial model was estimating science achievement fairly well within and between countries, future participation in science proved much more affected by factors not yet taken into consideration. The model proposed in Figure 6.5 seemed particularly inadequate for Australia (R^2=.08). and Hungary (R^2=.11), countries for which complete data were available so that the low explained variance for these two countries must obviously be due to other factors.

Before attempting to generalize the results it is considered useful to take a brief look at each country separately.

→*Australia* (Figure 6.6) At first glance it appears astonishing to find no relationship between achievement and expectations (ExpEd and ExpOcc) as well as between achievement and students' intentions to enrol in future science courses. It was hypothesized in Chapter 2 that science achievement influences students' decision to participate in future science courses. It was further thought likely that better achieving students would more preferably choose to enrol in science courses where this became optional at the upper secondary level. This was not at all the case in Australia and, as will be seen when cross-comparing the 10 countries, there was generally no significant effect from achievement to future enrolment.

Like educational expectations, attitudes and values towards science, home background and aptitude had only small direct effects on future enrolment, which, in turn, led to the overall low variance of eight percent. School- and teacher-related factors might have been more dominant in shaping students' decisions to proceed with studying science.

Though considerable gender differences were found in Australia at the terminal secondary level across all science subjects, and although many more boys enrolled in science courses in the upper secondary forms, gender did not appear to have a direct influence on FutEnrol at the 14-year-old level. But, and this was in contrast with the averaged model (see Figure 6.16), gender tended to affect science values, which themselves largely shaped attitudes. The latter, however, only had a minor effect on FutEnrol. Thus, it appears that in Australia gender effects are operating through SciValue and indirectly through SciAtt and ExpEd. This, in turn, is evidence for gender to affect science achievement through a network of influences and rather than directly in Australia.

The effects that HomeBack and Aptitude had on the students' expected education did not interfere with students' interest to enrol in future science courses. This was in line with the comparatively low correlation between ExpEd and FutEnrol (r=.22).

The estimated path model using student data only seemed inadequate to explain future enrolment in Australia. This came as a surprise, insofar as the model was well suited to explain science achievement. It remains to be seen whether school or teacher-related factors determined participation in science to a larger extent.

➞*England* (Figure 6.7) In the absence of the LVs ExpEd and ExpOcc students' home background (p=.14) and motivations (p=.10) affected FutEnrol directly. But the strongest influence on future participation in science courses came from SciValue with the direct path coefficient amounting to p=.25. Working also through students' attitudes the total effect of SciValue on FutEnrol even rose to a very strong t=.35. Though values and attitudes seemed to influence interest in future science courses, the overall variance accounted for by the path model was only 16 percent.

It should also be noted that gender had a direct impact on students' values (p=-.14). This was to be expected after having found significant gender differences with respect to the attitudinal scales (see Chapter 5) at the 14-year-old level.

In contrast to the general overall linkage between values, attitudes, and motivations no such relationship was observed in England. Motivations were not related to science values or attitudes. Students' efforts were only marginally effective in influencing students' future enrolment in science courses.

Similar to the Australian results, students' achievement in science seemed to have no influence on stimulating the decision to participate in science courses. Possibly, it is too premature for the students to make up their mind at the age of 14 in choosing to study science at the terminal secondary school level.

In England attitudes and values towards science apparently played a more important role both upon science achievement and future enrolment. Values and attitudes should be regarded as fairly independent composites which strongly accounted for success in school science.

The main effects of gender were through SciValue and then on to FutEnrol as well as indirectly through SciAtt. The role of gender effects operating through science values and attitudes in England was very similar to the effects of gender observed in Australia. Moreover, it is obviously not the effect of gender on science achievement that is of consequence in influencing the decision to enrol in future science courses, but the effect on science values which, in turn, have considerable impact on FutEnrol.

➞*Finland* (Figure 6.8) With no data for students' aptitude and efforts, only three LVs had a direct effect on future enrolment, namely SciAtt, Achieve, and ExpEd. Both attitudes towards science (p=.24) and educational expectations (p=.28) had the strongest influence on FutEnrol. While the direct effect of science achievement was small (p=.13) its total effect, working through

ExpEd, was much stronger (t=.22). There were also noticeable total effects from HomeBack (t=.16) and SciValue (t=.18). The overall variance accounted for was 17 percent.

Though educational expectations had a marked effect on future participation in science courses (p=.28), the path coefficient was smaller than the average regression weight ($\bar{p}$=.31). This might have been caused by the exceptionally strong causal relationship that gender had on students' educational expectations (p=.26) which seemed to interfere with the impact of the LV ExpEd. In Finland, it were the girls who showed considerably higher educational expectations. These, in turn, did not have to be oriented towards science in particular, thus weakening the usual impact of ExpEd on FutEnrol. This finding would be supported by the fact that many more girls aspired to an academic career in the upper secondary forms in Finland compared with male students who more commonly left school to follow a practical- and technical-oriented training (Leimu, 1988; Keeves, 1991).

Thus, in Finland student gender had a positive rather than a negative indirect effect on future enrolment in science courses, operating through expected education. In contrast to other countries, the effects of gender operated negatively through science achievement. It was only the male students who had a higher level of achievement in science who sought to proceed with enrolment in future science courses.

→*Hungary* (Figure 6.9) Despite the appropriate explanatory power of the Hungarian path model to predict science achievement, it accounted for only 11 percent of variance for future enrolment. As with Finland, this might have been due to the relatively low effect of ExpEd on FutEnrol (p=.20).

It was interesting to see students' aptitude having a total effect on future participation in science courses (t=.20) with a direct effect coming through (p=.12). Also HomeBack had a moderate total effect on FutEnrol (t=.14) while the impact of values towards science were comparatively small (p=.16). Moreover, science attitudes did not have a sizeable effect on either science achievement or FutEnrol.

This leaves room for the hypothesis that contrary to Australia, England and Finland, where attitudes had an impact on the decision to enrol in future science courses, they were less influential in Hungary. Furthermore, the home environment and individual student aptitude became much more important for future science enrolment. This finding was somewhat unexpected because Hungary was the only socialist country in the sample and such dominance of the students' home and personal characteristics appeared not to reflect the Marxist-Leninist doctrines of equality.

So, while students' values showed comparatively little overall variability among the 14-year-olds and also little differences between the sexes (see Tables 6.1 and 6.2), which, indeed, could have been a result of systematic political effort in the Hungarian society and educational system, this might explain the limited impact science-related values and attitudes had on future enrolment. Furthermore, it might account for the fact that gender had no effect on values and, while influencing science attitudes, such attitudes had no effect on achievement, educational and occupational expectations on future enrolment in science courses.

Hungary, like Finland, is a country where the participation and enrolment of boys in academic schooling at the terminal secondary school level is visibly reduced (Báthory, 1988; Keeves, 1991). Policies in these two countries directed towards greater involvement of girls and women in further science training including science-based occupations are strong (for Hungary see Báthory, 1989). However, in Hungary and Finland gender still has its effects on science achievement.

→*Italy* (Figure 6.10) Educational expectations had the strongest influence on students' decisions to take up further science courses in the upper grades of formal schooling (p=.43).

Though not coming through as direct effects, HomeBack (i=.21) and Aptitude (i=.11) excised measurable indirect effects on FutEnrol, too.

While SciValue only had a moderate influence on FutEnrol (p=.11), Gender affected this outcome directly (p=-.11). In Italy, boys (all other variables kept constant), were more likely to opt for future science courses than girls. This might have well been a reflection of the socialization and education process, which in Italy was considered to emphasize conservative ideals and was, therefore, bound to stimulate a more rigid gender-role identity among boys and girls (Visalberghi, 1988; Rosier and Keeves, 1991).

Basically as a result of the strong causal link of educational expectations with future enrolment in science, the explained variance of the Italian path model was 21 percent.

It should be noted, however, that the effect of gender on future enrolment in science courses was not indirect through attitudes and values but rather directly operating in unspecified ways to lead boys to enrol more frequently in future science courses than girls.

→*Japan* (Figure 6.11) As data were available neither for ExpOcc nor FutEnrol the outcome measure in the Japanese path model became ExpEd. Despite this flaw it was still considered appropriate to present the 'incomplete' model here because it shows congruence with those of other countries in the path structure as far as ExpEd is concerned.

Parallelly to what was seen in the countries previously discussed, the explanatory power of the model decreases significantly once expectations and future aspirations are introduced when compared with those analyses where the outcome measure was science achievement. By having ExpEd defined as the dependent variable in Japan the overall explained variance was 21 percent.

HomeBack (t=.35) and Aptitude (t=.30) turned out to be the two most powerful overall predictors for ExpEd. However, even Gender had a sizeable total effect on students' expectations (t=-.14) with Japanese boys expressing interest in achieving higher educational goals than girls.

It was also interesting to observe that students' efforts (t=.23) and science achievement (t=.17) had a clear positive impact on expectations though this also reflected both their home background and aptitude.

Thus, in Japan the effects of Gender on ExpEd are both direct (p=-.10) and indirect (i=-.04) acting through values and attitudes to influence science achievement which, in turn, acts to influence educational expectations.

→*The Netherlands* (Figure 6.12) Of all 10 countries values and attitudes towards science had the biggest effect in predicting students' plans for future enrolment in science courses in the Netherlands. While the direct effect of SciValue was already quite strong (p=.27), its total effect, operating through the related LV SciAtt, reached a very high value (t=.43). Rather than home-related factors students' values and attitudes had a most impressive impact on future enrolment. In turn, values and attitudes towards science were shaped by students' gender. This finding leads to the conclusion that in the Netherlands values were of much more importance in building expectations than in any other country in the study.

Thus, when looking for variables that had a total effect on FutEnrol it was noticeable that, although not marked, gender differences were found with respect to FutEnrol in the Netherlands. Gender affected both SciValue (p=-.14) and SciAtt (p=-.16) directly which in other countries was only the case in Japan and Sweden. This could be taken as a possible explanation for the gender specifity of values and attitudes resulting in the significant gender differences observed in the Netherlands (see Table 6.1).

Educational expectations, which across all countries were mainly affected by students' home background, aptitude, effort, and achievement, scarcely contributed to future enrolment (t=.13) in the Netherlands.

→*Sweden* (Figure 6.13) In a similar way to the path model for the Netherlands, Gender had a strong total impact on both the values (t=-.14) and attitudes towards science (t=-.29). But, in contrast to the Dutch model, both students' values and attitudes affected educational expectations directly (p=.21). This should be taken as the reason for the very high causal influence ExpEd exerted on FutEnrol (p=.64). In other words, contrary to the Netherlands, where the impact of educational expectations was reported to be small, students' expectations were found to be of much greater importance for the decision to enrol in future science courses.

That students' educational expectations had effects that were very different from those of all countries (with the possible exception of Italy which, as was seen, appeared to be the consequence of the different societal setting assumed to be fairly conservative) was underlined by the absence of paths from Aptitude and Effort to ExpEd which were found in most other countries.

The high overall proportion of variance accounted for in the Swedish path model of 41 percent ought to be seen largely as the result of the strong effect that ExpEd had on FutEnrol. Moreover, it is through educational expectations that attitudes and values operate to transmit the effects of Gender to ExpOcc and FutEnrol.

Science achievement in Sweden has only a relatively minor effect on educational expectations and thus on future occupation and future enrolment in science courses. The absence of the effects of Aptitude are, of course, the result of the lack of an effective measure of aptitude in the Swedish data.

→*Thailand* (Figure 6.14) In the absence of the LV Aptitude and with the weak explanatory power of the HomeBack construct only two LVs exercised a direct effect on FutEnrol, namely SciAtt (p=.20) and ExpEd (p=.35). The incompleteness of the Thai path model should also be seen as the reason for the comparatively low overall variance accounted for (16 percent).

Nevertheless, it is noteworthy that neither of the attitudinal constructs (SciValue or SciAtt) were directly influenced by HomeBack or Gender. Hence, the effect going from SciAtt to FutEnrol carried mainly the impact that Effort had on SciAtt (p=.22). This was to be expected as a similarly strong influence was reported when explaining science achievement for the Thai data where Effort had exactly the same effect on SciAtt.

In general, it should be suggested that the Thai model suffered markedly from a lack of data related to the latent variables in question. Nevertheless, it is clear that while Gender has a noticeable effect on science achievement it has little effect on any of the educational outcomes involving ExpEd.

➞*The United States* (Figure 6.15) The path model is characterized by the relatively large number of paths inherent in the model. This is even more true when looking at the total effects, where the majority of theoretically possible paths were found to have a regression weight of $|\beta| \geq .10$. However, the model for future enrolment explained only 24 percent of variance in the United States.

The effect of Gender on FutEnrol was largely through the attitudes of the students towards science and not through their values. There were also small indirect effects through science achievement and ExpEd, but surprisingly SciValue did not appear to be gender-related in the United States.

The strongest total influence on FutEnrol stemmed from SciValue (t=.37), whereas HomeBack (t=.25) and Aptitude (t=.14) showed considerably lower total effects. This might indicate that values and attitudes as well as background effects apparently had a nearly equally strong impact on the students' decision to continue studying science. However, it has to be considered that HomeBack and Aptitude have effects on FutEnrol that are mediated by SciValue and SciAtt. In addition, SciValue has effects that are mediated by SciAtt which accounts for the difference between direct (p=.28) and total effects (t=.37) on FutEnrol.

But, as indicated by the large number of relevant paths, the model proved to be very stable during PLS analyses. This fact should be taken into consideration when speculating about the 'meaningfulness' of the proposed path model in reflecting the 'real' processes responsible for influencing students' decisions on future science courses.

Having discussed the differences between countries with respect to the model of future enrolment it appears rather questionable that any additional information can be obtained from an averaged model which reflects a mean 'artificial' country as was done when explaining science achievement on the basis of student-level data.

However, for comparative purposes only, where data for direct and total effects were available, these were then averaged. Table 6.10 presents the effects, again with paths $|\beta| < .10$ omitted. In addition, as already briefly indicated, these mean direct effects have been concocted to form the averaged path model with future enrolment as the outcome variable represented in Figure 6.16.

TABLE 6.10 *Path coefficients (standardized regression coefficients) for direct and total inner model effects derived from PLS analyses with future enrolment in science courses as outcome measure for between-students data averaged across 10 countries at the 14-year-old level, 1983/84*

Direct Inner Model Effects averaged	HomeBack	Aptitude	Gender	SciValue	Effort	SciAtt	Achieve	ExpEd	ExpOcc
HomeBack									
Aptitude	.34								
Gender									
SciValue	...	.12	...						
Effort	.16	.20	.10	.17					
SciAtt	...	...	...	.37	.16				
Achieve	.16	.47	-.15	...	...	...			
ExpEd	.26	.13▼	...	...	.14	...	.10		
ExpOcc	...	...	...	...	...	...	...	.36	
FutEnrol	...	...	...	.13	...	.16	...	.35	...

Total Inner Model Effects averaged	HomeBack	Aptitude	Gender	SciValue	Effort	SciAtt	Achieve	ExpEd	ExpOcc
HomeBack									
Aptitude	.34								
Gender									
SciValue	.11	.12	...						
Effort	.24	.22	...	.17					
SciAtt	...	...	-.11	.40	.16				
Achieve	.31	.48	-.17	.11	...	...			
ExpEd	.36	.21▼	...	.11	.16	...	.10		
ExpOcc	.20	.12◆	...	...	...	...	...	.36	
FutEnrol	.16	.12▼	...	.23	...	.14	...	.31	...

Notes:
all figures are based on data from at least seven countries
◆ data were available from four countries only
▼ data were available from six countries only

FIG. 6.16 *Path model with future enrolment in science courses as outcome measure averaged across 10 countries; 14-year-old level, 1983/84*

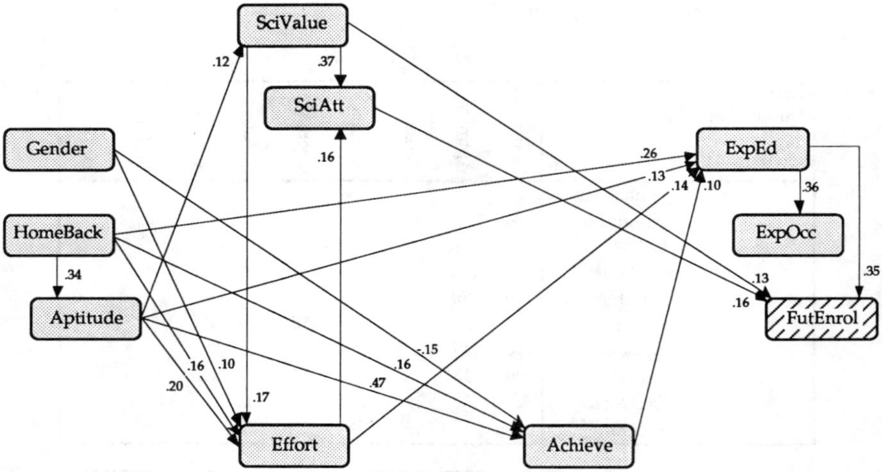

Notes:

paths with $|\beta|<.10$ have been omitted; for further details see legend for Figure 6.5

As can be seen from Table 6.9 and from the path models for each country, substantial differences between countries were detected. These differences not only occurred with respect to the strength of certain paths but also involved the structural relationships shown by the paths within the model. This variability was less evident when devising a model for estimating effects with respect to science achievement where, as was shown, substantial agreement existed among the 10 countries. Why then did the prediction of future enrolment in science courses change the general consistency of the path model so radically?

Several reasons could have accounted for this change.

➞*Measurement error* Though this point cannot be taken up here in detail, it is recognized that it was statistically not possible to estimate each single error component inherent in the data for each country as too many LVs were involved. However, statistical fit indices suggested the general appropriateness of the path models employed (see Appendix 6.2).

➞*Missing data* Australia, Hungary, Sweden, and the United States were the only countries which managed to collect sufficient data for all of the LVs forming the inner model. In all other countries at least one of the constructs was missing.

→*School and teacher effects* Although for Sweden the path structure suggested seemed appropriate with 41 percent of variance explained, other countries showed a rather low proportion of variance accounted for by the model. This would suggest that school- and teacher-related factors which had not been taken into consideration had a substantial impact on FutEnrol.

→*Adequacy of measurement* When asking 14-year-old students about their expectations regarding personal, educational, and occupational goals many of the answers given were believed to contain some degree of wishful thinking. At the age of 14 students may not be able to make a 'reasonable' decision whether to enrol in upper secondary science courses or not. Students of that age group may aspire to reach an educational level or occupation which has little to do with their achievement in class. This assumption is supported by the absence of the expected path from Achieve to ExpOcc and the weak path leading from Achieve to ExpEd ($|\beta|<.10$) when looking at the averaged model.

→*Structural problems* The path model applied to estimate achievement in science had a single, distinct outcome measure (Achieve). When extending the model to predict FutEnrol introducing also the intermediate constructs of ExpEd and ExpOcc, this created the situation of heterogeneous outcome variables namely Achieve (Australia, England, Italy, Thailand), ExpOcc (Australia, Finland, the Netherlands, Sweden, Thailand, the United States) and FutEnrol (all countries except Japan). Japan, as mentioned beforehand, did not collect data for ExpOcc and FutEnrol which bore another conceptual problem for averaging the path structure across countries. This structural problem becomes evident when comparing the percentages of variance accounted for by the path models for Achieve and FutEnrol in each country. While in Australia, England, Hungary, Japan, and the United States a considerable proportion of the variance was lost when extending the model to estimate FutEnrol, the path model proposed for Sweden became more accurate (gained variance: 20 percent). In the remaining countries (Finland, Italy, the Netherlands, Thailand) extending the model to incorporate expectations and future enrolment did little change to the structure of the earlier model. Additional analyses, not reported here, have also shown that it would not be advisable to omit ExpEd and ExpOcc completely from the extended path model just predicting FutEnrol by the LVs included in the basic model for science achievement (Figure 6.2). With ExpEd and ExpOcc dropped, the overall explained variance was considerably lower in each of the 10 countries than with ExpEd and ExpOcc included.

Although grave restrictions apply to a generalization across the path models for each of the countries, an attempt has been made to do so in order to show the main results found across the 10 countries. However, it was not considered appropriate to carry out a separate PLS analysis using data pooled between all 10 countries. This idea was rejected on grounds that the path models with FutEnrol as the outcome did not appear to produce generalizable patterns across the countries.

Data from Table 6.9 exhibits that at the between-student level different factors (LVs) influence students' decisions to enrol in future science courses in each of the 10 countries. But it became apparent that there were five major constructs affecting FutEnrol which were HomeBack, Aptitude, SciValue, SciAtt, and ExpEd. In contrast, Gender, Effort, Achieve, and ExpOcc did not have effects directly or indirectly in most countries. Where these constructs had an influence this was generally very low and most likely due to the absence of one or more other LVs (e.g. Japan).

The students' home background and ability level each influenced future enrolment positively which was to be expected in that more able students or students from higher status homes were more likely to select future science courses when this became optional. Nevertheless, HomeBack and Aptitude both did not work directly to affect FutEnrol. Rather these LVs stimulated the educational expectations to such an extent that these, in turn, stimulated most significantly the interest to study more science.

However, there seemed to emerge another major set of factors influencing future enrolment: student values and attitudes. Though values and attitudes were partly results of antecedent factors, which in many countries also included gender, and student motivation or effort, the data suggest that attitudes and especially values towards science are of considerable importance in shaping students' decisions to enrol in future science courses. It could be inferred that the two major forces shown to affect FutEnrol, 'student background' (comprising HomeBack, Aptitude, and Effort) and 'student attitudes' (comprising SciValue and SciAtt) worked in all countries. Just **how** they had an impact on future enrolment varied from country to country.

As a conclusion, it must be recognized that the decision to enrol in science courses in the upper secondary grades was very much correlated with the students' general educational expectations and attitudes (attitudes here understood in the broader sense comprising attitudes and values). Phrased differently, future enrolment did not directly depend on either HomeBack, Gender, or Aptitude. It obviously was a function of values, attitudes, and aspirations. Interesting enough, FutEnrol was not linked to occupational expectations at all. This, however, suggests that it was probably too premature to put 14-year-olds to the test about their future careers. It is probable that students of that age could not develop concise ideas about their future careers. Questions regarding educational expectations were more likely to be responded to as these were more closely related to classroom and school affairs.

Nevertheless, it remains to be seen in which way the school or teaching context excised an influence on future enrolment.

6.3 Summary and Conclusions

In this chapter student-level factors which influenced science achievement were identified and their effects evaluated. The first two sections examined gender differences on the attitudinal (Section 6.1.1) and effort scales (Section 6.1.2). Also, consideration was given to the relationship between attitudes and science achievement. It became apparent that despite country-specific differences marked gender differences were found on the attitudinal and effort scales. Boys and girls seemed to have and express a different 'gender identity' with respect to science. These gender differences not only varied from country to country but also across the three grade levels tested. Further to this, the results showed that it was favourable for subsequent multivariate analyses to operate with two distinctly different attitudinal constructs: science attitudes and science values.

In Section 6.2.1 a general path model with science achievement as the outcome was developed for the between-student analyses and tested for each of the 10 countries using PLS (see Appendix 6.1). The results presented in Section 6.2.2 underlined the appropriateness of the general path model for most of the countries. However, the comparatively low amount of variance accounted for by this path model in Finland, Japan, Sweden, and Thailand gave rise to take into consideration also teacher and school variables to explain science achievement. Despite country-specific differences in the direct and total effects of the LVs in the model it was possible to make some important generalizations between the countries. Though HomeBack and Aptitude turned out to be the strongest overall predictors of science achievement at the 14-year-old level, throughout all countries Gender had a remarkably strong effect on science achievement of $\bar{t}=-.18$. As the average direct effect of Gender of $\bar{p}=-.16$ was only minimally lower, this led to the conclusion that, even when taking into account the effects of other LVs in the model, gender differences favouring the boys were quite substantial. Obviously, Gender was strongly associated with science achievement. It is, of course, misleading to interpret this finding in that boys were achieving better in science than girls because of their sex. A statement like this would be superficial as it disregards the gender-typical socialization and pre-school science education of boys and girls (see, for example, Levy, 1989; Duncan, 1989).

In Section 6.2.3, the relationships among factors at work at the student level which lead to the development of a student's interest to obtain further science training, to take up a science-related occupation, or to enrol in future science courses at the terminal secondary level were also examined. For each country an extended path model incorporating ExpEd, ExpOcc, and FutEnrol was developed and tested (see Appendix 6.2). The cross-country differences for the path models having FutEnrol as the outcome measure increased and did not promote the development of an averaged path model as was possible when operating with science achievement as the independent variable. These path models for FutEnrol were, with the exception of Sweden where the amount of variance explained by the model could be raised when compared with the path model for science achievement, less appropriate to explain the overall variance. Nevertheless, it became obvious that science-related values and attitudes as well as educational expectations played an important role to stimulate a student's decision to study science at the upper secondary school level.

However, the discussions of the country-specific path models for future enrolment suggested, once again, to give special emphasis to teacher and school variables when predicting science achievement and, moreover, examining the relevant factors which influence the interest of students to continue studying science at the terminal secondary school level where this becomes optional.

The effects of teacher and school-level variables on students' science achievement and participation of students in science courses will be discussed in detail in the following chapter.

7. Teaching and School Factors

In the previous chapter the effects of gender on science-related values and attitudes, science achievement and future participation in science courses were examined using between-student data. As was shown, cross-national comparisons of the path models revealed substantial similarities in the factors and mechanisms influencing science achievement, but this kind of equivalence disappeared when extending the path model to predict the effects of factors promoting future enrolment in science courses.

It should be recalled that the path models presented in Chapter 6 were based on between-student data only and did not include any teacher or school variables. Thus, it was possible to examine the effects of gender across all students at the 14-year-old level within each of the 10 countries separately. This allowed careful analyses of the student-related factors on science achievement while focussing particularly on the role gender effects played. However, the proportions of variance accounted for by the path models for science achievement (see Table 6.7) and future enrolment (see Table 6.9) suggest that further to student-level variables teaching and school-related factors should also be taken into consideration in order to understand better the network of gender effects on the outcome measure of science achievement in each of the countries.

This chapter, therefore, presents the results of an extended path model which incorporates, over and above the student-level constructs, latent variables from the teachers and schools across the 10 countries. Combining data from students, teachers, and schools into one analytical approach requires theoretically either the disaggregation of teacher and school information to the student level or the aggregation of student data to the class or school level. Without going into detail here about the problems of aggregation (or disaggregation) bias, which have already been outlined in Section 4.2, it was considered appropriate to aggregate the relevant student-level data to the school level for each of the 10 countries (see Cheung et al., 1990). While Chapter 6 exhibited gender effects on the student level, thereby allowing statements about each individual student in the sample with respect to his or her achievement in science, the results presented in this chapter are based exclusively on between-school analyses. In this chapter it is of primary concern to examine gender differences between classrooms or schools rather than students when taking into account teacher- and school-related factors.

The results discussed below underline the importance and necessity to integrate student, teaching, and school variables analytically in order to understand fully the complexity of gender effects on science achievement. It is made clear that gender differences cannot be examined isolated from all other factors which are believed to influence achievement in science. As will be seen, gender differences are indeed related to the teaching processes in the classroom as well as to the school-specific context in which teaching and learning occurs. Though the students' home background, attitudes and values towards science, and motivation play a cardinal role in establishing differences in science achievement between boys and girls it is also the setting in the classroom and school which helps to increase or decrease gender differences. This is the focus of this chapter.

7.1 An Extended Path Model for Science Achievement

Following the design of a path model for science achievement outlined in Figure 6.2, first adequate blocks of variables representing those teaching and school level constructs assumed to influence science achievement at the 14-year-old level have to be selected. In analyzing SISS data from several countries with PLS, Postlethwaite and Wiley (1991, pp. 129-135) successfully employed a number of manifest variables collected at the school level. Keeves (1991) using a different conceptual approach proposed another set of school level variables. Depending on the analytical questions raised it becomes necessary to operate with different constructs and thus, selecting a different set of manifest variables representing the latent variables. For further details on the manifest and latent variables considered for this study the reader is referred back to Section 4.1 (see also Appendices 4.1 and 4.2).

The teacher- and school-level variables finally selected for the partial least squares analyses in this study emerged after careful preliminary analyses, not presented here, were conducted (see also the discussion on the selection of MVs in Section 4.1). Finally, three teacher-related constructs and one school-related construct were incorporated into an extended path model for science achievement when examining data at the 14-year-old level across the 10 countries.

These four school-level constructs were:

TchBack	*Teacher's educational background*
TchGend	*Sex of teacher*
SchFacil	*Facilities available for science teaching at school*
TchEmph	*Emphasis on science content taught*

An overview of the MVs constituting each of these four LVs can be found in Table 4.1 with additional information on the coding and specificities of the MVs given in Appendix 4.1. While TchBack and TchGend were exogenous LVs SchFacil was believed to be a function of the characteristics of the students enrolled at school (HomeBack, Gender) as well as the kind of teachers teaching science (TchBack, TchGend). But before discussing each of the constructs used for between-school analyses the extended path model which incorporates, additionally to the LVs already described in Chapter 6, the four new constructs is presented in Figure 7.1.

The antecedent constructs constituting the inner model are HomeBack, Gender, TchBack, and TchGend. It is contended that these four constructs are not influenced by any other construct in the model, hence their placement to the far left.

Of the mediating latent variables SchFacil and TchEmph are hypothesized to precede student's aptitude, values, attitudes, and motivation chronologically. From the research review on gender differences in science achievement presented in Chapter 2 it emerged that the student-related constructs (Aptitude, SciValue, Effort, and SciAtt) appear to be influenced by the facilities available at school (SchFacil) and by what happens in the classroom (TchEmph). For a more detailed discussion see, for example, Saario et al. (1973), Leinhard et al. (1979), Kahle and Lakes (1983), Duncan (1989), and Kahle (1989).

FIG. 7.1 *Between-school within-country inner path model with science achievement as the outcome measure for the 14-year-old level, 1983/84*

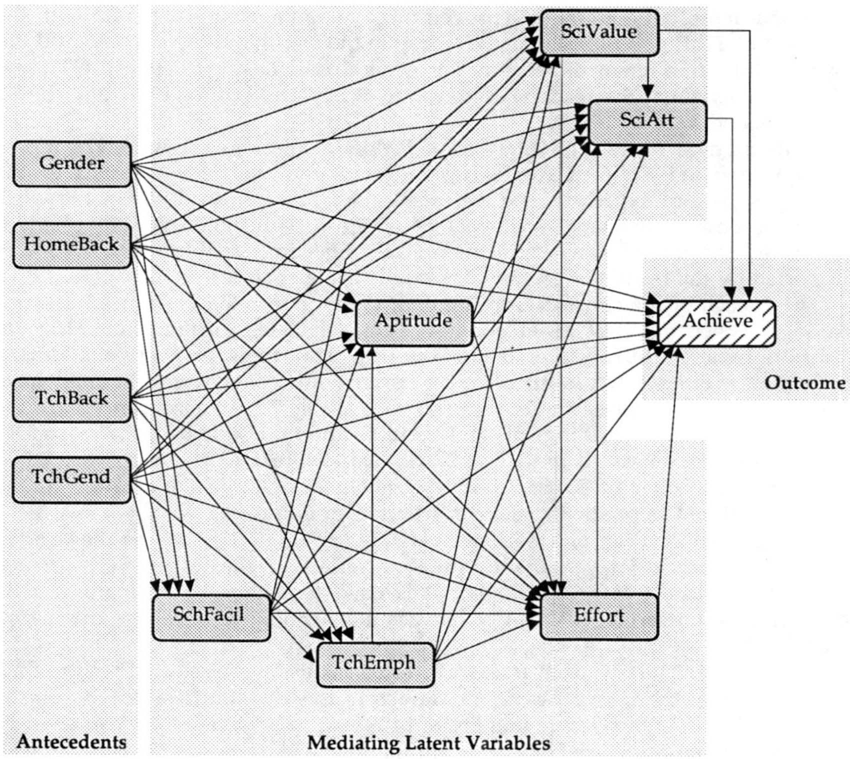

Notes:

construct, latent variable (LV)

construct of outcome measure

path, with arrow showing the direction of the effect

The sole outcome measure, again, is science achievement (Achieve) comprising the biology, chemistry, and physics subtest scores.

Figure 7.1 shows the fully recursive path model with all theoretically possible paths among the constructs. It is, of course, understood that several paths will turn out to be trivial or non-significant, i.e. $|\beta|<.10$, when analyzing the country data. Having included them here is not just a theoretical precaution it also serves the purposes of an instructive example of how complex the relationships of even single-level path models become. After all, for simplicity of the graphical presentation all manifest variables have not been displayed.

Originally, it was intended to include a construct of opportunity to learn (OTL) into the model. However, during preliminary exploratory analyses it turned out that this would reduce the stability of the path model proposed in Figure 7.1 across all countries. It proved more advantageous to operate with the construct of TchEmph rather than OTL despite the fact that these two LVs did not measure the same latent dimension. For general relationships between OTL and science achievement the reader is referred to Wiley and Wolfe (1991, pp. 91-107) or Keeves and Muller (1991).

In contrast to the PLS analyses of Postlethwaite and Wiley (1991), it was also not deemed appropriate to incorporate latent variables related to the teaching style and experiments done in science mainly on grounds of the weak predictive power on science achievement within a path model when looking at gender effects on science achievement. The rejection of these two LVs was supported by the results of preliminary multivariate analyses.

For each of the constructs in the path model presented in Figure 7.1 factor loadings of their manifest variable(s) have been calculated for the 10 countries. In addition, where appropriate, mean loadings and their respective standard deviations averaged across all countries from which data were obtained have been provided, too. These figures are shown in Table 7.1. In accordance with the presentation of factor loadings of the LVs in Chapter 6 (Table 6.6), the respective estimation modes (*inward mode* ↑; *outward mode* ↓) are also given for each MV.

From the data in Table 7.1 it is possible to obtain an indication of the strength of the MVs predicting the constructs in question.

The mean loading of a manifest variable can be understood as the relative strength of a particular variable 'forming' or 'reflecting' the respective construct. The resultant mean standard deviation is used to obtain an estimate of how much variation a certain MV showed across the 10 countries (see also discussion in Chapter 6).

All in all, 33 MVs taken to explain 11 constructs were selected for the final PLS analyses of the between-school data across the 10 countries. But, as can be seen from Table 7.1, a sizeable number of MVs had to be deleted prior to analysis so that the actual number of MVs used in the final PLS runs was, in general, well below 33. The reason for deleting several of the originally selected MVs in each country was necessary to improve either the predictive power of the construct a set of MVs was estimating, or the overall stability of the inner model could be significantly raised when dropping certain MVs. It might be argued that operating with different sets of manifest variables in different countries would imply a lack of comparability of the path models for science achievement across countries. To some extent, this is, indeed, the case. On the other hand, PLS is basically used for exploratory purposes, and in this respect, the information which MVs contribute to the explanatory pattern and which do not is still very useful.

TABLE 7.1 *Factor loadings for manifest variables for between-school within-country PLS analyses for 14-year-olds, 1983/84*

Latent Variable / Manifest Variable	Australia	England	Finland	Hungary	Italy	Japan	Netherlands	Sweden	Thailand	United States	average mean▲	average sd
❑ **HomeBack**												
↑ Father's Occupation	.89	d	d	d	d	nc	.90	.81	.66	d	.82	.11
↑ Mother's Occupation	d	d	d	d	d	nc	d	.64	d	d	na	na
↑ Father's Education	.92	nc	.75	1.00	.84	nc	d	.69	d	.92	.85	.12
↑ Mother's Education	d	nc	d	d	d	nc	d	.64	.68	d	.66	.03
↑ Books in Home	.89	1.00	d	nc	.94	1.00	.90	d	.96	d	.95	.05
↑ Use of Dictionary	d	nc	.80	nc	.64	nc	.72	d	d	.74	.73	.07
❑ **Gender**												
↕ Sex of Student	1.00	1.00	1.00	1.00	1.00	1.00	1.00	1.00	1.00	1.00	1.00	.00
❑ **TchBack**												
↑ Age of Teacher	.82	d	d	d	1.00	d	d	d	d	d	.91	.13
↑ Teacher Post-Sec. Education	d	d	d	1.00	d	d	d	nc	.71	d	.86	.21
↑ Teacher Post-Sec. Sci. Educ.	d	nc	1.00	d	d	1.00	1.00	1.00	d	d	1.00	.00
↑ Years of Teaching Experience	.99	d	d	d	d	d	d	nc	.90	.57	.82	.22
↑ Time Spent Inservice Training	d	1.00	d	d	d	d	d	d	d	.85	.93	.11
❑ **TchGend**												
↕ Sex of Teacher	1.00	1.00	1.00	1.00	1.00	1.00	1.00	1.00	1.00	1.00	1.00	.00
❑ **SchFacil**												
↕ No. of Science Laboratories	1.00	1.00	1.00	1.00	1.00	1.00	1.00	nc	1.00	1.00	1.00	.00
❑ **TchEmph**												
↑ Proportion Teaching Science	1.00	nc	.81	.72	nc	d	.94	1.00	.67	d	.86	.14
↑ Emphasis on Knowledge	d	nc	d	d	d	d	.69	nc	d	.70	.70	.01
↑ Emphasis on Student Needs	d	nc	d	d	d	.88	.72	nc	d	.78	.79	.08
↑ Emph. on Scientific Investig.	d	1.00	.68	.86	1.00	.75	d	nc	.75	d	.84	.14

to be continued

Notes:
- ❑ construct, variate, latent variable
- ↑ manifest variable estimated by inward mode
- ↓ manifest variable estimated by outward mode
- ↕ single manifest variable; value becomes unity
- ▲ mean loadings calculated as the sum of loadings across all countries divided by the number of countries for which data were obtained
- d deleted prior to or during exploratory analysis
- nc data not collected
- na not applicable

TABLE 7.1 (ctd.) Factor loadings for manifest variables for between-school within-country PLS analyses for 14-year-olds, 1983/84

Latent Variable / Manifest Variable	Australia	England	Finland	Hungary	Italy	Japan	Netherlands	Sweden	Thailand	United States	average mean▲	average sd
❑ **Aptitude**												
↑ Year Level or Age	d	d	d	d	d	d	d	1.00	d	d	na	na
↑ Language	d	d	d	d	nc	nc	d	d	d	d	na	na
↑ Word Knowledge	.96	.98	nc	nc	1.00	nc	nc	nc	nc	1.00	.99	.02
↑ Mathematics	.94	.90	nc	1.00	nc	1.00	nc	nc	nc	nc	.96	.05
❑ **SciValue**												
↑ Career Interest	d	d	.76	.97	.65	.96	.87	.76	1.00	.95	.87	.13
↑ Beneficial Aspects	.98	.98	.99	.69	.93	.85	.88	.91	d	d	.90	.10
↑ Non-Harmful Aspects	.56	.57	d	d	.67	d	.81	d	d	.70	.66	.10
❑ **Effort**												
↑ Homework Effort	nc	nc	nc	.94	d	nc	d	.85	d	d	.90	.06
↑ Classroom Effort	nc	nc	nc	.75	.76	.92	.60	.82	d	.67	.75	.11
↑ Hours Homework	1.00	1.00	1.00	.61	.78	.72	.98	.71	1.00	.94	.87	.15
❑ **SciAtt**												
↑ Interest in Science	.97	1.00	.96	.84	nc	1.00	.73	.97	.70	.97	.90	.12
↑ Ease of Learning Science	.67	d	.76	.58	1.00	nc	.96	.63	.84	.75	.77	.15
❑ **Achieve**												
↓ Biology Score	.92	.93	.88	.94	.91	.86	.94	.85	.93	.95	.91	.04
↓ Chemistry Score	.94	.93	.86	.90	.94	.89	.96	.87	.97	.96	.92	.04
↓ Physics Score	.92	.94	.74	.91	.94	.90	.95	.83	.96	.96	.91	.07

Notes:

❑ construct, variate, latent variable
↑ manifest variable estimated by inward mode
↓ manifest variable estimated by outward mode
↕ single manifest variable; value becomes unity
▲ mean loadings calculated as the sum of loadings across all countries divided by the number of countries for which data were obtained
d deleted prior to or during exploratory analysis
nc data not collected
na not applicable

It should, however, be pointed out that similar to the analysis of the data on the between-student level several MVs were not collected in one or more countries. This, indeed, places a limit on any cross-national comparison. Thus, emphasis will put on these cases when discussing each LV and when analyzing the path model for each country.

In the following, each construct and the manifest variables it comprises will be described.

❑ **HomeBack** The average loadings of each of the six MVs reflecting HomeBack were either high or very high. While *Books in Home* turned out to be the most potent predictor ($\bar{a}$=.95) and was finally included in the analyses in six countries, *Mother's Occupation* had to be deleted in all but one country (Sweden), due to its high correlation with *Father's Occupation* in most countries which, would it have remained to constitute HomeBack, would have caused effects of multicollinearity. Nearly the same could be said for *Mother's Education* which was appropriate to be used only in Sweden and Thailand. The standard deviations of the mean factor loadings were low for each of the MVs.

Unfortunately, only one MV, namely *Books in Home* could be used to estimate HomeBack in England and Japan. But as this MV proved the overall most powerful indicator for the construct, it was still considered sufficient to operate with just this single MV within the path models for England and Japan.

The construct HomeBack should be interpreted as the socio-economic situation of the class or school because the data, though originally collected on the student level, had been aggregated to the school level. Therefore, the LV reflects, at least to some extent, the social and economic status of the school as it can be inferred that due to the sampling frames applied, in general, just one class had been sampled per school. It is hypothesized that parallel classes or streams would not vary considerably in their socio-economic status and if slight variations would occur these types of bias would probably cancel out each other. Hence, it seems appropriate to assume that HomeBack does represent the school's social and economic climate.

❑ **Gender** Data were available for each country. The sole variable predicting Gender was *Sex of Student*, hence a value of unity for each country. As the data were aggregated to the school level, Gender should in this case be interpreted as the 'sex of class'. While the MV was coded '1' for boys and '2' for girls it should be stressed that due to the aggregation values between '1' and '2' were now possible. Thus, a *Sex of Student* (or, better 'Sex of Class' or 'Sex of School') value of 1.5 would indicate that there were exactly equal numbers of boys and girls in that class. Likewise, a value of 1.9 would signal that the class in question was predominantly female.

❑ **TchBack** The LV TchBack estimating the science class teacher's background was comprised of five MVs which generally predicted the construct equally well. These variables were the *Age of Teacher* ($\bar{a}$=.91), the *Teacher's Post-Secondary Education* ($\bar{a}$=.86), the *Teacher's Post-Secondary Science Education* ($\bar{a}$=1.00), the number of *Years of Teaching Experience* ($\bar{a}$=.82), and the *Time Spent on Inservice Training* ($\bar{a}$=.93). With the exception of England (no information on the *Teacher's Post-Secondary Science Education*) and Sweden (no information on the *Teacher's Post-Secondary Education* and *Years of Inservice Training*) data were available for each of the manifest variables for each country. Further information on the construct and its MVs has already been presented in Section 4.1.

Despite this, it is evident from Table 7.1 that in all but three countries (Australia, Thailand, and the United States) only one of the five MVs proved adequate to be used in the final PLS run to predict science achievement. This is partly a function of high correlations among these MVs which would have caused multicollinearity effects within the path models. Thus, numerous MVs were removed in the process of refining the path model for each of the countries. The fact that in all but three countries just one manifest variable was entered in the final PLS runs does not limit the meaningfulness and interpretability of the construct TchBack.

However, it is interesting to note several differences between the countries. While it is the *Time Spent on Inservice Training* which becomes the key predictor for TchBack in England and the United States, *Years of Teaching Experience* mainly constitutes TchBack in Australia and Thailand. In contrast to this, the key predictor for TchBack in Italy is the *Age of the Teacher*. In the remaining five countries, TchBack is reflected by the years of general post-secondary education (Hungary) and post-secondary science education (Finland, Japan, Netherlands, and Sweden). But it should be reemphasized that this pattern of key predictors is simply the result of the optimizing process in arriving at a 'final' path model most suitable to predict science achievement at the between-school level in the 10 countries. Yet, that this optimizing process leads to the use of different key variables to constitute TchBack in different countries seems noteworthy.

❑ **TchGend** Taken from the Teacher Questionnaire, the *Sex of Teacher* was used to form this construct. Similarly to the coding of the *Sex of Student*, a male teacher was coded '1' and a female teacher was coded '2'. It should be noted that where more than one teacher was sampled for a school or classroom group, the average sex of teacher was calculated and used in analyses.

❑ **SchFacil** The only appropriate variable providing information about the science teaching facilities available for the science teacher at school was the *Number of Science Laboratories* a school had. This variable showed sufficient variation across countries as can be seen from findings presented by Postlethwaite and Wiley (1991, p. 31). None of the other school-related variables was found appropriate for inclusion into a path model for cross-national comparisons of science achievement using PLS.

❑ **TchEmph** Four different manifest variables formed this LV, namely the *Proportion of Time Teaching Science*, the teacher's *Emphasis on Knowledge*, *Emphasis on Student Needs*, and *Emphasis on Scientific Investigation*. The overall factor loadings varied from $\bar{a}$=.70 (*Emphasis on Knowledge*) to $\bar{a}$=.86 (*Proportion of Time Teaching Science*). Similarly to the stepwise elimination of MVs, as described for TchBack in order to optimize and stabilize the path model for each of the countries, one or more MVs were removed to predict TchEmph in the final PLS runs due to the results of the exploratory analyses.

Emphasis on Knowledge turned out to be the least potent predictor but the loadings were still sufficiently high. Though for the United States (a=.64) the loading was comparatively low, it was still considered to be included in the final path model. For England and Sweden data for only one MV were available (*Emphasis on Scientific Investigation* in England; *Proportion of Time Teaching*

Science in Sweden). But as these MVs showed a high overall loading in forming TchEmph it was believed that this would not restrict the interpretability of the construct.

Aptitude This construct is meant to represent the general preparation received by the class or school group for learning science. It should be interpreted with caution though, especially because data from two predictors, the *Word Knowledge* and the *Mathematics* subtests, were available for Australia and England only (see also discussion in Sections 4.1 and 6.2.1). Further to this, data for one of these two subtests had been collected only in Hungary, Italy, Japan, and the United States. The mean loadings for the Word Knowledge and Mathematics subtests showed equally high values of $\bar{a}$=.99 and $\bar{a}$=.96.

For Sweden an attempt was made to include the *Language* of the students in class as a substitute to provide an estimate of the aptitude of the class, but it must be borne in mind that this, of course, has severe restrictions on the interpretability and meaning of 'aptitude'.

No data were available for Finland and the Netherlands, thus the LV was not incorporated into the path model for science achievement in these two countries.

SciValue The average science-related values of the students in a class could be estimated very satisfactorily by the attitudinal scales *Career Interest in Science* ($\bar{a}$=.87) and *Beneficial Aspects of Science* ($\bar{a}$=.90). This was underlined by the low standard deviations of the mean loadings. Non-Harmful Aspects of Science turned out to be a less powerful predictor in the five countries for which this MV was included in the final PLS runs ($\bar{a}$=.70). The measurement problems related to the latter scale have already been discussed in Chapters 5 and 6, so this slightly lower mean loading in comparison with the other two science values scales was to be expected.

Effort At the school level Effort was intended to describe the general degree of motivation of class or school. This effort was estimated using the two scales *Homework Effort* ($\bar{a}$=.90) and *Classroom Effort*, ($\bar{a}$=.75) and the *Hours Homework* the class spent on average per week. The latter MV was a suitable predictor in each of the countries with a high average loading of $\bar{a}$=.87. *Homework Effort* could be used to estimate Effort in Hungary (a=.94) and Sweden (a=.85) only. Data for the effort scales were not available for Australia, England, and Finland but as *Hours Homework* showed high factor loadings in all other countries it was considered appropriate to use this MV as the sole predictor for Effort in these three countries. Further information on the LV Effort has been given in Sections 4.1 and 6.1.2.

SciAtt High to very high factor loadings were found with the two attitudinal scales reflecting the construct SciAtt ($\bar{a}$=.90 for *Interest in Science;* $\bar{a}$=.77 for *Ease of Learning Science*) but differences occurred which of these MVs became the stronger predictor. Hence, it is safe to assume that the attitudes towards science of a class were sufficiently represented by the set of attitudinal items.

❑ **Achieve** In general, each of the three science subtests scores for a class of the subsets of items for biology ($\bar{a}$=.91), chemistry ($\bar{a}$=.92), and physics ($\bar{a}$=.91) worked equally well in predicting science achievement. However, two interesting observations should be mentioned.

In Finland, the factor loading for physics (a=.74) was somewhat lower than that of biology (a=.88) or chemistry (a=.86). This finding could not be explained by differences in achievement in the three subtests (see Postlethwaite and Wiley, 1991, p. 200) but it may be hypothesized that the lower loading of the physics subscore was the result of the heterogeneity between the groups of students studying biology, chemistry, and physics. As Keeves and Kotte (1991, Table 6.4) have shown very large effect sizes for physics achievement with respect to gender have been found for Finland's physics classes when compared with biology or chemistry classes at the 14-year-old level (see also the discussion in Section 5.2.1). The slightly erratic 'behaviour' of the factor loading of physics in Finland was not visible when looking at factor loadings at the between-student level. This leads to the speculation that there may be similar heterogeneity within the biology, chemistry, and physics classes in science achievement but more homogeneity in physics achievement between schools while the heterogeneity still seems to prevail for biology and chemistry when comparing schools in Finland. This explanation would be supported by the factor loadings for the science subtests in Finland found at the between-student level already discussed in Section 6.2.1 (see also Keeves and Kotte, 1991, Table 6.10).

Even with these minor inadequacies found in Finland the factor loadings for the biology, chemistry, and physics subscores suggest a justification of combining them into a single outcome measure of science achievement.

The factor loadings presented in Table 7.1 for the 33 manifest variables which were entered into a the PLS analyses to estimate the 11 constructs of the inner model for science achievement on the school level were satisfactorily high and the deviations between countries appear to be of minor degree. Thus, it seems appropriate to devise similar path models for each of the 10 countries for a cross-national comparison of gender differences in science achievement when using data aggregated to the between-school level.

As a summary, the extended recursive path model for science achievement showing all MVs and LVs is depicted in Figure 7.2 For clarity, inner model paths have been omitted as the figure is intended to illustrate the relationships of the outer model between the manifest variables and their respective constructs rather than the inner model relationships which have already been presented in Figure 7.1.

7.2 Student, Teaching and School Effects on Science Achievement

To estimate gender differences in student, teacher, and school-related effects based on the model developed in Figure 7.1, and using the MVs just described, separate PLS analyses were run for each of the 10 countries. Appendices 7.1 to 7.10 show selected summary statistics of the final run for each country.

FIG. 7.2 *Between-school within-country outer path model with science achievement as the outcome measure for the 14-year-old level, 1983/84*

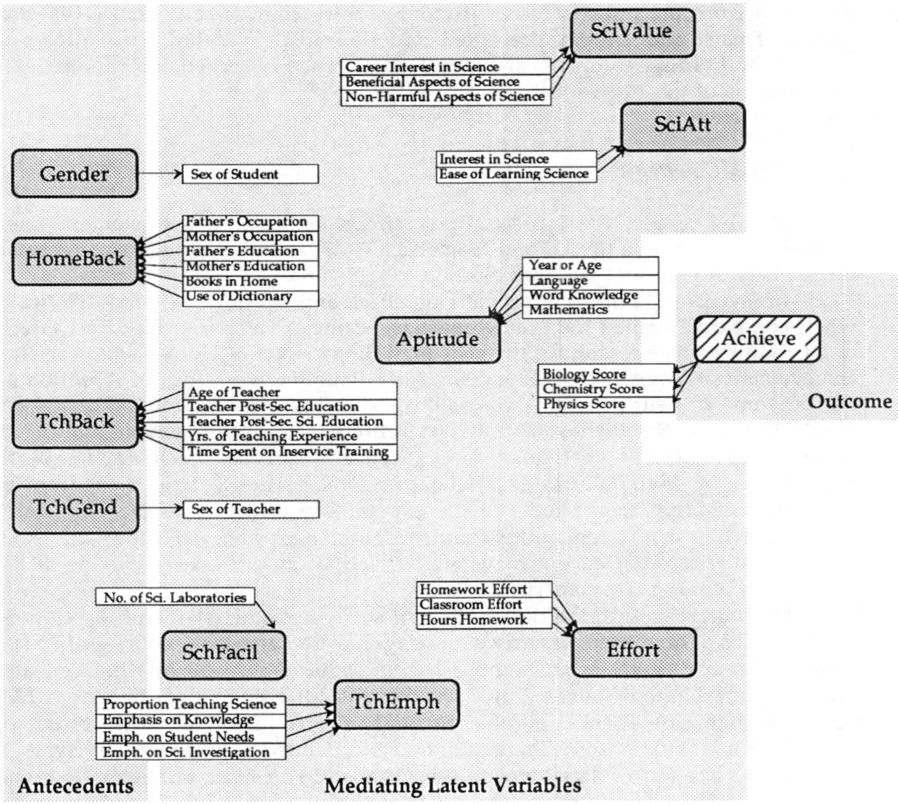

Notes:
For clarity, all paths between constructs (inner model paths) have been omitted

construct, latent variable (LV)

construct of outcome measure

manifest variable

outer model path, with arrow showing inward or outward estimation mode

Table 7.2 gives an overview of the direct and total effects of the inner model latent variables estimating science achievement in each country. Additionally, average direct and total effects, i.e. the grand unweighted means as well as their respective standard deviations have been calculated to express between-country variations for each LV. Also, the percent of variance explained by the model is presented for each country in the bottom row of Table 7.2. All figures are derived from between-school within-country data at the 14-year-old level obtained from the 1983/84 data collection.

In the following, direct and total effects on science achievement will be discussed from the point of view of cross-country comparisons before the path model of each countries is presented and described in detail. This allows to compare individual countries against general trends observed in cross-national comparisons of the 10 path models.

7.2.1 Cross-Country Relationships

On the whole, the largest direct effects on science achievement were exerted from Aptitude ($\bar{p}$=.62) and HomeBack ($\bar{p}$=.30) similar to what was reported for the between-student analyses (Section 6.2.2).

In Sweden, the direct effects of HomeBack and Aptitude differed drastically when compared against the grand unweighted mean values of the direct effects of all countries. The reason for the very high direct effect of HomeBack on science achievement in Sweden (p=.52) and the considerably lower effect of Aptitude on the outcome measure (p=.27) is possibly indicating the inadequacy of the LV Aptitude for the Swedish data as it was represented by the average age of the students in class rather than a more accurate measure such as the Word Knowledge or Mathematics test (see previous section). This measurement problem seemingly forced most of the explanatory power onto the HomeBack construct while in the remaining countries HomeBack and Aptitude had about similar influence on science achievement which is underlined by looking at the total effects of these two constructs.

In contrast to this, the direct effect of socio-economic status of the class as expressed by HomeBack was trivial in England and the United States (p<.10). Here, the major direct effect stemmed from Aptitude (p=.86 for England and p=.91 for the United States). But a look at the total effects of these two LVs suggests that HomeBack operates through Aptitude as the total effect of HomeBack amounts to t=.64 (England) and t=.66 (United States) respectively.

In the absence of data for Aptitude, Thailand showed a relatively low total effect of HomeBack on science achievement. This was partly the result of a comparatively low variance with respect to the socio-economic background of the school. This finding was also in line with the results of Keeves (1991) and Postlethwaite and Wiley (1991, p. 129) who stated the low influence of HomeBack on science achievement at the between-student level.

Gender had a moderate overall direct effect on science achievement favouring the boys ($\bar{p}$=-.18) coming through in seven countries. This gender effect was larger in the Netherlands (p=-.30) and Thailand (p=-.28). No sizeable direct gender effects were found in Finland, Japan, and Sweden. As the total effects of Gender were, on average, smaller than the direct effects a careful examination of gender differences in the path models of each of the 10 countries is required to locate the different ways Gender seems to exert (and lose) its influence on achievement. So, the total gender effect is down by about half in Thailand in comparison with the direct effect gender had.

TABLE 7.2 Direct and total effects derived from PLS analyses on science achievement as outcome measure for between-school within-country data for the 14-year-olds, 1983/84

Relationship to Science Achievement	Australia	England	Finland	Hungary	Italy	Japan	Netherlands	Sweden	Thailand	United States	average mean▲	average sd
Direct Effects												
HomeBack	.19	...	.34	.23	.16	.20	.44	.52	...	...	.30	.14
Gender	-.16	-.12	...	-.12	-.11	...	-.30	...	-.28	-.17	-.18	.08
TchBack	...	...	...	...	...	...	...	...	...	...	na	na
TchGend	-.14	...	...	...	.10	...	nd	-.14	nd	...	-.06	.14
SchFacil	...	...	.12	...	...	...	...	nd	.29	...	.21	.12
TchEmph	...	...	...	...	.16	...	.27	...	.23	...	.22	.06
Aptitude	.71	.86	nd	.65	.47	.49	nd	.27	nd	.91	.62	.23
SciValue	...	...	.16	...	...	...	.13	.22	...	...	.17	.05
Effort	...	...	.19	-.15	.14	.13	.24	...	.66	...	.20	.26
SciAtt	...	...	-.16	.23	...	.14	...	...	...	...	.07	.20
Total Effects												
HomeBack	.72	.64	.41	.53	.53	.43	.72	.56	.39	.66	.56	.12
Gender	...	-.14	...	...	-.15	...	-.26	...	-.16	...	-.18	.06
TchBack	...	...	...	...	...	...	...	...	.29	...	na	na
TchGend	-.21	...	...	...	.13	...	nd	-.14	nd	...	-.07	.18
SchFacil	...	.11	.12	...	...	...	...	nd	.26	-.11	.10	.15
TchEmph	...	...	...	.13	.26	...	.35	...	.35	...	.27	.10
Aptitude	.71	.86	nd	.64	.51	.53	nd	.23	nd	.91	.63	.23
SciValue	...	...	.12	...	...	.12	.22	.22	...	...	.17	.06
Effort	...	...	.15	...	.14	.19	.24	...	.66	...	.28	.22
SciAtt	...	...	-.16	.23	...	.14	...	...	...	...	.07	.20
% Variance Accounted for Science Achievement	**67**	**76**	**23**	**59**	**42**	**39**	**58**	**42**	**64**	**83**	**55◆**	**19■**

Notes:

▲ mean effects calculated as the sum of effects across all countries divided by the number of countries for which data were obtained (effects with |β|<.10 have been excluded)

◆ mean variance accounted for science achievement across 10 countries

■ standard deviation of mean variance

... the effect was discarded with |β|<.10

nd data not collected, or deleted prior to analysis, or all manifest variables not significant

na not applicable

It should also be noted that in Australia and Hungary the direct effect Gender had on achievement was cancelled out such that the total effect in both countries was non-significant, i.e. $|t|<.10$. Here, obviously gender-typical effects favouring the boys (directly) and the girls (indirectly) annihilated when looking at the size of the total effect HomeBack had.

A very interesting finding is the virtually non-existent influence of TchBack across the 10 countries. Only in Thailand, an experienced teacher makes the difference in class achievement with respect to science (p=.29). Though this result needs to be examined more thoroughly, it supports the observation of Postlethwaite and Wiley who reported an average direct effect of a construct labelled *Teacher Training and Experience* on science achievement of $\bar{p}$=.05 (1991, p. 127).

Fairly heterogeneous results were found with the LV TchGend. While a male teacher appeared to raise the achievement level of the class in Australia and Sweden - in both countries the direct effect of TchGend was p=-.14 - female teachers seemed to have higher achieving classes in Italy. But in all other countries there was no sizeable gender effect of the science teacher on the achievement level of the class.

Science facilities at school were of importance for science achievement only in Finland (p=.12) and Thailand (p=.29). This should not be misinterpreted in that way that science laboratories do not help or enrich teaching science at the 14-year-old level in the other countries. The result allows the conclusion that either not much variation occurred in the availability of school science laboratories within the other countries or that the use of school laboratories for science teaching did not have a big impact on the achievement level. The latter hypothesis could be attributed to the fact that the science subtests administered were testing students' theoretical knowledge rather than their practical skills, for which learning and experimenting in a school lab would be of assistance. Hence, school science facilities might become of importance mainly when testing practical skills and not for paper-and-pencil science tests. However, in Finland and Thailand differences seemed to exist in the number of laboratories per school and, indirectly, in the type of equipment. It could be speculated that not all rural schools in both countries were sufficiently equipped in their science laboratories if they possessed a science laboratory at all. Further to this, emphasis on practical work was perceived as low by the Finnish students in comparison with other countries (Oon Chye, 1991, p. 25). The finding that the number of laboratories was related directly to achievement in science in Thailand contrasts earlier observations by Fuller and Heyneman who stated that "science laboratories...in Third World countries...do not consistently boost student achievement and may not be relevant to teaching basic science scientific topics" (1989, p. 17). At least in the case of Thailand the presence of school laboratories appeared to influence science achievement very positively.

The slight negative overall effect SchFacil had on science achievement in the United States (t=-.11) may be taken as an alarming sign for an adequate use of the equipment of science laboratories at certain schools. Though no direct relationship could be detected it cannot be rejected that inadequate school facilities, in turn, had a negative influence on the quality of teaching science, an assumption underlined by students reporting an unsatisfactory opportunity to learn science in the United States (Keeves, 1991, Table 4.8).

In Italy (p=.16), the Netherlands (p=.27), and Thailand (p=.23) the emphasis a teacher put on science teaching had a moderate positive direct effect on science achievement. The remaining countries showed no effect except for Hungary where a low total effect of t=.13 was found.

Except for Australia, England, Sweden, and the United States higher classroom and homework efforts increased the level of achievement in science. The average direct effect amounted to $\bar{p}$=.20. Its high standard deviation was mainly due to the very pronounced direct effect of Effort in Thailand which led to the skewed distribution. In Thailand, the motivation of the class was indeed the key factor influencing science achievement (p=.66).

Science-related values which, on the between-student level had an effect on science achievement only in Finland, the Netherlands, and Sweden, also had a direct effect in Japan (p=.12). Overall, the effect of SciValue was small ($\bar{p}$=.17).

At the between-school level the effects of science-related attitudes on achievement occurred in only three of the 10 countries, namely in Finland, Hungary, and Japan. As with SciValue the effect of SciAtt within the path model of science achievement was also low ($\bar{p}$=.07). Of some interest was the negative impact SciAtt had in Finland (p=-.16). This was not necessarily to be expected because at the between-student level the direction of the effect of science-related attitudes was found to be positive (p=.14). An explanation for this phenomenon might be the organizational differences between the Finnish schools.

As a consequence of aggregating data to the class or school level, a higher proportion of variance explained was to be expected (see discussion about aggregation bias in Section 4.2; see Cheung and Tsoi, 1990, pp. 245-256; Sellin, 1990 [2], pp. 267-268). However, due to the introduction of four additional school-level constructs (TchBack, TchGend, SchFacil, and TchEmph) the percentage of variance explained by the path model for science achievement in each country could be still be increased. It is, of course, not possible to exactly determine the amount of aggregation bias which means that the actual contribution of school-level variables to the country-specific path models cannot be judged precisely.

Despite these procedural shortcomings some interesting observations should be mentioned. In adding three school-level variables (TchGend was missing in both countries) the adequacy of the path model could be increased markedly for the Netherlands and Sweden even despite the fact that no data were available for Aptitude. In contrast to this gain, the introduction of school-related variables hardly contributed to the adequacy of the models for science achievement in Finland and Japan. In these countries factors not yet considered seemed to account for a larger portion of achievement in science. An indication of which type of factors seemed to affect science achievement in these two countries is given by Postlethwaite and Wiley:

> "Substantively, the above findings imply that, compared with other societies, the Nordic countries and Japan have strong societal factors which homogenize science achievement across classes and schools, but which leave within school variation similar to that over most of the rest of the world. ...these must be educational rather than background effects." (1991, p. 61)

Overall, the proportion of variance explained is high and seems to underline the adequacy of the basic path model shown in Figures 7.1 and 7.2. It is, of course, understood that the higher amount of variance explained by this between-school path model may, to a large extent, be the result of the aggregation of student data to the school level.

While it appeared justified to propose and describe an 'average' path model which explained the relationships of the latent variables on science achievement at the between-student level the results of Table 7.2 suggest that this might be undesirable at the between-school level since there was greater heterogeneity at the between-school level when comparing the path models of the 10 countries. Thus, in the following section, the path models for each country will be presented and discussed separately in terms of gender effects on science achievement. Also, for each country a short general statement relevant for policy-makers and educators will be given.

7.2.2 Comparison of Country-specific Path Models

➞*Australia* (Figure 7.3) This model explains 68 percent of variance of science achievement, therefore constituting a fairly accurate picture of the relationships influencing achievement at the 14-year-old level.

Both science-related attitudes (p=-.19) and achievement (p=-.16) are affected negatively by Gender which indicates that single-sex boy schools and classes where there are more boys than girls showed a higher performance in science and more favourable attitudes towards science. On the other hand, single-sex girl schools and classes with a minority of girls, on average, had a lower achievement level and a less favourable attitude (Keeves, 1991).

Nevertheless, though science-related values strongly influenced science attitudes as well as did Effort, there were no measurable effects from either SciValue or SciAtt on science achievement, a finding already observed at the between-student level.

The most important predictor for achievement in Australia was HomeBack operating directly (p=.14) and indirectly through Aptitude (p=.71). It is interesting to note that although Gender affects science achievement directly it also has an indirect effect through Aptitude which looking at the total effects cancels out its direct influence (p=-.16 versus t=-.03; see Appendix 7.1). In other words, while the sex of a class (school) was directly influencing achievement in science girls gained superiority in science achievement by showing a higher score with respect the Word Knowledge and Mathematics skills. As can be seen further, the effect exerted by Aptitude appeared to be fairly gender-biassed.

Moreover, Aptitude became also gender-biassed by TchGend. Interestingly enough, TchGend showed a direct (p=-.14) and indirect effect operating through Aptitude on science achievement which caused its total effect (t=-.21) to be even higher. This suggests that it did make a difference for the achievement of a class in science whether the teacher was male or female: classes of male teachers performed better, however difficult this may be to interpret. Also, the possibility of artefacts due to aggregation bias should be taken into consideration.

Whether these gender effects can be further attributed to single-sex schools or whether they are also found in coeducational schooling, which would imply that the gender proportion in class becomes of importance remains to be seen. The reader is referred to the analyses presented in Chapter 8.

FIG. 7.3 *Between-school path model for Australia with science achievement as outcome measure; 14-year-old level, 1983/84; N=233*

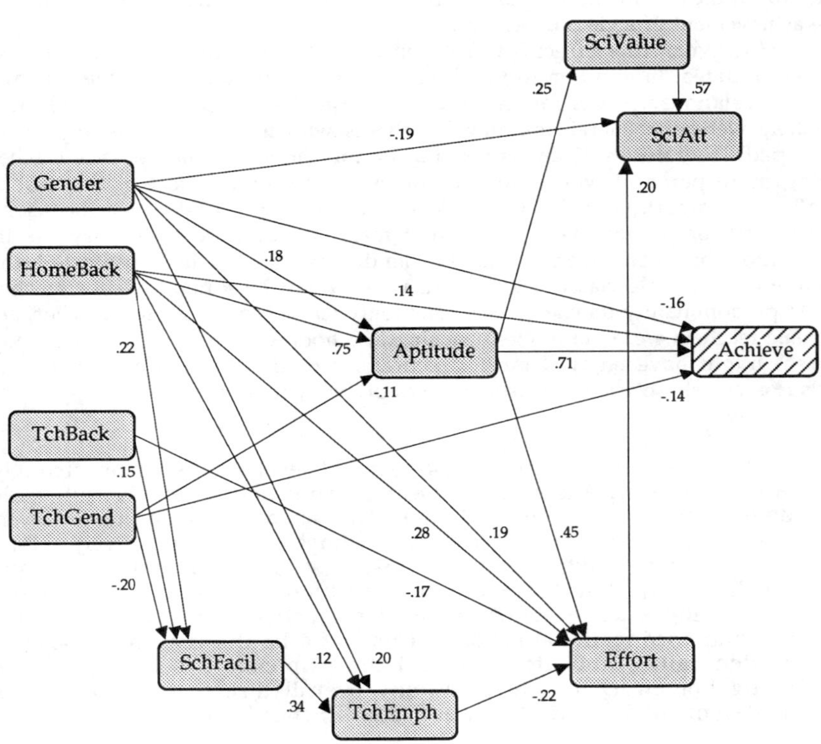

Notes:

paths with |β|<.10 have been omitted; for further details see legend for Figure 7.1

Two other direct gender of class effects should be noted. Gender influenced Effort positively (p=.19), in that the motivation of predominantly female classes was reported as higher. Yet, as Effort appeared not to influence achievement directly or indirectly, a result also found on the between-student level, this would have only boosted the attitudes a class has towards science but it was not related to science achievement.

The other class-related gender effect found was a positive influence on TchEmph (p=.20). The gender-type of a science class appeared to have led to a different emphasis on what was taught in science. Why did Australian teachers view girls as learning science differently than boys? As it was the proportion of teaching science, i.e. the proportion of the time a teacher spent teaching science to the time spent on teaching science and other subjects, which constituted the LV TchEmph, predominantly girls classes were taught by teachers who were more specialized in science instruction. Noting the negative direct effect TchEmph had on Effort (p=-.22), indicating that the higher the proportion of teaching science was the lower the amount of homework given became (Homework Effort was the only indicator in the LV Effort). But this had, surprisingly, no influence on the achievement of a class in science.

However, the effect of TchGend on achievement should not be underestimated. In addition to the direct effects on Aptitude and achievement, a marked relationship was found between TchGend and SchFacil (p=-.20). Male teachers were, in general, teaching in schools with more (and, possibly, better equipped) laboratories. If this is taken as an indication for male science teachers doing more practical work and experiments - the direct effect of SchFacil on TchEmph (Proportion of Teaching Science) was strong (p=.34) - then the science instruction for predominantly male or female classes appears to vary. As the correlation between Gender and TchGend was r=.40, which indicated that predominantly male classes more often were taught by a male science teachers while predominantly female classes commonly had a female science teacher, and as male teachers were more often teaching in school with more laboratories, boys classes might have enjoyed more practical instructions and experiments than girls classes, whose science instruction was probably more theoretical.

Science teaching at the 14-year-old level seems to be strongly-gender typed in Australia at the between-school level. Overall, male-dominated classes achieve better. The socio-economic status of the parents in the catchment area of a school is highly positively related to the general ability level and, thus, to science achievement. It could be, however, that there are major differences among the factors influencing science achievement at the school level between single-sex and coeducational schools. This could be the reason for science-related values and attitudes and the emphasis a school places on doing homework not to show any overall direct effect on science achievement. This will be investigated further below.

➞*England* (Figure 7.4) Similar to the Australian results, science-related values and attitudes did not exercise any influence on the class achievement in science in England. Thus, the small direct effect SciAtt had on achievement at the between-student level (p=.12) vanished completely.

In England, achievement at the between-school level was influenced mainly by the socio-economic status of the class operating indirectly through Aptitude and SchFacil, where the latter might be interpreted broadly as an indicator for the 'richness' or status of the school. The total effect of HomeBack on science achievement amounted to t=.64.

But the only other direct effect on achievement was that of Gender. As in Australia, predominantly boys classes showed a higher performance than classes with predominantly girls. Again, it is unclear at this stage whether this finding can be attributed to the different gender types of schooling, i.e. single-sex versus coeducational schooling. Further clarity is expected from additional analyses with HLM reported in Chapter 8.

The sex of class had two other direct effects. On the one hand female-dominated classes seemed to put slightly greater efforts in learning science (p=.16) than male-dominated classes did. On the other hand, male-dominated classes were found in schools equipped with more laboratory facilities (p=-.16).

No school level construct (TchBack, TchGend, SchFacil, TchEmph) influenced science achievement. The small direct effect of SchFacil on Aptitude (p=.13) was associated with the effect of HomeBack on SchFacil (p=.19). Nevertheless, the proportion of variance explained for the path model in England was very high and with 76 percent significantly higher than that of the path model for the between-student level (43 percent). This seemed to support the assumption that the within-class variance with respect to science achievement was lower than the between-school variance, a result in retrospect justifying the high intraclass correlation coefficient reported for sampling in England (Keeves, 1991). This finding was also in accordance with the results of the FISS study reported by Comber and Keeves (1973).

As in Australia the correlation between student gender and sex of teacher (TchGend) was high (r=.53) indicating that male-dominated classes were taught most often by male teachers and, likewise, that female dominated-classes were commonly taught by female teachers. Yet, TchGend had no direct or indirect influence on the achievement level of a class. Male teachers affected the manifestation of science-related values more favourably than did their female colleagues (p=-.12), and male teachers also were teaching in schools which were better equipped (p=-.13). Yet, female science teachers were putting a slightly higher emphasis on scientific investigation in their instruction. This did not, however, lead to higher efforts or achievement of the class but the effect of TchEmph on SciValue (p=.19) and SciAtt (p=.17) should not be overlooked. It should be added, though, that TchEmph was mainly influenced by TchBack in that teachers with less inservice training focussed more on principles of scientific investigation when teaching science.

Another key finding in England at the between-school level is that the amount of homework given seemed very much dependent on the status of the class. Classes, whose students had a higher home background or status, had to do much more homework (p=.38) underscored by the high direct effect coming from Aptitude (p=.48) through which HomeBack operated indirectly on Effort, too. Noting the positive gender effect of p=.16, this suggests that single-sex girls schools stressed the concept of homework in science to a much greater extent than other did schools. This result will be elaborated further in Chapter 8.

FIG. 7.4 *Between-school path model for England with science achievement as outcome measure; 14-year-old level, 1983/84; N=146*

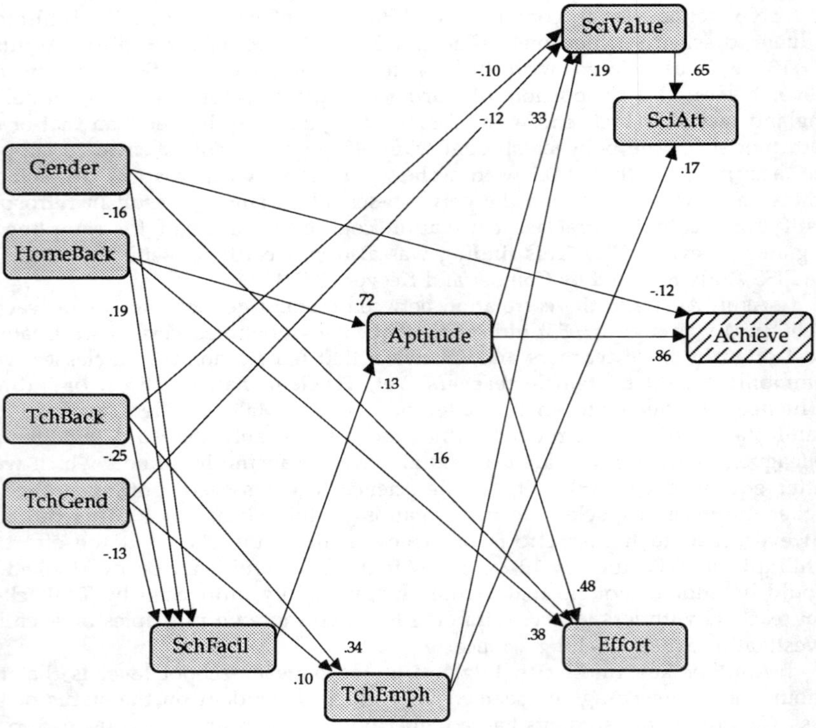

Notes:
paths with |β|<.10 have been omitted; for further details see legend for Figure 7.1

In general, it must be seen that the path model for England reflects a conservative, status- and achievement-oriented educational policy with respect to science instruction. It was home background which was, most likely in conjunction with the gender-types of schooling (single-sex versus coeducational) to determine the achievement of a school in science. Science-related values and attitudes as well as teacher-related effects played only a very minor role in influencing the outcome measure. Clearly, further research is necessary to shed light on teacher and teaching related effects.

FIG. 7.5 *Between-school path model for Finland with science achievement as outcome measure; 14-year-old level, 1983/84; N=90*

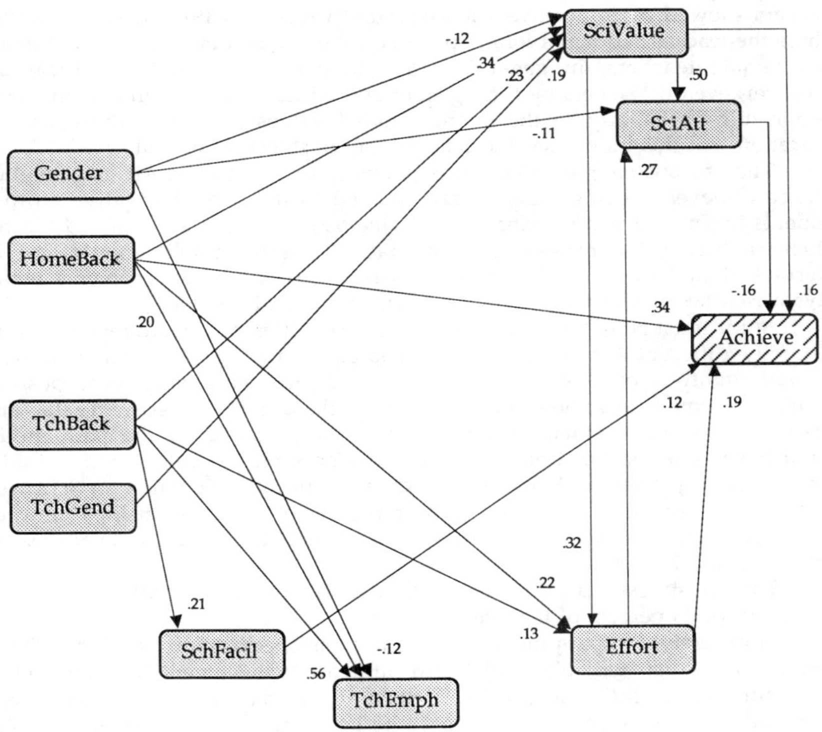

Notes:
paths with |β|<.10 have been omitted; for further details see legend for Figure 7.1

➞*Finland* (Figure 7.5) In contrast to Australia and England, gender differences were operating through science-related values and attitudes. With both constructs, SciValue and SciAtt, low negative gender effects were identified (with p=-.12 on SciValue and p=-.11 on SciAtt respectively). This would indicate that male-dominated classes - there were no single-sex schools in Finland - stimulated the manifestation of science values and attitudes in class. Though gender influenced these two constructs, the effect did not carry through to affect science achievement as can be seen by the total effects of SciValue (t=.12) and SciAtt (t=-.16) (see Table 7.2). This could also explain why SciValue had a positive and SciAtt had a negative effect on achievement: the gender effect was in the same direction for both constructs so values and attitudes were influenced differently by other LVs of the inner path model. It might further be concluded that although a class-related gender effect on science values seemed to exist in

Finland this effect was virtually annihilated when analyzing the impact SciValue had on science achievement.

While the socio-economic status of the class apparently turned out to be the biggest predictor of SciValue (p=.34) it should be noted that classes of female teachers showed more positive science-related values (p=.19) and so did classes where the teacher had had a longer post-secondary education (which is not to say that female teachers, in general, had a longer post-secondary education). However, even if teachers by their gender and educational background might be responsible for modifying the science-related values of a class there was no indication that this influenced the achievement of the class as well.

The obvious counter-effects of SciValue (positive) and SciAtt (negative) on science achievement might be, in part, related to the overall negative attitude students in Finland had in science (see Table 6.2), an observation already stated when analyzing the between-student data. Nevertheless, it should not be overlooked that these results might also reflect specification problems which have arisen because of multicollinearity within the path model.

But it must be hypothesized that as gender differences were reported to be distinctly different for SciValue when compared with SciAtt such that boys showed slightly more favourable attitudes and girls marginally more positive values this would, to some extent, account for the different effects SciValue and SciAtt had on achievement. However, this interpretation of the path model might have gone too far. Gender effects on science-related values and attitudes were reportedly small and so were the diverse effects of SciValue an SciAtt on science achievement. To obtain a clearer picture of the relationships between gender, values, attitudes and science achievement more specific analyses are certainly needed.

This is stressed also by the fact that the proposed path model only accounted for 23 percent of the overall variance.

Most likely, the path model for Finland has specification problems. Those manifest and latent variables which **did** influence achievement in science were not incorporated into the model. Indeed, they may not even have been measured. At this stage, there can only be speculation about which variables might have accounted for the variance of achievement between classes. There may be regional differences between rural and urban areas but it could have also been the absence of a suitable measure for aptitude in Finland. Another possibility in explaining the structural problems inherent in the path model presented could have been the comparatively low variability on the outcome measure which made it more difficult to identify clear-cut factors which influenced science achievement. However, Postlethwaite and Wiley (1991, p. 129) could also only unsatisfactorily explain science achievement in Finland although they had taken care of splitting the HomeBack effect into status-oriented and literacy-oriented effects and at the same time introducing two further constructs reflecting the teaching style and amount of experimental work done in science classes. Their path model, though applied to between-student data, was only able to account for 19 percent of the variance. Keeves, in his attempt to analyze science achievement at the 14-year-old level using between-student data, could not explain more than 16 percent of variance either (1991, Table 6.11). This underscores the difficulties in arriving at a detailed picture of how gender differences really influence science achievement in Finland and, indeed, further analyses which are beyond the scope of this work are required.

One further suggestion of how gender affected achievement in science can probably be obtained by the fact that predominantly male classes were taught differently than female-dominated ones. Whether this or other methodological differences in teaching boys or girls played some importance in creating gender differences with respect to science achievement was unclear.

Finally, it should be pointed out that there were achievement differences between the sexes with respect to the science subjects, a finding underlined by the somewhat lower loading of the physics score on achievement in comparison with the biology or chemistry scores (see Table 7.1). Major gender differences in physics had already been reported at the between-student level. However, additional PLS analyses at the between-school level run separately for biology, chemistry, and physics (the results are not shown here) indicated only very marginal differences in the path model presented in Figure 7.5. In no cases, the explained variance of the model could be raised.

In Finland, the science-related values and attitudes played an important role in influencing the achievement level of a school in science. While attitudes were negatively related to achievement the effect of values was found to be positive. Though this could have been a sign of gender differences it is highly likely that misspecifications within the path model, missing data and/or multicollinearity problems had accounted for it. Certainly, further analyses are required to find out to which degree instructional differences in science teaching contribute to gender differences.

→*Hungary* (Figure 7.6) The path model for Hungary is characterized by a multitude of paths linking the constructs in the inner model. The model explained 59 percent of the variance of science achievement.

Though the direct gender effect on achievement was found to be p=-.12 favouring male-dominated classes, its total effect of t=.08 (see Appendix 7.4) indicated that gender effects operated in different ways for classes which were dominated in numbers by either boys or girls.

Similar to what was observed in Australia, Gender (p=.32) and HomeBack (p=.39) showed a marked positive effect on Aptitude. While female-dominated classes showed higher Aptitude they, at the same time, enjoyed a stronger emphasis on scientific investigation during science instruction. The latter finding combines with the fact that the science teachers perceived, wrongly or rightly, girls were more receptive to theoretical-oriented instructional strategies such as the teaching of scientific principles.

FIG. 7.6 *Between-school path model for Hungary with science achievement as outcome measure; 14-year-old level, 1983/84; N=99*

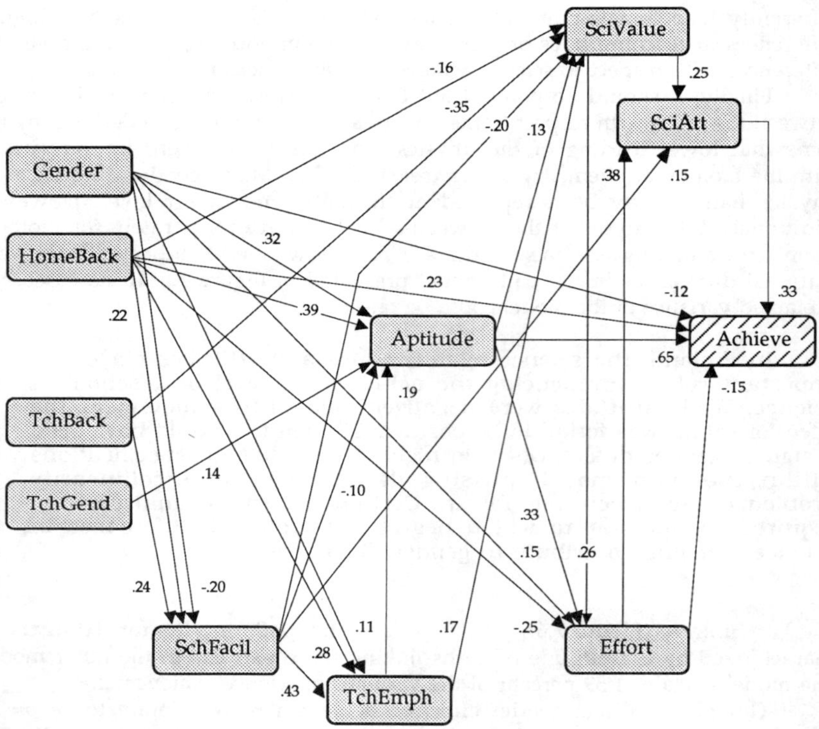

Notes:

paths with |β|<.10 have been omitted; for further details see legend for Figure 7.1

Yet, gender differences between classes seemed to have been caused by the equipment available for science teaching at school, too. Despite a centralized Hungarian school system, embedded in a socialist, political setting, it **did** matter which school a boy or girls went to for science instruction. More often, male-dominated classes were in schools with more and, presumably, better science equipment (p=-.20). These classes were also taught more commonly by experienced science teachers as indicated by the direct effect from TchBack on SchFacil (p=.24). It may be concluded that this finding was in reality a reflection of the differences existing between rural and urban areas, where in urban area schools were often better equipped and had more experienced science teachers (Keeves, 1991). If this was so, it could also explain the negative direct effect found between SchFacil and Aptitude (p=-.10): in less-equipped schools (meaning smaller schools located in rural areas) classes had slightly higher Aptitude. If, at the same time, it were female-dominated classes which showed higher Aptitude, girls profited from studying science in smaller classes, probably due to a more gender-identical environment in which male domination as observed with practical, experimental work was less likely (Kahle and Lakes, 1983; Smail, 1984 [2]; Kahle, 1989).

Another direct effect coming from the sex of the class positively influenced the motivation and effort of the class. This indicated that, in general, where girls outnumbered the boys in class the amount of effort spent on science at home and in class was higher than that of classes where boys were in the majority. Extending the assumption that in rural schools the ratio of boys to girls was, more commonly, in favour of girls these rural classes showed stronger efforts with respect to science (p=.15) than urban schools where boys were dominating the classes in terms of numbers. This explanation would be supported by the medium-sized negative path coefficient between HomeBack and Effort (p=-.25) which could as well be seen as a regional difference between towns and countryside in Hungary: rural homes showed a lower socio-economic status.

Despite the somewhat deprived educational and economic situation of rural areas, where about half of the population lived, the Hungarian educational system offered a considerable amount of equality with respect to girls studying science. Hence, the motivation to study science observed in rural classrooms was higher than in town schools. In studying harder and spending more time on science girls 'accepted' the educational opportunities offered. As reported earlier, the perception of learning science as being easy showed the smallest gender differences in Hungary (see Table 6.1). Likewise, overall gender differences in science achievement were smallest in Hungary when compared with the other nine countries (see Table 5.1).

At this stage it should be recalled that in comparison with other countries taking part in the SISS study, Hungarian students performed highest at the 14-year-old level (Postlethwaite and Wiley, 1991, p. 60). Interest in science was highest, too, while gender differences were found to be small both in achievement as well as in science-related attitudes and values.

This would also explain why no direct gender effects were found in the path model leading from Gender to SciValue and SciAtt: basically gender differences were non-existent. Nevertheless, there was some indication that regional differences enforced the manifestation of different science-related values. Classes in town schools seemed to show lower science values if HomeBack could be interpreted as being parallel to the geographical location of the class or school (p=-.16). Provided, less experienced teachers were also more likely to teach in rural areas (the correlation between HomeBack and TchBack was r=.22) these teachers were able to increase the science-related values within their classes, in contrast to experienced (urban) teachers whose classes showed a lower career interest and more often saw less beneficial aspects in science (p=-.35).

It should be noted, however, that science-related attitudes, although hardly different between boys or girls, and male- or female-dominated classes, were much less directly influenced by the students values (p=.25) than in other countries. A possible explanation for this finding could, once again, be a discrepancy between rural and urban students: in order to escape the relatively poor employment chances in rural areas migrating to towns and taking up a career in science seemed a promising possibility for many. Despite the socialist rhetoric educational and occupational opportunities were clearly favourable in urban areas (Budapest, Miscolc, Debrecin, Szeged, Pécs) than the possibilities offered when working in an agricultural field (mainly the only available type of job in rural Hungary).

Similarly, it was surprising to see the strong total effect HomeBack had on science achievement (t=.53). This was presumably, in part, another reflection of the socio-economic differences between rural and urban schools.

Gender differences in science achievement were small at the between-school level in Hungary when compared with other countries. However, there were some indications for urban/rural differences both in terms of gender and science teaching. Keeping in mind that Hungarian students performed well in science at the 14-year-old level in comparison with other countries, it may be still advisable to educational planners to improve the science laboratories and put more emphasis on practical work especially in school located in rural areas.

➞*Italy* (Figure 7.7) As was the case in Hungary, Gender neither affected SciValue nor SciAtt. At the same time, science-related values and attitudes had no direct or indirect influence on the outcome measure. This would seem to suggest that values and attitudes towards science played a remarkably lower role in influencing achievement in science than in other countries, such as Finland or the Netherlands.

The main effect on science achievement was exercised by the socio-economic status of the class (t=.53) where the classes of students from higher status homes had higher test scores. As with Hungary, this finding might represent the latent geographical difference both between the 'rich' North of Italy (Milan, Turin, Genoa) and the 'poor' South (Naples, Sicily) as well as between rural and urban areas, the latter areas having, on average, higher socio-economic status (Visalberghi, 1988).

This would explain the marked positive relationship between HomeBack and SchFacil (p=.38) and, likewise, the strong positive effect HomeBack had on Effort (p=.34). In the 'richer' Northern provinces of Italy science classes spent more time on learning science and doing homework than in the rural schools in the South. In addition, rural schools were equipped with fewer laboratories than schools of urban areas where the students were better off at home.

The direct gender effect on science achievement was low (p=-.11), the total effect was just marginally larger (t=-.15). In both instances, predominantly male classes achieved higher.

FIG. 7.7 *Between-school path model for Italy with science achievement as outcome measure; 14-year-old level, 1983/84; N=224*

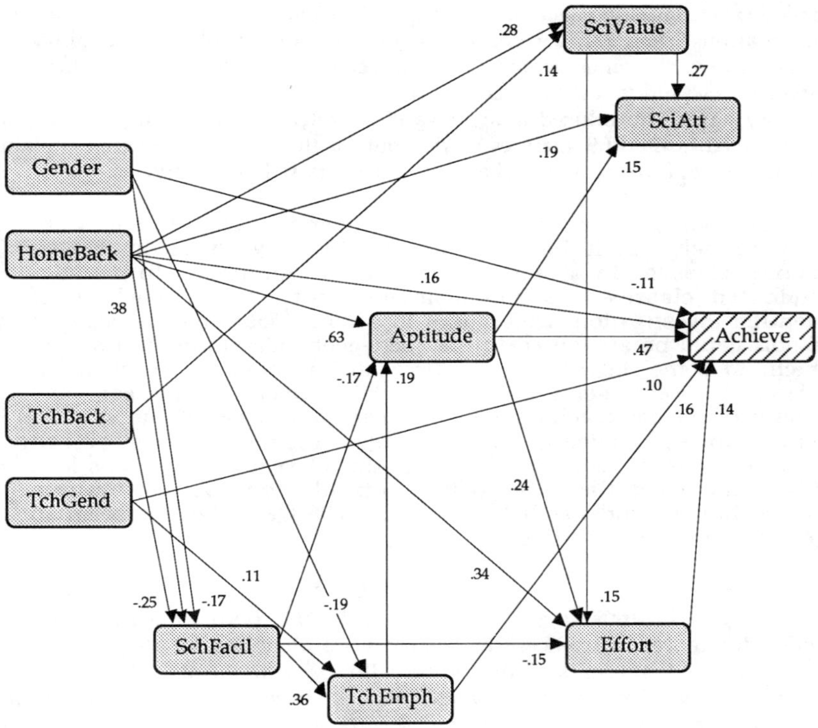

Notes:
paths with |β|<.10 have been omitted; for further details see legend for Figure 7.1

More frequently female-dominated classes were taught in schools with less laboratory equipment available. In addition, girls were more commonly found in classes where teachers put less emphasis on scientific investigation. This different emphasis for boys- and girls-dominated classes was aggravated by the poorer laboratory equipment found in these schools as the relationship between SchFacil and TchEmph seemed to suggest (p=.36). Given these preconditions of poorer science equipment, which could indicate that predominantly female classes undertook fewer practical exercises and experiments in science lessons compared with male-dominated classes, and a lower emphasis put on the teaching of scientific principles, then these could help to explain the marked achievement differences found between boys and girls at the 14-year-old level (Table 5.1).

TchGend showed little effect on science achievement (p=.10) and the experience of the science teacher was of virtually no effect at all (t=.01). The total effects on science achievement stemming from TchEmph (t=.26) and Effort (t=.14) were, most likely, resulting from the socio-economic situation of the class/school as indicated by the correlation between HomeBack and TchEmph (r=.20) and the direct effect from HomeBack on Effort (p=.34). Hence, achievement differences in science at the *Scuola Media* were to a large extent the result of status differences of the students' homes which were reflected as socio-economic differences between 'rich' and 'poor' schools.

By having introduced a latent variable reflecting 'regionality', which, was not included in the PLS analyses carried out for this study, the overall explained variance of 42 percent could probably have been raised decisively.

Though gender differences in science achievement were comparatively small between schools in Italy, male-dominated science classes, in general, were better equipped. In female-dominated classes less emphasis was put on the teaching of scientific investigation. Differences in science achievement between schools could be attributed to the socio-economic status of a school which, to some degree, was a surrogate measure for rural-urban differences. Nevertheless, it is considered useful to increase the amount of science teaching and practical work, especially in schools located in economically disadvantaged regions as the overall performance of students in science at the 14-year-old level was low. This would meet the very positive attitudes and values towards science Italian students held which, in 1983/84, did not seem to promote achievement in science.

→*Japan* (Figure 7.8) At the between-school level no gender differences in science achievement were identified for the age group of the 14-year-old students. This would allow for the conclusion that class performance in science was very homogeneous across schools in Japan in terms of gender. These differences, though observed at the between-student level (see Table 5.1), vanished completely when investigating class performance in science (t=-.01).

The only gender effect traced at the between-school level was on SchFacil. Its extent was minor indicating that some predominantly male classes enjoyed slightly better school facilities for studying science (p=-.11). In reference to the overall path model this gender effect might be arbitrary.

Though considerable gender differences were noted within classrooms in terms of different science-related values and attitudes (see Table 6.1) these disappeared when comparing classrooms. As boys and girls were nearly equally distributed across all classes and as no single-sex schools existed in Japan's educational system this result was not surprising.

Japan's educational policy, in particular related to science subjects, offers equal opportunities to boys and girls (NIER, 1988). Science plays a very important role throughout formal schooling and is widely accepted as one of the key sources for Japan's economic success which is largely technologically based (Rosier and Keeves, 1991). The science curriculum is centrally developed for all '*kens*'(districts) and planning within the educational system occurs 'top down' (Kanaya, 1988).

FIG. 7.8 *Between-school path model for Japan with science achievement as outcome measure; 14-year-old level, 1983/84; N=199*

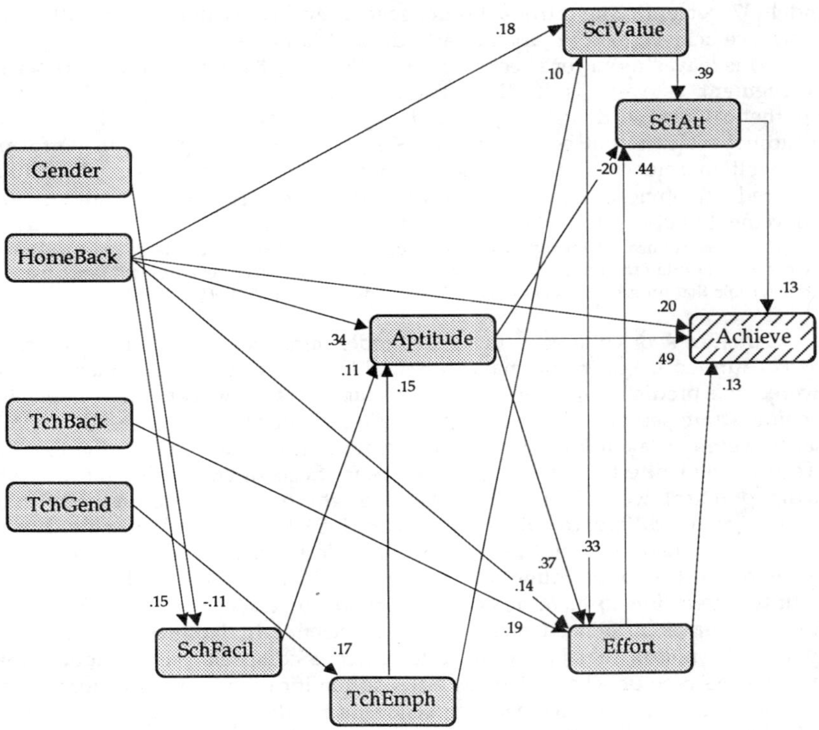

Notes:
paths with $|\beta|<.10$ have been omitted; for further details see legend for Figure 7.1

This centralized educational policy can be regarded as a major force behind the obvious homogeneity in science achievement between boys and girls (NIER, 1988).

Consequently, science achievement at the between-school level was largely a function of the socio-economic background of the school (HomeBack; t=.49) and the aptitude of the class (t=.53). Although HomeBack operated through Aptitude and also through Effort, Aptitude turned out to be the strongest overall predictor for science achievement. This would suggest that socio-economic differences still existed between schools even when taking into account that neither TchBack nor SchFacil nor TchEmph had a total effect of t≥.10 on science achievement. Thus, it was not the influence of the teacher- and school-related variables that promoted differences in performance between classes. Rather, it was the socio-economic situation of the students' homes which was responsible for better or worse achievement in science at the 14-year-old level. The relatively small influence of school- or teacher-related variables was underlined by the fact that the variance accounted for by the path model presented in Figure 7.8 was only two percent higher (39 percent) than the variance explained by the between-student path model (37 percent). The introduction of school level constructs into a path model for science achievement in Japan clearly did not improve it.

This leaves room for speculation why the model still fails to predict science achievement as well as it did for other countries. Although this may be hypothetical, the students' neutral and negative responses to the items in the attitudinal scales might be one reason (Keeves, 1991). Though Japan performed very well in science at the 14-year-old level, the generally neutral attitudes indicated an obvious mismatch between the interest in science and science achievement of students. Kida suggested that...

> "...science education for the student at school as well as the public at large must make clear the multilateral relationship between science and social development, and the crucial role that progress in science plays in social development." (1991, p. 195)

Another shortcoming of the path model might have been that HomeBack was constituted by only one manifest variable (Books in Home) which, in turn, lowered the predictive power of the construct. It is, however, more likely that teaching strategies and methodological or instructional differences in teaching science were not taken into proper consideration in the path model due to a lack of respective manifest variables. Group or peer effects of boys being more 'pushy' during practical work in science lessons might have also accounted for the observed gender differences (Kahle and Lakes, 1983; Smail, 1984 [1]; Kahle, 1989).

At this stage, it is not possible to provide further information to explain the gender differences found at the between-student level. However, it is certainly interesting to have detected no school- or teacher-related effect strong enough to have influenced achievement in science in Japanese schools. This supports the notion that Japan had established a comprehensive school system which by the type of school, courses offered, qualification and sex of teacher did not foster gender differences between classes, a result supporting the findings of the FISS study (Comber and Keeves, 1973, pp. 280-284; Kojima, 1974).

> At the between-school level basically no gender differences were found in science achievement in Japan. Science achievement of schools was influenced by the socio-economic status of the catchment area of a school and, related to this, the general ability level of the students. Teacher- and school-related variables apparently did not affect science achievement noticeably. Though classroom efforts and the amount of homework done very positively stimulated the attitudes towards science, the effect of science-related values and attitudes on science achievement was minimal. The obvious mismatch between science achievement and science attitudes appeared to reflect a general problem of the rapid technological changes introduced to the rather closed-up Japanese society, a problem educational planners should take care of.

→*The Netherlands* (Figure 7.9) As hypothesized in Section 6.2.2, the introduction of school-level variables increased the variance explained by the model from a mere 18 percent on the between-student level to 58 percent on the between-school level. School- and teacher-related variables seemed to have influenced science achievement to a large extent in the Netherlands. This became even more true as Aptitude and TchGend had to be removed from the path model due to the lack of data.

The direct negative gender effect was large (p=-.30), a finding already made at the between-student level where the Netherlands produced the highest gender differences of all 10 countries. Male-dominated classes - there were no single-sex schools in the sample - had much higher science test scores than classes in which girls outnumbered boys. But this direct effect was partially countered by the indirect gender effects operating through TchEmph and SciValue as could be seen by the lower total effect that gender had on achievement (t=-.26). While predominantly female classes were classes where the proportion of science taught was higher than in comparable male-dominated classes (p=.20) 'girls classes' reported both higher science-related attitudes (p=-.23) and values (p=-.11). SciAtt had no influence on science achievement whereas the total effect from SciValue on achievement amounted to p=.22. However, the indirect effect of Gender working through SciValue was small. This indicates that although in predominantly male classes more favourable science-related values were observed, it was largely attributable to the HomeBack component which very strongly affected SciValue (p=.54).

Thus, the relationship of Gender and TchEmph becomes important to be looked at more carefully as it is through the proportion of teaching science where girls apparently profited in terms of achievement compared with male-dominated classes.

TchEmph, in the Dutch educational system's context, was suggested to be a construct partly reflecting the various school types students were enrolled in at the 14-year-old level. The relative complexity of the educational system at the junior secondary level can be nicely illustrated by the description of the target population for the Netherlands, which consisted of:

> "All students in Year 9 of *Gymnasium, Atheneum, VWO* (general secondary, pre-university), *HAVO* (general secondary, senior), *MAVO* (general secondary, junior), *LTO* (lower technical vocational), *LHNO* (lower vocational: domestic science), *LEAO* (lower vocational: administrative and business), and *LAO* (lower vocational: agricultural) schools on 31 May, 1984. Students in schools which were various combinations of the above as well as students in the *Middenschool, LNO* and *LMO* were excluded. The excluded group represented about 2 percent of the population." (Postlethwaite and Wiley, 1991, p. 174)

Without going into detail here, it could be inferred that *Gymnasium* and *Atheneum* offered more academically-oriented science instruction when compared with the other school types (Rosier and Keeves, 1991). Hence, science teachers were more specialized in these two school types to teach either biology, chemistry, or physics. The reader should be reminded that it was in physics where the biggest gender differences occurred (see Table 5.3). If TchEmph, in part, could be considered as a surrogate for school type, this would explain also the sizeable effects it had on SciValue (p=.36) and science achievement (p=.27) which, in both cases, was positive indicating that male-dominated classes held higher values towards science and had superior test scores.

It should, however, be kept in mind that the specification of the Dutch path model suffered from the lack of a suitable LV which should have encompassed the type of a school or the science program offered at school.

FIG. 7.9 *Between-school path model for the Netherlands with science achievement as outcome measure; 14-year-old level, 1983/84; N=224*

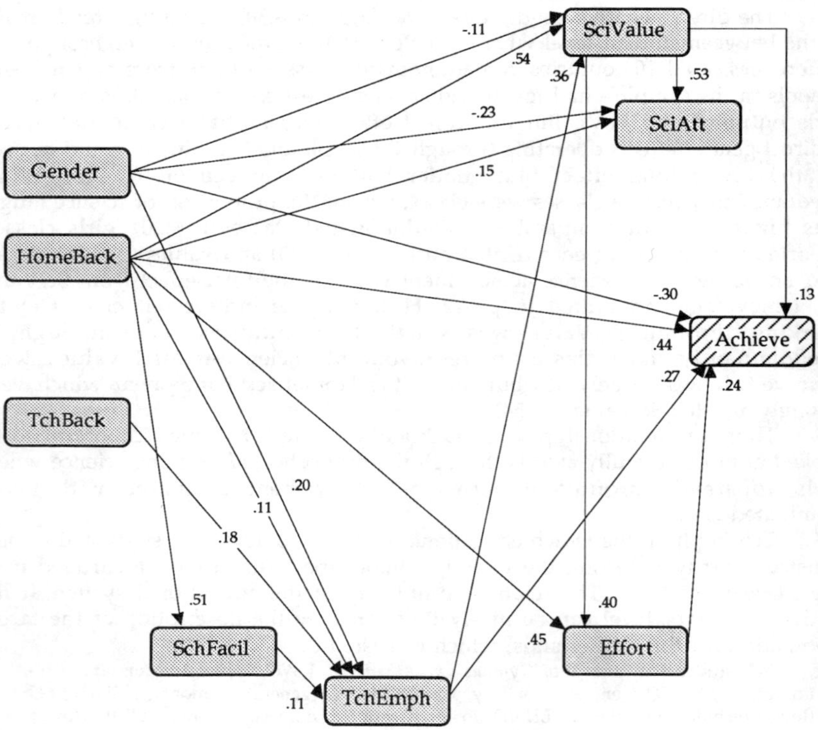

Notes:

paths with |β|<.10 have been omitted; for further details see legend for Figure 7.1

Apart from the impressive gender and school type effects, it was the socio-economic status of the students in school which influenced achievement in science most (t=.72). The impact of HomeBack became even more pronounced in the absence of the LV Aptitude. Thus, HomeBack showed strong positive direct effects on SchFacil (p=.51), SciValue (p=.54), Effort (p=.45), and science achievement (p=.44). This would seem to indicate large (status) differences between the schools which, indeed, were another reason for the drastically increased proportion of variance explained by the between-school path model in comparison with the model for the between-student analysis where the HomeBack effect was found to be relatively limited (t=.22).

It was interesting to note that TchBack and SchFacil had no significant effect on science achievement (|p|≤.10) which might be taken as an indicator for the 'balance' of the Dutch policy of teacher training (Nijhof and Streumer, 1988).

This might, however, be an artefact only showing that there was little variation within certain school types in relation to the qualifications of teachers and larger variations between the qualifications of teachers from different types of schools. This assumption was further supported by the fact that TchGend exercised no effect whatsoever within the inner path model though data were obtained for sex of teacher. As a consequence, TchGend was removed from the path model completely.

The interpretation of the Dutch between-school path model suggests that in order to determine specific school type effects additional analyses are needed. It exceeds the scope and limit of this study to do so.

Large gender differences in science achievement were found at the between-school level for the 14-year-olds in the Netherlands. To some degree, these differences were reflecting the various organizational and instructional forms of schooling. It would be highly advisable to increase the proportion of science instruction in relation to other subjects a teacher is supposed to teach as classes of more specialized teachers seemed to perform better. More emphasis should be put on classwork and homework in science especially in school types which offer less exposure to science at the 14-year-old level. However, school-type related differences need to be investigated further.

➞*Sweden* (Figure 7.10) The path model explained 42 percent of variance which was twice as much as the variance accounted for by the model at the between-student level. Data for SchFacil were not available.

Gender effects on science achievement could not be observed (t=-.04), contrary to what was found when analyzing between-student data (see Table 6.7). Still, Gender had a positive effect on TchEmph (p=.18) and appeared to influence directly both SciValue (p=-.19) and SciAtt (p=-.10) negatively. However, the effects of SciValue (p=.22) but not SciAtt carried through to affect the achievement level of a class. Thus, it seems justified to conclude that school at the lower secondary level were not gender-biassed as far as achievement in science was concerned. As pointed out earlier, the Swedish educational system, in particular at the primary and secondary level, was perceived as fairly homogeneous and effective in science (Postlethwaite and Wiley, 1991, p. 153-154). This, in turn, could be attributed to the common type of comprehensive schools in which students were free to choose from a variety of subjects and topics even at lower grades additional to the compulsory set of lessons (Marklund, 1988; Riis, 1988).

Although sex of class did not contribute greatly to the variance of science achievement, the effects Gender had on science-related values and attitudes were noteworthy. Classes with predominantly male students held higher values and attitudes towards science than female-dominated classes. As there were no single-sex schools in Sweden this result was even more astonishing. A possible explanation would be that boys and girls were, indeed, educated following a different gender-role perception of the parents and perhaps, if not necessarily consciously, the teachers (Morse and Handley, 1982; Kroeger, 1990; Keeves and Kotte, 1991).

FIG. 7.10 *Between-school path model for Sweden with science achievement as outcome measure; 14-year-old level, 1983/84; N=137*

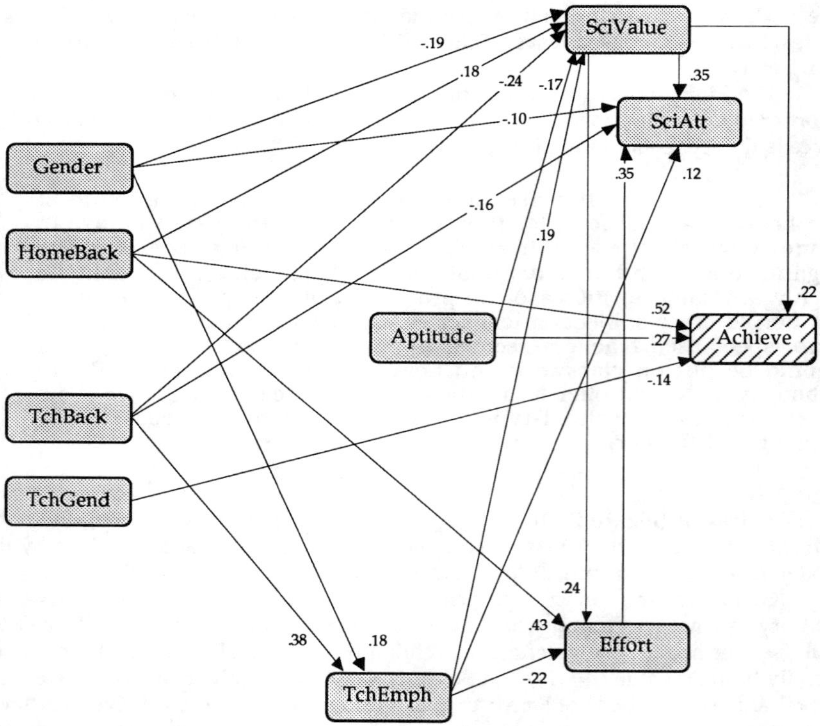

Notes:

paths with |β|<.10 have been omitted; for further details see legend for Figure 7.1

As already stated when analyzing the Finnish and Dutch path models, SciValue had a direct effect on achievement in science. But in contrast to the other two countries, the effect SciValue exercised could be largely attributed to the socio-economic status of the class (p=.18). Gender effects, as mentioned, tended to cancel out: while there was a direct negative influence, this was countered by a positive gender effect operating through TchEmph (p=.19). Similar counter-effects came from the teachers educational background: the direct effect from TchBack on SciValue was negative (p=-.24), the effects on TchEmph (p=.38) and onwards on SciValue were positive. The emphasis of a science teacher promoted higher science-related values in the students if the teacher was teaching a proportionately higher amount of science (science specialist teacher). In contrast, teachers less experienced in teaching science (non-specialist science teacher), had classes which showed lower science-related values. However, the effect of

TchBack on SciValue was independent of the sex of the teacher (|t|<.10). This finding could not necessarily be expected as about three quarters of the science teachers at the 14-year-old level were male (Postlethwaite and Wiley, 1991, p. 153). Even more, the effect from TchGend on science achievement (p=-.14) came as a surprise indicating that female science teachers were, on average, teaching slightly lower achieving classes. Whether this result reflected the performance difference between male and female science teachers, the outcome of an educational policy, or a filter effect in that those females who decided for a career as science teacher were, in general, more able than their male colleagues and probably higher motivated could not be determined from the data available. Here, further research would be required. It may be ruled out that the relationship between TchGend and science achievement was spurious or incidental as the path coefficient (p=-.14) was noticeably higher than the correlation between the constructs (r=-.02).

Despite the general uniformity of the Swedish educational system it could not compensate for very pronounced background effects (see also Riis, 1988). HomeBack had a high positive effect on science achievement (p=.52) which had classes with higher status of the student homes score considerably higher. This would suggest that the largest predictor of achievement in science in Sweden was the socio-economic status of the home, a finding shared by Keeves (1991) as well as Postlethwaite and Wiley (1991, p. 129).

Another interesting observation was that Effort had no influence on class achievement in science. Though the amount of classroom and homework effort of a class was heavily affected by the status of the class, i.e. students of higher or lower status homes (p=.43), as well as TchEmph, which, unexpectedly, had a negative effect (p=-.22) and SciValue (p=.24), this high motivation 'only' led to an increased attitude towards science in general but not to an improvement of the achievement level of the class. This might be taken as another support for the conclusion that HomeBack had a large influence on science achievement.

Finally, it should be mentioned that the age of the student as measured by the construct Aptitude, though negatively influencing science-related values, where classes with older students showed less favourable values towards science (p=-.17), had a moderate positive effect on science achievement (p=.27). This came not unexpected as it could be reasoned that older students would tend to score higher. Hence, the negative effect of Aptitude on SciValue seemed to emphasize the finding already discussed that with increasing age students in Sweden showed less favourable values and attitudes towards science (see Section 6.1.1).

In Sweden science-related values considerably influenced science achievement at the between-school level. However, attitudes towards science did not have any effect on science achievement for 14-year-old students. While science instruction as well as science-related attitudes and values were apparently strongly gender-typed the overall effect of gender on science achievement was small. Socio-economic effects dominated the observed between-school differences in Sweden. It is recommended that the negative attitudes and values Swedish students held towards science, and also to a less degree, to schooling in general be analyzed further. Additional investigation is also necessary to find out why classes of male teachers did not perform significantly higher.

→*Thailand* (Figure 7.11) Some of the findings for Thailand, the only developing country in the sample of countries tested, were different from other countries. These might be, in part, the result of two key constructs, namely TchGend and Aptitude, missing but it might also have something to do with the cultural and economic differences (e.g. lower overall enrolment in school; lower GNP) between Thailand and the other countries (Chantavanich and Fry, 1988).

Possibly due to its economic situation with its low GNP and per capita income, which was illustrated by Oon Chye (1991, pp. 24-31) when comparing economic and educational indicators for the countries participating in the SISS study, the comparatively poor socio-economic background of the parents as measured by HomeBack had the lowest total effect of all 10 countries (t=.39). In Thailand, HomeBack had no direct effect on science achievement. These findings could certainly be attributed to the relatively low variance of HomeBack between schools and the low value of HomeBack per se when compared with other countries. It should also be recalled that only about 32 percent of the age group were in the target population in Thailand (Postlethwaite and Wiley, 1991, p. 156). However, HomeBack affected SciValue (p=-.30) and SciAtt (p=-.11) negatively, though science-related values and attitudes did not influence science achievement at the between-school level.

Despite the relatively small impact HomeBack had on achievement it should be noted that it operated indirectly through SchFacil, TchEmph, and Effort. Yet these indirect effects cannot be understood correctly if the effects of other antecedents, namely Gender, are not considered simultaneously.

The direct effect Gender had on science achievement was quite remarkable (p=-.28) as it indicated that simply due to its gender distribution, predominantly male classes, in general, scored consistently higher on the science tests. This, seen from the perspective of most developing countries, was not surprising. Education, in particular science education, is viewed as a more male 'domain' with the result that the enrolment in secondary schooling is higher for boys in many developing countries (Kelly, 1978; Jimenez et al., 1988; Duncan, 1989; Walberg, 1991). This finding was certainly true for science education at the terminal secondary school level in Thailand. It should, however, not be overlooked that the enrolment ratio for girls at the 14-year-old level in science was slightly higher than that of boys (Comber and Keeves, 1973; Keeves, 1991). This comparatively balanced enrolment ratio between boys and girls in science classes at the lower secondary level might the result of many boys from economically poorer families leaving school at an early age to help earn a living (see also discussion of the relationship between achievement and enrolment in science in Section 5.4).

The strong direct negative effect of Gender on science achievement was, however, to some extent countered by the indirect positive effects of Gender operating through SchFacil (p=.14) and TchEmph respectively through Effort (p=.12). Classes, in which there were more girls than boys, were more commonly found in better equipped schools. This observation coincided with the fact that the better equipped schools had students with higher socio-economic status. It can be hypothesized that, most likely, these were private schools in Bangkok.

FIG. 7.11 *Between-school path model for Thailand with science achievement as outcome measure; 14-year-old level, 1983/84; N=96*

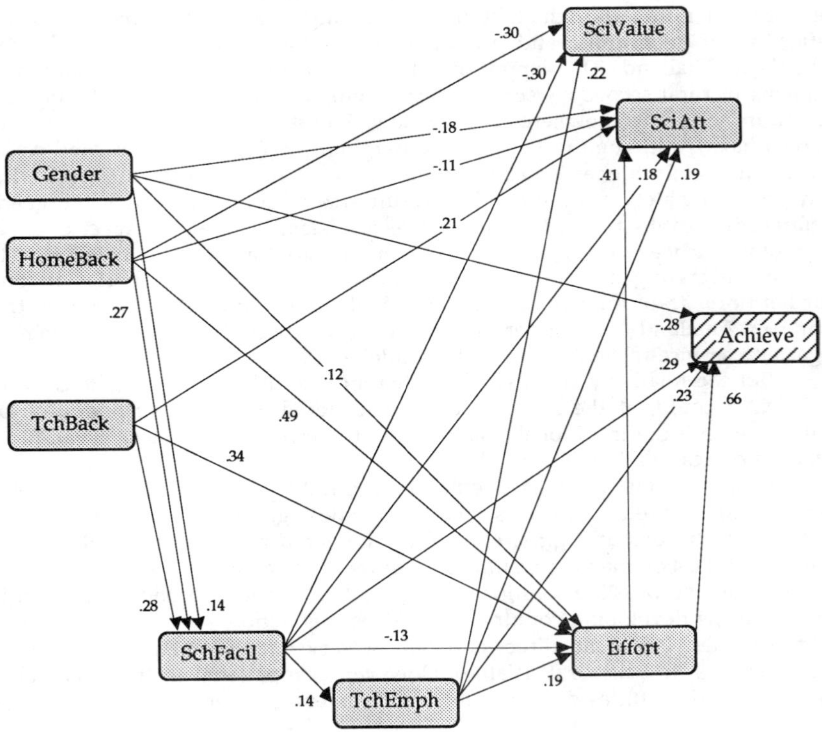

Notes:
paths with $|\beta|<.10$ have been omitted; for further details see legend for Figure 7.1

Surprisingly, the better equipped schools did 'produce' less favourable values towards science although the classes, in general, held more favourable science-related attitudes. Obviously, there was much 'ambiguity', or variance, between science-related values and attitudes among Thai classrooms, an assumption supported by the fact that SciValue did not at all influence SciAtt. In this, Thailand was unique. One possible explanation would be that SciValue was constituted by only one scale, namely Career Interest in Science. It could, therefore, be speculated that occupational perspectives were very different from science-related attitudes among Thai students. This seemed to be even more valid for classes with a low socio-economic background in poorly equipped lower secondary schools as was indicated by the effects from HomeBack (t=-.37) and SchFacil (t=-.26).

Clearly, students from these low status classes in low status schools showed higher science career interests than classes in better equipped schools where the average socio-economic status of the parents was, on average, higher. This might cast some doubt on the validity of the Career Interest in Science Scale. It should be borne in mind, though, that this result might, at least in part, also be a reflection of geographic and economic differences between rural and urban schools in Thailand. The aspirations, hopes, and occupational expectations of students in rural secondary schools more commonly centred around taking up a career in science somewhere in town (e.g. Bangkok, Thon Buri, Chiang Mai, Khon Khaen). Although the effect was marginal (t=-.04) male-dominated classes showed a slightly higher interest to follow a career in science than classes which had proportionately more girls. This result was in accordance with the gender differences identified already at the between-student level with respect to Career Interest in Science (see Table 6.1). This was to be expected as more males leave the villages migrating to the cities in search of a job and a better income. Furthermore, tradition and culture had males look after the income of their family in Thailand - a role perception probably reflected also in the statements regarding science-related values and attitudes.

But the relatively small gender differences found with respect to SciValue and SciAtt could, at the same time, be considered as a sign of liberality and equality within the traditional Thai culture if one rules out a lack of validity of the attitude scales administered.

Despite its fast growing population, about 2.1 percent in the years 1980 to 1985, Thailand's educational system undertook great efforts to have more students, both boys and girls, enrol for a longer period at school (Rosier and Keeves, 1991). For many, parents and students alike, sufficient education was regarded as one possible solution to surpass the everyday economic difficulties present in the developing country. It could be this notion which, with respect to science values, led to no direct and only a very minor total gender effect on Career Interest in Science (SciValue). However, occupational interests as well as science-related attitudes did not affect achievement in science at the 14-year-old level.

The indirect positive effect Gender had on achievement operated basically through SchFacil and Effort. Although female-dominated classes tended to spend more time on homework these were also classes which had a significantly higher socio-economic background (p=.49) and these classes were also taught by more experienced teachers (p=.34). Though the findings presented by Jimenez et al. (1988) were based on mathematics achievement their conclusion of major differences between public and private schools can be reasonably transferred to explaining achievement differences in science. Thus, more homework would be given in private schools (idem, p. 156) which would suggest that the strong direct effect of Effort on science achievement (p=.66) represents, to some extent, the differences between private and public secondary schools, too.

Here, indeed were the shortcomings of the Thai path model. Though it clearly illustrates the different gender effects on science achievement, which worked directly for predominantly male classes and indirectly for female-dominated classes, it falls short in taking into account influences and differences caused by the type of school, public versus private, as well as caused by the 'regionality' of the school, whether the school was located in Central Bangkok and other urban areas compared with rural areas.

Finally, the importance of the teacher deserves attention. While there was no direct effect of TchBack on science achievement, its total effect amounted to t=.29. TchBack seemed to operate through SchFacil (p=.28), TchEmph, and Effort (p=.34). Experienced teachers had better achieving students partly because they were able to teach in better equipped school laboratories and partly because they put more stress on homework. Further to this, classes in which the teacher taught a proportionately higher amount of science showed higher achievement. These teacher-related results advocate the need for more qualified science teacher specialists if Thailand wishes to improve its science education at the secondary level.

Above all these findings, it should not be overlooked that the between-school path model, although it could explain 64 percent of the variance, which was more than five times as high as the variance accounted for by the between-student model, was applicable to a minority of the age group only. According to the sampling frame, only 32 percent of the age group of the 14-years-olds were in the target population of Thailand (Postlethwaite and Wiley, 1991, p. 156). This meant that two thirds of the age group did not go to school any more or, in many cases, had never gone to any school. Even if school-related factors, such as school type and regionality were to improve the adequacy of the PLS model presented above, the model cannot offer much to overcome the basic problems the Thai educational system faces. Gender differences in science achievement were certainly not among the gravest problems.

There appeared to be marked gender differences in science achievement at the between-school level for the 14-year-old students in Thailand. Gender showed a direct positive effect on science achievement for male-dominated classes. This was countered by an indirect gender effect operating through the available school facilities and classroom and homework efforts undertaken by the school which in both cases favoured predominantly female classes. These may represent underlying differences between rural and urban areas. While Thai schools held particularly positive attitudes and values towards science, this seemed to reflect the need for more education rather than that it was in correspondence with the real level of science achievement which, in cross-national comparison, was to be regarded as low. In order to improve the overall performance level in science it is recommended to investigate regional and rural-urban differences in relation to the science equipment available in schools, the teacher pre-service and inservice training, and the amount of classwork and homework given to the students. It is, however, acknowledged that the more basic problem of increasing the overall enrolment ratio is of foremost interest to educational planners. Here, differences in science achievement between rural and urban areas as well as between public and private schooling have also to be taken into consideration.

➞*The United States* (Figure 7.12) At first sight, the model, with which 83 percent of the overall variance could be explained, appears comparatively complex with data available for each of the 11 constructs and with a network of inner model relationships.

The overall effect of Gender on science achievement was small (t=-.04) but predominantly male classes scored slightly higher on the science tests than did female-dominated classes (p=-.17). However, this direct effect was partly balanced out by the indirect gender effect operating through Aptitude which favoured female-dominated classes. This could be inferred from the direct effects from Gender on Aptitude where the sign of the path coefficient was positive (p=.15). Likewise, Aptitude showed an extremely strong positive influence on science achievement (p=.91). This indicated that classes which had more female than male students - only a mere six percent of schools were single-sex (Postlethwaite and Wiley, 1991, p. 155) - had a higher 'aptitude'. In addition, these classes were achieving better on the science tests. However, the very pronounced direct effect of Aptitude on science achievement must be interpreted with caution as it was certainly not just the aptitude level of a class which was positively related to science achievement (r=.71). Much more, as can be seen by the strong positive effect HomeBack had on Aptitude (p=.75), the 'aptitude' of a school was largely dependent on the socio-economic status of the school's catchment area.

It is difficult to interpret the interrelationships among the latent variables of the inner model properly without taking into account the structure of the United States educational system, in particular the structure of science education offered at the lower secondary school level (Staver and Walberg, 1986; Eckstein, 1988; Humrich, 1988 [1]; Walberg, 1991). This becomes apparent when examining the various effects HomeBack and Gender had on the other LVs in the model (see also Haertel et al., 1981). Combining this observation with the structure of the educational system offered at lower secondary schools, it could be concluded that there appeared to be a tendency of having science classes of quite imbalanced gender proportions at the 14-year-old level in the United States. In other words, due to mechanisms not yet discussed, it seemed likely to have science classes with considerable 'gender bias' such that these classes were either predominantly male or predominantly female, a finding shared by Humrich (1988 [1]) in her analysis of the United States SISS data.

Leaving the mechanisms leading to this gender bias aside for a moment, one possible explanation for the existence of these 'typically male' and 'typically female' science classes could be seen in the course offerings of most American High Schools. In many schools, 14-year-olds had the choice for enrolling into one of a few parallel science courses offered which sometimes were teaching the same, sometimes different syllabi. This form of comprehensive schooling undoubtingly advocated a gender-typical selectivity of science courses at the lower secondary school level. Phrased differently, boys, quite often, seemed to enrol in different science courses than girls of the same school. This was underlined by the findings that Gender showed positive direct effects both on TchEmph (p=12) and Effort (p=.14). In female-dominated classes more emphasis was put on student needs, or, phrased differently, girls seemed to favour to enrol in classes where more emphasis was put on student needs (see Saario et al., 1973; Leinhardt et al., 1979; Good and Findley, 1985). The influence Gender had on

Effort demonstrated that classrooms with predominantly female students were more likely to do homework and also put more effort into classwork.

What were the reasons behind this gender-biassed selection process? At this stage, only speculations can be brought forth. Opting for a science course considered 'gender-typical', i.e. either 'typically male' or 'typically female', could have been dependent on the gender self-concept and perception of a gender role identity of the student as suggested by Morse and Handley (1988) as well as parental and teacher influences (Kelly, 1978; Comber and Keeves, 1973). It could also have been the result of internal school structure arrangements (Kelly, 1978; Petersen, 1983; Becker and Chang, 1986).

In order to throw some light on the complexity of the link between gender, socio-economic status of the school, and science achievement regional, ethnic, and cultural aspects also need to be taken into consideration. According to the United States Department of Education, National Center for Education Statistics, total minority enrolment in elementary and secondary education rose to almost 30 percent in 1986, the year of data collection of Phase 2 in the United States (NCES, 1990). Strong gender differences in science achievement had already been noted at the between-student level which, at each population level tested, were in favour of the boys (see Chapter 5). In conjunction to the results presented by Humrich (1988 [1]), it could be reasoned that among the White majority of students proportionately more girls enrolled in some sort of science course more frequently than did female students of minority groups, such as Blacks or Hispanics (the enrolment ratio for boys and girls of the Asian minority in lower secondary science courses should be about equal but, unfortunately, no detailed statistics could be obtained for the sample in question).

Even if ethnicity and school structure seemed to influence gender differences in science achievement at the between-school level indirectly neither TchBack nor SchFacil affected achievement ($|t|<.10$). This would suggest that science achievement was invariant against the experience and qualification of science teachers as well as the quality of school science equipment. It should be kept in mind, though, that this observation did not allow any inferences to be made about the quality of the science teacher training in the United States which, according to several indications, would have to be improved (NCES, 1990; Keeves, 1991; Postlethwaite and Wiley, 1991). It was, however, interesting to find TchBack related positively to SchFacil ($p=.22$). As HomeBack was also showing a positive effect on SchFacil, it appears likely to conclude that, again, this could be taken as another indication for the large variation between schools in the United States: more qualified teachers happened to teach in schools where the students were economically better off and in which more laboratory equipment was found.

On the other hand, SchFacil probably did not have any influence on science achievement because only 19 percent of all science instruction took place within science laboratories at the 14-year-old level "despite the fact that the schools stated that they had an average of five laboratories or equipped classrooms per school" (Postlethwaite and Wiley, 1991, p. 155). Thus, it was not surprising that the effect of SchFacil on Aptitude turned out to be slightly negative ($p=-.12$).

FIG. 7.12 *Between-school path model for the United States with science achievement as outcome measure; 14-year-old level, 1983/84; N=88*

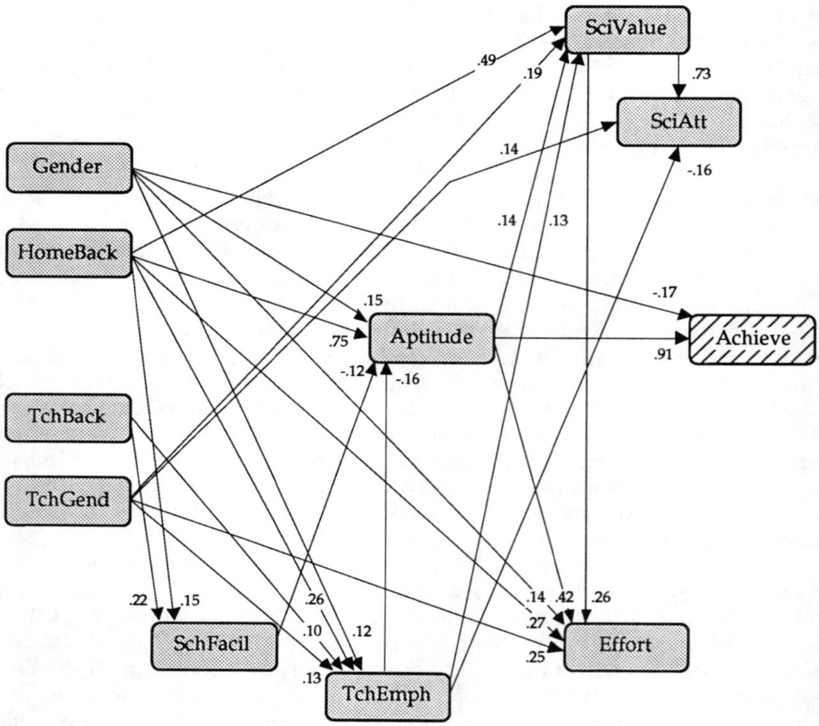

Notes:

paths with |β|<.10 have been omitted; for further details see legend for Figure 7.1

Nevertheless, this finding and the fact that neither SchFacil nor TchEmph influenced science achievement at all should be also taken as a warning as one would expect that better equipped laboratories and an increased emphasis on teaching scientific knowledge should lead to higher achievement in science. At least in the United States at the 14-year-old level this could not be observed. Most likely, these findings have to be seen in connection to the overall poor performance of the United States compared with other countries in the SISS study (see Keeves, 1991; Postlethwaite and Wiley, 1991).

Another interesting result was that sex of teacher had direct effects on four other LVs, namely Effort (p=.25), SciValue (p=.19), SciAtt (p=.14), and TchEmph (p=.13). Moreover, all these effects were positive meaning that female teachers stimulated the classes to take more efforts, contributed to more positive attitudes and values towards science, and put more emphasis on students' needs. Despite these obvious gender differences in teaching the achievement level classes taught by male and female teachers was the same.

Finally, it should be pointed out that science-related values and attitudes had no influence on science achievement at the between-school level. Also, the amount of classwork and homework done did not contribute to higher achievement. This was, indeed, surprising especially because science-related values showed a positive effect on the efforts of a class. These findings and the fact that Aptitude influenced Effort quite strongly (p=.42) seemed to illustrate one of the cardinal problems of science instruction in the United States: what was taught and how this was taught was apparently insufficient to increase the achievement level of schools in science. Consequently, also the quality of the pre- and inservice training should be put to the test.

It should have become clear that despite the many new and useful aspects the path model for the United States still lacked several key factors such as ethnicity, type of science course, or school program. Without these data, gender effects and effects related to the socio-economic status of the school could not be explained fully at the between-school level. However, the relations found to be effective in influencing the achievement in science at the school level should have illustrated not just the complexity of the network of major latent variables but also the need for more refined multivariate analyses.

In the United States gender differences were an important factor influencing science achievement at the school level. Though it could not be concluded that male-dominated classes achieved significantly better than classes in which the majority of students were female, it is possible that a kind of 'gender-typical' selection process existed at the lower secondary school level in terms of what type of science course to enrol in. The results clearly illustrated the enormous variation between schools in science achievement in 1983/84 which, at least in part, reflected regional and rural-urban differences. Despite major structural, curricular, and instructional differences between the schools the findings point to major deficits in science teaching, only partly related to gender differences.

7.2.3 Implications

After having presented in detail the path models for science achievement at the between-school level for each of the 10 countries, some of the major implications should be discussed.

The average overall amount of variance accounted for by the between-school path models across the 10 countries was with 55 percent nearly double the figure for the between-student path models (29 percent). Though this might have been largely due to possible aggregation bias is might also be a justification for the inclusion of school- and teacher-related variables in a path model which was meant to trace possible gender differences in science achievement.

It has become obvious that the school organization, school status and school setting played a more or less important role in fostering gender differences. Likewise, the teacher's sex and teaching background as well as the effort and the emphasis that was placed on teaching science inside and outside the classroom were of considerable importance in influencing science achievement for boys and girls. But it should also have become clear that the socio-economic status of the school, class and student was **the** key factor in shaping science achievement. However, the socio-economic background of the student's home seemed to be unrelated to gender differences in any country. Although in five of the 10 countries HomeBack was found to have either a direct or indirect positive effect on science achievement while at the same time the effect of Gender was negative, this only symbolized the existence of gender differences but allowed no speculation about the 'link' between HomeBack and Gender. The reader should be reminded that *a priori* no causal link was anticipated between these two antecedents. Students of higher status homes scored better and so did, in general, boys. The sign of the effects of HomeBack (+) and Gender (-) on science achievement could be explained legitimately such that 'boys from higher status home achieved best'.

In fact, in most countries male-dominated classes had an overall advantage in science achievement compared with female-dominated classes. Yet, this result just reinforced, at an aggregated level, the finding already stated at the between-student level (see discussions in Chapters 5 and 6). Nevertheless, and this could be illustrated impressively in the path model for the United States, gender differences at the school level were the result of different influential patterns and networks of effects favouring either boys or girls. Hence, the factors which helped to establish gender differences in science achievement varied considerably between countries.

Unlike the average path model presented in Chapter 6 for the between-student level (see Figure 6.3), no similar path model could be advanced for the between-school level. Where possible, similarities between the countries with respect to gender differences in science achievement were already pointed out. It is, however, understood that if sufficient data would have been available for each MV in each country this could have increased the probability of arriving at an 'average' path model for between-school analyses.

The issue of gender differences found in coeducational schools will be dealt with and further elaborated in the following chapter by analyzing data from Australia and England, the only two countries which still had sizeable numbers of coeducational and single-sex schools in operation at all population levels.

7.3 An Extended Path Model for Participation in Science

To find out about the factors influencing future participation in science at the school level separate PLS analyses have been carried out for each of the 10 countries. These results and the major differences found in the path models for science achievement at the between-school level suggested not to discuss in detail the extended path models for each country explaining participation in science. Indeed, a similarly diffuse picture was obtained when comparing the path models with FutEnrol as the outcome measure using data aggregated to the school level.

However, several interesting findings emerged which need to be analyzed carefully as they might throw some light on major problems related to gender differences in science education and possible structural deficiencies within an educational system. Therefore, Table 7.3 has been prepared which shows the direct and total effects of each of the constructs of the inner model on participation in science represented by the LV FutEnrol. Aggregated to the school level, FutEnrol expressed the degree to which a class described the intention to continue taking science at the upper secondary school level. Again, grand unweighted means and standard deviations across the 10 countries have been calculated to give an impression of the cross-national homogeneity of an effect. Additionally, the amount of variance accounted for by each path model is given.

By extending the general path model estimating science achievement at the between-school level to predict future enrolment the path model arrived at is presented in Figure 7.13. This path model was tested for all 10 countries. For completeness, the final PLS statistics for each country were included as Appendices 7.11 to 7.20.

The extended path model for participation in science as depicted in Figure 7.13 consists of 14 inner model LVs estimated by 36 MVs. As could be seen from the discussion of the extended path model for science achievement not all MVs from preliminary PLS analyses warranted inclusion for predicting FutEnrol. In a similar way different MVs had to be excluded from the outer model in different countries. For details the reader is referred to Appendices 7.11 to 7.20.

The inner model was arranged according to the path model for science achievement on the between-school level as shown in Figure 7.1 enhanced by the constructs ExpEd, ExpOcc, and FutEnrol with FutEnrol becoming the outcome measure. Theoretically, there were 85 inner model paths to be estimated which would give an idea of the relative complexity of the model for FutEnrol.

In order not to overload the graphical representation of the path model by showing all theoretically possible inner model effects, it was decided to draw only the chronological placement of the LVs plus the direct paths between ExpEd, ExpOcc, and FutEnrol. However, Figure 7.13 makes clear the complexity of the inner model effects.

Keeping in mind the manifold differences between the path models for FutEnrol at the between-student level for the 10 countries, it came as no surprise to detect an even greater disparity within the network of effects influencing participation in science at the between-school level.

Although an average of 49 percent of variance for FutEnrol could be explained, the large standard deviation of 21 percent can be taken as an indication for the applicability of the path model cross-nationally. Though aggregation of data, in general, would raise the amount of variance explained of a path model the introduction of school- and teacher-related variables in forecasting participation in science of a class X in country Y added to the 'precision' of the estimation in five of the 10 countries only. These countries were Hungary (explained variance: 75 percent), the Netherlands (62 percent), Sweden (72 percent), Thailand (53 percent), and the United States (62 percent). The proportionately biggest gain in the amount of variance accounted for by the extended path model occurred in the case of Hungary.

Overall, HomeBack was the only predictor effective throughout all countries. Except for Sweden, which showed a very minor HomeBack effect on FutEnrol ($t=.10$) - the direct effect was even negative ($p=-.12$) - the effect of HomeBack on FutEnrol was, on average, moderately positive ($\bar{p}=.39$). Hence, in most countries classes which had a higher socio-economic status intended more frequently to participate in future science courses.

It was, indeed, unexpected to realize that in the United States HomeBack had no direct influence on FutEnrol. However, the indirect effect was enormous (t=.76). This should be interpreted in light of the results presented for the path model for science achievement at the between-school level (see Section 7.2.2) as the large indirect effect of HomeBack on FutEnrol might have, in part, been an artefact illustrating the effect of ethnicity, regionality, or different school types and science programs.

In five of the 10 countries male-dominated classes showed a slightly higher interest to enrol in future science courses (Australia, England, Finland, Italy, and Sweden). With the exception of Italy this result was not found when comparing all students in a country (see Table 6.9). It could therefore be reasoned that in these five countries the gender-typing of a class played a role, though not a big one, in promoting future participation in science. This finding was applicable also for Hungary with the difference that there it were female-dominated classes which had expressed a higher interest in future science enrolment. It would exceed the framework of this study to scrutinize the way Gender influenced FutEnrol in each of the countries. Certainly, more detailed analyses will be necessary to disentangle the complex network of class-level gender effects on expectations (ExpEd and ExpOcc) and future enrolment. It should be borne in mind, though, that the gender-typing of a class can very well be of importance to stimulate or inhibit students' interest to study any further science at upper secondary levels. Grouping of students, streaming, or the offering of different science programs at the lower secondary level all can add, positively or negatively, to the students' decision to enrol in upper secondary science courses. However, more country-specific analyses are needed.

While TchBack was found to be effective to influence FutEnrol directly only in Sweden (p=.21) and the United States, where the effect was found to be negative (p=-.11), sex of the teacher hardly made the difference. It were in Australia (p=.17) and England (p=.14) where TchGend had a small direct effect on FutEnrol. As mentioned beforehand, these countries will be discussed in more detail in the following chapter. It was interesting to see that in the United States TchGend happened to have a marked total effect on FutEnrol (t=.27). This finding was in line with the observation that with science achievement as the outcome TchGend also showed positive effects on SciValue (p=.19) and SciAtt (p=.14). It can, therefore, be hypothesized that female teachers were more likely to stimulate students to enrol in future science courses at the upper secondary level.

No common trend could be obtained for the effects of SchFacil and TchEmph on FutEnrol. While it were the better equipped schools which promoted to take further science courses in Finland (p=.13), Hungary (p=.21), Japan (p=.20), and Thailand (p=.10) schools with poorer science laboratory equipment had classes with higher FutEnrol in England (p=-.11) and the Netherlands (p=-.20). A similarly heterogeneous picture was found with respect to the effect of TchEmph as indicated by the high standard deviation.

TABLE 7.3 *Direct and total effects derived from PLS analyses on future enrolment as outcome measure for between-school within-country data for the 14-year-olds, 1983/84*

Relationship to Future Enrolment	Australia	England	Finland	Hungary	Italy	Japan	Netherlands	Sweden	Thailand	United States	average mean▲	average sd
Direct Effects												
HomeBack	...	...	.14	.43	...	.47	.16	-.12	.11	...	.20	.22
Gender	-.19	-.18	-.13	...	-.18	...	...	-.19	...	...	-.17	.03
TchBack	...	...	...	...	...	...	...	.21	...	-.11	.05	.23
TchGend	.17	.14	...	...	...	...	nd	...	nd	...	.16	.02
SchFacil	...	-.11	.13	.21	...	.20	-.20	nd	.10	...	.06	.17
TchEmph	.25	...	...	-.15	...	...	.25	-.33	...	...	.01	.29
Aptitude	.26	...	nd	.52	...	.27	nd	-.13	nd	-.21	.14	.30
SciValue	...	.17	.12	.42	.18	...	.46	...	.22	.22	.26	.13
Effort	...	.35	...	-.11	...	...	.22	-.11	-.20	.24	.06	.23
SciAtt	.13	.24	...	.22	...	...	.22	...	.31	...	.22	.06
Achieve	...	nd	...	-.55	...	...	...	...	.26	.13	-.05	.44
ExpEd	...	nd	.41	...	.58	...	-.22	.82	.14	.60	.39	.37
ExpOcc	...	...	...	...	nd	nd	-.15	...	.38	...	.12	.37
Total Effects												
HomeBack	.26	.32	.42	.36	.46	.59	.36	.10	.31	.76	.39	.18
Gender	-.10	-.11	-.10	.10	-.16	...	...	-.21	...	...	-.10	.11
TchBack	...	...	...	-.11	.10	...	...	...	.17	-.11	.01	.14
TchGend	.12	.12	...	...	...	...	nd	.10	nd	.27	.15	.08
SchFacil	...	...	.15	...	...	.23	-.15	nd	...	...	.08	.20
TchEmph	.25	.10	...	...	...	...	.36	...	.24	...	.24	.11
Aptitude	.28	.28	nd	.15	...	.27	nd	...	nd	.28	.25	.06
SciValue	...	.33	.16	.45	.19	...	.62	...	.29	.37	.34	.16
Effort	...	.35	...	...	...	...	.19	...	.16	.34	.26	.10
SciAtt	.13	.24	...	.10	...	...	.22	...	.36	...	.21	.10
Achieve	...	...	.16	-.55	...	...	...	...	.35	.22	.05	.40
ExpEd	...	nd	.41	...	.58	▼	-.22	.82	.37	.60	.43	.35
ExpOcc	...	nd	...	...	nd	▼	-.15	...	.38	...	.12	.37
% Variance Accounted for Future Enrolment	19	27	28	75	40	40▼	62	72	53	62	49◆	21■

Notes:

▲ mean effects calculated as the sum of effects across all countries divided by the number of countries for which data were obtained (effects with |β|<.10 have been excluded)

◆ mean variance accounted for FutEnrol across 9 countries (Japan has been excluded)

■ standard deviation of mean variance

▼ the outcome variable for Japan was ExpEd; there were no data collected for ExpOcc or FutEnrol

... the effect was discarded as trivial with |β|<.10

nd data not collected, or deleted prior to analysis, or all manifest variables not significant

FIG. 7.13 Between-school within-country inner path model with future enrolment as the outcome measure for the 14-year-old level, 1983/84 (paths prior to ExpEd have not been shown)

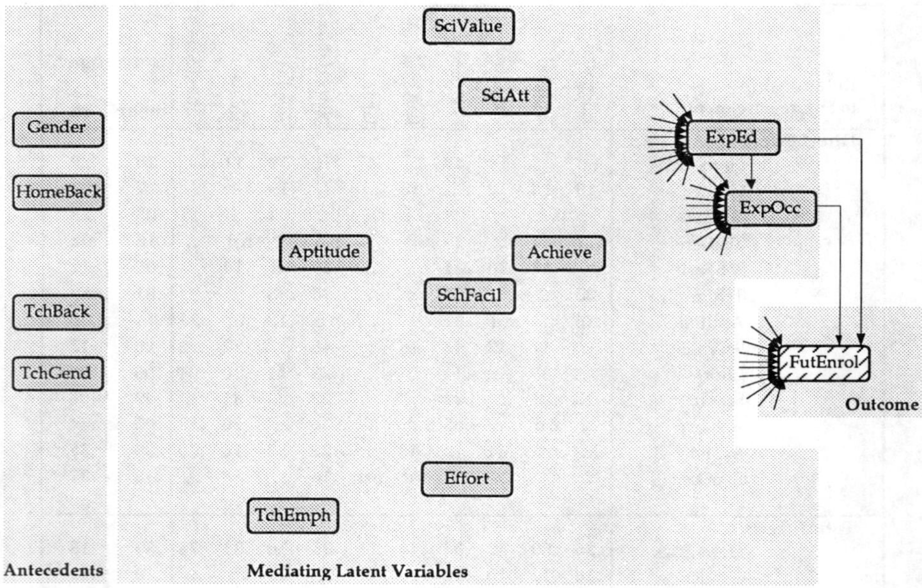

Notes:

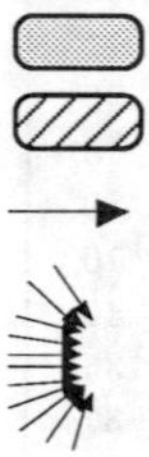

construct, latent variable (LV)

construct of outcome measure

path, with arrow showing the direction of the effect

paths leading to construct originating chronologically prior to ExpEd; the left-hand part of the model is identical to the path model which had science achievement as the outcome measure (see Figure 7.1), for clarity these path have not been shown

It seems justified to conclude that cultural, ethnic, regional, and structural differences in the educational systems of the countries led to the heterogeneity in the effectiveness of the four school-level constructs TchBack, TchGend, SchFacil, and TchEmph to predict FutEnrol.

Aptitude positively influenced FutEnrol in four countries, namely Australia (t=.28), England (t=.28), Hungary (t=.15), Japan (t=.27), and the United States (p=.28). But it was even more interesting to note that the direct effect of Aptitude on FutEnrol found in the United States was negative (p=-.21) while the total effect was strongly positive. Most likely, this sign reversal could be ascribed to school type, science program, and ethnic differences existing among United States schools pointed to earlier if multicollinearity effects are ruled out.

As would be expected SciValue showed, in general, a positive effect on FutEnrol ($\bar{p}$=.26). This could be taken as a supportive finding for the validity of the model outlined in Figure 6.17 in which it was hypothesized that background factors and science-related values and attitudes were responsible for stimulating or inhibiting participation in science courses by operating either directly or indirectly through ExpEd. However, it was the Netherlands which had negative total effects of ExpEd (t=-.22) and ExpOcc (t=-.15) on participation in future science courses. This might reflect school type differences likely to be found in a very diversified educational system such as the Dutch one.

Finally, it should be noted that achievement in science was the most powerful predictor for FutEnrol in Hungary, but possibly much to the reader's surprise, its impact on participation was negative (p=-.55). This meant that the higher the achievement level of a class was, the more likely this class did not have an interest to enrol in any future science course. The reasons for this disinterest in future science course can only be speculated about and more refined analyses need to be undertaken. Nevertheless, this result has also to be seen in relation to the fact that neither students' educational nor occupational expectations were apparently related to FutEnrol. No direct or indirect effects were found. It might therefore be concluded that the disinterest in future science found with high-achieving classes and those of higher status (see table for direct and total effects for Hungary in Appendix 7.14) reflected some kind of hesitation to keep on studying science but it did not seem to have led to a general disinterest with further science-related tertiary education or with taking up a science-related job in the future.

Overall, the most influential factors in promoting future participation in science courses at the between-school level for the 14-year-olds were the socio-economic status of the class, the science-related values, and the educational expectations. However, much ambiguity and country-specific differences made it difficult to trace any further similarities between the countries. Thus, sex of class had an impact on FutEnrol in only six of the 10 countries. In general, its effect was very small.

Despite these shortcomings the inclusion of school-level variables led to the improvement of the between-student path model for future enrolment which became certainly true for Hungary, the Netherlands, Sweden, Thailand, and the United States. It would certainly be useful to undertake more country-specific analyses with respect to gender differences in participation in future science courses. This should be in the interest of the educational planners in each of the countries.

7.4 Summary and Conclusions

In this chapter different effects of student, teacher, school variables on gender differences in science achievement were identified for the 10 countries. It became obvious that the gender-typing of a class, i.e. a more female- or male-dominated class, still played an important role in affecting achievement in science in most countries. Aggregating data to the school-level made it possible to trace a number of relationships in the countries which had considerable impact in creating variability in science achievement between the schools. The results shown suggested that, additional to the gender differences identified at the between-student level, emphasis has to be put also on teacher- and school-related variables in understanding the complex network of effects promoting gender differences with respect to achievement in science at the 14-year-old level.

The results presented should have made it clear that the issue of gender differences should not be looked at solely by conducting between-student analyses. In most countries, there was found to exist school-level factors which accounted for gender differences in science achievement in addition to the factors shown to be effective at the student level. But at the same time it became evident that by operating with aggregated data further latent variables would be required to understand better the network of factors influencing science achievement. These additional factors could be ethnicity, regionality, school type or science program. Further to this, emphasis has to be given to gender differences arising from coeducational or single-sex schooling (see Chapter 8).

Though the effects of HomeBack and Aptitude (with all their interpretative limitations) were found to be most influential on science achievement, it should be kept in mind that several other effects were of importance in each of the countries. However, these effects varied quite remarkably between the 10 countries. It could be argued, though, that science-related values and attitudes as well as the school- and homework-related efforts also accounted for much of the variance of the outcome measure.

In contrast, the effects of TchBack, TchGend, SchFacil, and TchEmph were of little effect on science achievement except, perhaps, in Thailand. This could lead to the conclusion that school- and teacher-related variables become less influential the higher the general educational standard (or economic status) of a country is. Once certain standards have been achieved in relation to school science equipment and science teacher training, other factors, some of them not yet identified, stimulate or inhibit students' achievement in science at the lower secondary school level. The results of this chapter presented some evidence that teaching practices in the classroom or laboratory (science curriculum) and how this was done (teaching strategies) had considerable influence on both increasing or decreasing science achievement and gender differences in science achievement. However, it would be ignorant not to accept that possibly the 'wrong' or inappropriate MVs and LVs have been employed to detect gender differences in science achievement at the school level. Possibly, information regarding the teaching style, instructional method, type of science program, or form of streaming, to name just a few alternative variables, would have proven useful.

Irrespective of this technical and conceptual limitation, it was the socio-economic status of a class which accounted for most of the differences in science achievement. In comparison with the other countries this was particularly true in Australia, England, the Netherlands, and the United States. There was some evidence that the educational systems of these countries were unable to counter or did not intend to counter the strong HomeBack effect in science achievement. In the cases of the Netherlands and the United States many of the between-school differences in science achievement were believed to be caused by the various school types or science programs offered.

Though not applicable in each of the 10 countries, the notion that more homework would improve the achievement in science, would seem true. Again, Effort was found to be most effective in the only developing country of the sample, Thailand. This country was also the only one where the background of the science teacher influenced achievement noticeably. It this stage it can just be speculated that this finding was a consequence of the science teacher training program in Thailand.

Comparing the between-school level path models with the models arrived at for the between-student level, it can be concluded that with an average of 55 percent of variance accounted for this was a sufficiently high value. It indicated that the path models proposed, with perhaps the exception of Finland, were capable of identifying some of the major student, teacher, and school factors which influenced science achievement at the 14-year-old level. However, it should be noted that much of the variance explained by the between-schools path models could be attributed to aggregation bias and this amount could not satisfactorily be separated out using PLS.

Another point of interest in the analyses presented in this chapter was the relationship between science achievement and future participation in science courses. It emerged that the close positive link between achievement in science and the students' interest to enrol in upper secondary science courses could not be verified in most of the countries. Moreover, science achievement had a very pronounced negative effect on future enrolment in Hungary, the country in which students scored top marks in the science tests. It should not be overlooked, though, that other factors not explicitly measured by any of the variables within the extended path model, such as ethnicity, regionality, school type, or science program, would also contribute to influence students' interest in future science courses. Although small gender differences were found in six of the 10 countries always favouring male-dominated or single-sex boy classes, gender-typing of the class turned out not to be the key factor in stimulating or depressing the students' interest to enrol in future science courses.

However, to obtain a more concise idea of how the gender-typing of a class would affect science achievement special multivariate analyses using HLM have been performed for Australia and England, the only two countries which had a sizeable proportion of single-sex schools in operation at both the data collection of the FISS and SISS studies (1970/71 and 1983/84). Thus, the following chapter will be devoted to the problem of coeducational versus single-sex schooling for science instruction.

8. Single-Sex Schooling in Australia and England

From the discussion in the previous chapter it emerged that gender differences in science achievement were influenced by student-, teacher-, and school-related factors. Although some similarities between the 10 countries in terms of variables that were increasing or decreasing the effect of gender in science achievement were found considerable differences were noted. Thus, it was deemed inappropriate to develop an 'average' path model having either science achievement or future enrolment as the outcome measure.

Taking into consideration technical, statistical, and methodological constraints, it could be concluded from the results presented in Chapters 6 and 7 that several key factors, such as socio-economic status (HomeBack), general aptitude level (Aptitude), or homework (Effort) affected science achievement both at the student and at the school level. Further to these, gender-typed effects on achievement were also found in several countries.

While all path models presented in Chapter 6 were based on student-level data, the results shown in Chapter 7 stemmed from analyses of between-school data. In order to integrate student-level variables into between-school analyses student data were aggregated to the school level. To evaluate the effects of student-, teacher-, and school-related variables jointly it would also have been possible to disaggregate data from the teacher and school background questionnaires to the student level. Nevertheless, both processes, aggregation and disaggregation of data, would introduce some sort of bias into the analyses and, subsequently, also into the interpretation of the results a fact which certainly limits the applicability of partial least squares techniques as was clearly to be seen in Chapter 7. The problem of aggregation and disaggregation bias has been elaborated in detail in Section 4.2.

From the researcher's point of view it would be advantageous to be able to estimate student, teacher, and school factors simultaneously within one path model without having to aggregate or disaggregate data. This underscores the desirability of analyzing the data using a two-level approach. As the educational setting of *student→class→school→country* clearly reflects an organizational hierarchy, it would seem to be advisable to develop a hierarchical linear model in order to predict gender differences on science achievement. The statistical procedure offering such type of modelling was HLM (Hierarchical Linear Modelling) developed by Raudenbush and Bryk (1986). The properties of this multivariate, multi-level technique were described in some length in Section 4.3.2 and need not be repeated here.

It would be interesting to develop separate HLM models for each of the 10 countries. However, given the complexity of the data and statistical procedure of HLM per se this would require an enormous amount of skill, effort and time far beyond the scope of this study. It should be recalled that the intention of this dissertation is to detect and analyze possible differences in science achievement due to the effects of gender thereby holding all other constructs in the path models constant.

The two countries having sizeable numbers of schools in operation that were sex-segregated were Australia and England. This chapter is concerned with gender-typed effects deriving from coeducational and single-sex schooling. Hence, this chapter presents the analyses for these two countries only. When discussing the path models for Australia and England in Chapter 7, several questions were raised about differences between single-sex and coeducational types of schools. In order to decompose the relationships of the various factors on science achievement into within- and between-school components for single-sex and coeducational schools special HLM analyses had been carried out exclusively for Australia and England.

8.1 Towards a Hierarchical Linear Model of Science Achievement

The multivariate analyses and results described in this chapter demand at least some familiarity of the reader with the statistical approach of HLM. Though the major ideas of this multivariate technique have already been discussed in Section 4.3.2, a short summary is given here with particular emphasis on the interpretation of the results furnished by HLM.

For the sake of a simultaneous estimation of student- and school-level data, it was decided to perform two-level statistical analyses using the program HLM-2 designed and distributed by Bryk and associates of the University of Chicago (1988). Though a three-level version of HLM was available, it was considered appropriate to restrict the analyses to using HLM-2 as data from just two hierarchically ordered levels, namely the student and the school level, needed to be computed (see also Section 4.3.2).

The aim of the HLM analyses was to shed some light on the factors fostering gender differences found in coeducational schooling in contrast to single-sex schooling in science achievement in Australia and England. Hence, separate analyses were to be performed for the sub-samples of coeducational, boys-only, and girls-only schools within each of the two countries.

Unlike PLS, which allows for the entry of all manifest and latent variables into one model at either level, HLM-2 can predict one outcome variable only taking into account the direct effects of each factor which was thought to influence the outcome measure. In other words, HLM-2 is not capable of differentiating the direct effects of LVs on the outcome measure **and** direct effects **among** the LVs in a particular path model simultaneously. To compensate for this shortcoming, a sequential or stepwise procedure had to be devised which estimated the key outcome measure of science achievement and then, in subsequent steps, regressed science achievement upon each of the chronologically earlier variables using a gradually more and more reduced part of the overall path model. It should be added that all subsequent 'outcome variables', i.e. all variables other than science achievement and other than antecedents, could originate from the student-level data only. This is a mathematical restriction in the conceptualization of HLM (Raudenbush and Bryk, 1986; Bryk and Raudenbush, 1988) of which the reader should be aware (see also the methodological discussion about HLM in Section 4.3.2).

It should be reiterated that the notion of linking several independent HLM analyses, carried out iteratively for each of the possible outcome measures, was to arrive at a complete path model rather than ending up with a two-level multivariate regression equation which, in simplified terms, was the statistical nature of a single HLM run. Employing this iterative technique results in obtaining separate 'quasi path models' for each of the three sub-samples of schools for Australia and England. These iterative HLM analyses should not be perceived as PLS-like multivariate analyses. But for the sake of being able to compare the complex structure of and effects within the path models for coeducational, single-sex boys and single-sex girls schools at the school **and** the student level simultaneously, this stepwise HLM technique was regarded as advantageous. Furthermore, the construction of separate path models for each gender type of schooling involved potentially more information to be analyzed and was expected to be helpful in tracing gender differences between students and schools over and above the information obtainable by one single HLM analysis for each of the three gender types of schooling having science achievement as the sole outcome measure.

This conceptualization has an advantage over the analytical approach of Lee and Bryk (1986) who conducted some interesting research into the effects of single-sex secondary schools on student achievement and attitudes in the United States. Lee and Bryk opted for the use of discriminant analyses to 'identify' personal, family, and background characteristics influencing student achievement and attitudes before proceeding to use these results in a multiple regression analysis for each of the gender types of schools (1986, pp. 383-387). Though they were able to show major differences between the coeducational and single-sex schools as well as between single-sex boys and girls schools they fell short of determining relationships **among** the relevant latent variables. In contrast, the stepwise linkage of separate HLM analyses applied in this study enabled the development of a comprehensive path model, one for each gender type of schooling in Australia and England.

The general between-school path model shown in Figure 7.1 served as a starting point for planning the sequential HLM analyses. However, preliminary HLM runs suggested the removal of the LVs TchGend and SchFacil from a two-level path model for Australia and England as TchGend had a small total effect on science achievement in Australia only while SchFacil showed a minor effect only in England.

A decision was taken to integrate four new school-level constructs which were:

Practice	*Practical and experimental work in class*
Region	*Region school was located in*
SchType	*Type of school* (Australia only)
RegType	*Combination of region and type of school* (England only)

In order to cast more light on students' motivation and effort regarding practical or experimental work done in science classes the new school-level construct with labelled 'Practice' was introduced (see also Postlethwaite and Wiley, 1991, and Keeves, 1991). Conceptually, Practice was placed directly before, respectively left of Achieve assuming that practical work had an effect on science achievement but was dependent on Effort as well as science-related values (SciValue) and attitudes (SciAtt).

FIG. 8.1 *Two-level path model with science achievement as the outcome measure for the 14-year-old level, 1983/84*

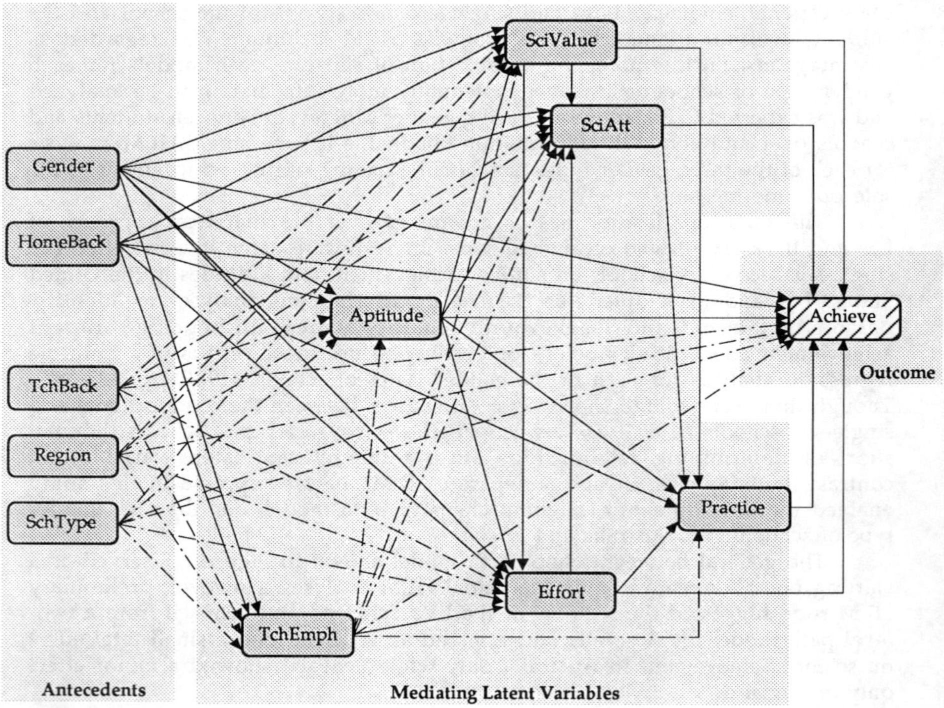

Notes:
In England, the antecedent **SchType** was substituted by **RegType** (see text for details).

construct, latent variable (LV)

construct of outcome measure

path, relevant on both student and school level

path, relevant on school level only

As the discussion of the path models for Australia and England in Chapters 6 and 7 suggested the effects of 'regionality' on science achievement, another LV, 'Region', thought to reflect regional or geographical differences was incorporated as an antecedent. However, 'Region' had a different meaning in Australia and England (see Appendix 4.3). Since it was hypothesized that the type of school played some importance in determining gender differences in science achievement, a school-type variable was also taken into account for the design of the two-level path model. The codes attached to the new variables were derived from criterion-scaling (Beaton, 1969) of the original variables with science achievement as the outcome. For further information regarding the coding the reader is referred to Appendices 4.1 and especially 4.3 as well as to the unpublished internal documents of the SISS study which may be obtained from the IEA Headquarters, The Hague.

Region and SchType/RegType were placed into the HLM path model as antecedents or exogenous LVs.

As it would have involved a great deal of extra work to compose each of the LVs anew by a set of MVs for each of the six HLM models, it was decided to use only the strongest predictor known from previous PLS analyses for each LV. Thus, in the HLM model presented in this chapter each LV is represented by a single MV (see Appendix 4.3).

The full model tested in sequentially linked HLM runs is presented in Figure 8.1. The outcome measure was science achievement (Achieve) again comprising the biology, chemistry, and physics subtest scores.

The mediating LVs employed in the model were TchEmph, Aptitude, SciValue, Effort, SciAtt, and Practice while HomeBack, Gender, TchBack, Region, and SchType/RegType were included as antecedents. Following the statistical properties of HLM, student- and school-level constructs were entered into the path model simultaneously. Data for TchBack, Region, SchType/RegType, and TchEmph were obtained at the school level only whereas all other LVs had been included in the model at both the student and school levels.

FIG. 8.2 *Three-dimensional representation of part of the HLM path model showing all possible effects between antecedents and Aptitude at and between student and school level*

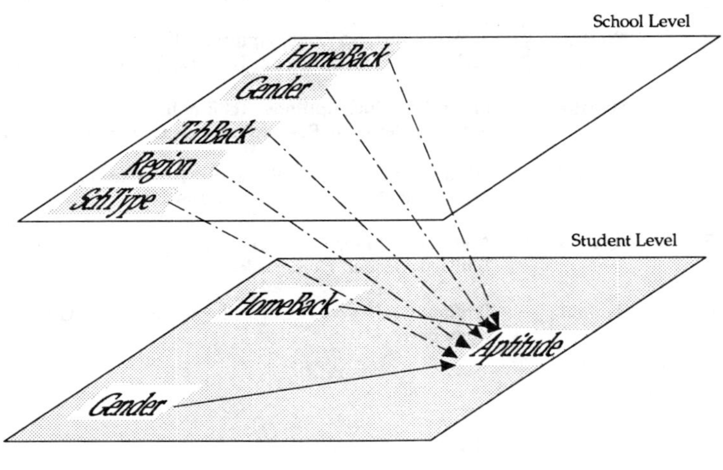

Notes:

——▶ path between student-level constructs

—·—▶ path leading from school-level construct to student-level construct

The two-level hierarchical structure of the model presented in Figure 8.1 can be seen by the two different types of paths assumed to exist among the LVs. Those paths drawn using a continuous line ——▶ symbolized possible causal effects measured at both student and school level. In addition, paths postulated among school-level constructs only were characterized by —·—▶.

It should be added, that for the simplicity and interpretability of the HLM path model a two-dimensional representation was preferred over a three-dimensional one. If the latter form of graphical design would have been used the paths between, for example, the antecedents and Aptitude would be as shown in Figure 8.2.

From Figure 8.2 one can easily imagine a loss in readability of the causal effects if to each single path within the whole model a path coefficient would also have been attached. Hence, preference was given to a two-dimensional representation of the two-level hierarchical path model.

It was pointed out earlier that in order to arrive at a 'final' two-level path model for each of the gender types of schooling for Australia and England multiple HLM runs needed to be performed. Due to the chronological order of the path model for science achievement, which has already been discussed in length in Chapters 6 and 7, each of the separate HLM runs was predefined. In a single HLM run only one outcome measure can be estimated by a set of student- and school-level predictors at a time. Direct or indirect relationships among these predictors could not be estimated. Hence, it was necessary to conduct several independent HLM runs. Each one was to predict another outcome chronologically prior to science achievement. Expressed differently, the following separate steps, identical to individual HLM runs, were performed:

Step	Outcome	Predictors
1	Achieve	Practice, SciAtt, Effort, SciValue, Aptitude, TchEmph, SchType/RegType, Region, TchBack, Gender, HomeBack
2	Practice	SciAtt, Effort, SciValue, Aptitude, TchEmph, SchType/RegType, Region, TchBack, Gender, HomeBack
3	SciAtt	Effort, SciValue, Aptitude, TchEmph, SchType/RegType, Region, TchBack, Gender, HomeBack
4	Effort	SciValue, Aptitude, TchEmph, SchType/RegType, Region, TchBack, Gender, HomeBack
5	SciValue	Aptitude, TchEmph, SchType/RegType, Region, TchBack, Gender, HomeBack
6	Aptitude	TchEmph, SchType/RegType, Region, TchBack, Gender, HomeBack

Initially, all possible within-unit (or student-level) variables as well as between-unit (or school-level) variables were entered into each single HLM run. The HLM-2 output was then checked, non-significant paths removed, and, where necessary, the run was repeated. In many instances this stepwise procedure needed to be repeated several times before a 'final' path structure emerged. As described in Section 4.3.2, paths were deemed non-significant or 'trivial' if the confidence level expressed by the *p-value* for a coefficient γ^* was larger than .01. Possible interaction effects were discarded if the p-value for the estimated parameter variance, or variance of the slope, of a latent variable was larger than .05. These cut-off criteria were considered conservative enough to arrive at paths which were meaningful and also represent the reality of the respective educational system (Raudenbush and Bryk, 1986; Cheung et al. 1990). It should be recalled that the standardized, non-metric γ^* coefficients have a similarly basic interpretation as β in multiple regression analysis.

Since the two-level HLM approach splits the whole variance of an outcome variable into a within-unit and a between-unit segment it is also of considerable importance to identify possible variation among the slopes of the within-unit variables. For this reason, HLM-2 provides an estimation of the slope or parameter variance calculating a χ^2-value alongside with a respective p-value. Should the p-value fall below the cut-off point of .05 it was assumed that the variance of the slopes of this particular LV was not incidental or a measurement error. Efforts were then undertaken to trace the LV or LVs which accounted for this variation. If the variation could be clearly attributed to one or more LVs, this constituted a so-called *interaction effect* I. Typically, interaction effects found in the path models for coeducational schools in both Australia and England did not account for more than 5 percent of the total variance. Hence, interaction effects should, as pointed out in Section 4.3.2, not be interpreted as major factors influencing an outcome variable.

In contrast to PLS, where only path coefficients of $p \geq .10$ were regarded as meaningful, a cut-off point of $\gamma^* \geq .05$ as applied for the HLM models (see Section 4.3.2) might not seem conservative for HLM runs. But it has to be kept in mind that the two-level HLM statistics were not flawed by possible aggregation bias due to its conceptualization of taking into account both variables from the student and the school level. For a more detailed statistical discussion of the cut-off criteria see Raudenbush (1988, 1989), Bryk and Raudenbush (1989), and Bryk et al. (1988).

Further information regarding the interpretation and 'reading' of the statistics produced by HLM-2 has already been given in Section 4.3.2 and need not be repeated here.

The outputs for all final runs for each of the six steps described above for each of the three path models for Australia and England may be obtained from the author on request. As an example, Appendix 8.1 shows the final HLM-2 run for estimating the effects on science achievement for coeducational schools in Australia.

In the following two sections the path models for the three gender types of schooling for Australia (Section 8.2) and England (Section 8.3) will be presented and discussed before a comparison of coeducational versus sex-segregated education is attempted (Section 8.4).

8.2 Two-level HLM Models for Australia

In 1984, at the age of 14 nearly all students in Australia were enrolled in so-called *Junior Secondary Schools*. Compulsory education ended at the age of 15, with the exception of Tasmania where it was 16 years. Basically, three different types of schools were found throughout all states and territories: *Government*, *Independent*, and *Catholic Schools*. In 1984 about 25 percent of all 14-year-olds were attending a non-governmental school, a proportion which had increased steadily since 1970. Moreover, of these 25 percent four-fifth of the students went to Catholic schools which were run and financed mainly by the Roman Catholic Church. Non-governmental schools were largely independent of governmental directives and had only to subscribe to a minimum educational standards of educational provision (McKenzie, 1988).

During the 1983/84 data collection about 84 percent of the 14-year-old students were enrolled in a coeducational junior secondary school, 10 percent went to single-sex boys schools and six percent attended single-sex girls schools (Oon Chye, 1991, p. 32). The number of single-sex schools had strongly decreased over the period since the first data collection in 1970/71 when a proportion of 36 percent single-sex schools was reported (Comber and Keeves, 1973, p. 151).

For each of the three gender types of schooling, coeducational, single-sex boys, and single-sex girls schools, separate HLM path models on the basis of Figure 8.1 were developed and tested. Of course, Gender had to be omitted for the models for single-sex schools due to zero variance on this construct.

→*Coeducational Schools in Australia* Figure 8.3 shows the 'final' two-level path model. The most important data necessary for interpreting the path model have been provided in Table 8.1. According to the analytical concept of HLM (see Section 4.3.2), the results have been split into those referring to the *fixed effects* and those referring to *random effects*.

Fixed effects can derive from either within-unit (i.e. student-level) or between-unit (i.e. school-level) constructs. As indicated the strength of these effects, or paths, is given by the respective γ^* coefficient. Recorded in Table 8.1 are also the standard error as well as the p-value for each fixed effect.

In addition, the HLM-2 output provides several statistics which allow an interpretation of the *random effects* inherent in the specified model. A χ^2-test is performed for each within-unit construct to examine the hypothesis that the residual differences among units, after taking into account the effect of between-unit predictors, is zero. Thus, Table 8.1 gives the amount of estimated parameter variance due to variation in the slopes of each of the within-unit constructs. The smaller the estimated parameter variance the more likely the fixed effects identified account for most of the variance of the outcome measure. If, on the other hand, the respective p-value for the estimated parameter variance fell below .05 it was necessary to test whether it was possible to introduce an interaction effect into the model by clearly identifying the particular construct(s) which caused a variation in the slopes of the within-unit path. As can be seen from Table 8.1 when predicting Practice the student-level LV SciAtt showed a direct, or fixed, effect of $\gamma^*=.17$. At the same time, the slopes of SciAtt appeared to be influenced, if only to a very minor extent (the estimated parameter variance was 1 percent), by some school-level effects as indicated by the small p-value of

.041. It was then decided to re-run HLM specifying possible school-level constructs said to show this interaction effect. As a result, it turned out that the social status of the class or school (HomeBack) accounted for the variation among the slopes of SciAtt with Practice as the criterion in the regression equation. This significant interaction effect (p-value .016), however, was not very strong (γ*=-.05) but it was interesting to note that it was bigger than the direct effect of Effort on Practice and that it had a negative sign. This could be interpreted as meaning that although the science-related attitudes of a student were positively related to the amount of practical work a student did during science lessons, this relationship was negatively influenced by the social status of the school: in schools where the students were of higher status, it was less likely that science-related attitudes had an impact on Practice.

As can be seen, all standard errors reported in Table 8.1 were small (≤.04) while the respective p-values for all fixed and interaction effects were ≤.05, in many cases they even amounted to ≤.000 which could be taken as an indication of the stability of the six separate HLM runs constituting the path model for science achievement in Australia's coeducational schools shown in Figure 8.3.

The data from Table 8.1 and the path model presented for coeducational schools in Australia gave rise to some interesting new observations. Science achievement was largely influenced by student-level effects, with the strongest one coming from Aptitude (γ*=.52). On the school-level, HomeBack (γ*=.12), Aptitude (γ*=.23), and Effort (γ*=.07) exercised an effect on achievement. Although these results confirmed previous findings (see Tables 6.7 and 7.2), it must be acknowledged that the interpretation of the between-school model developed for Australia (Figure 7.3) led to the assumption that it was the school-level effects which accounted for most of the variation on science achievement. That conclusion was underlined by the fact that the amount of variance explained by the between-school model (68 percent) was considerably higher than the amount of variance explained by the between-student model (47 percent). Here, the application of iterative HLM-2 runs succeeded in clarifying the conclusions drawn in Chapter 6 and 7. Putting it differently, the very strong total effects on science achievement found with the between-school model for HomeBack (t=.72) and Aptitude (t=.71) were artefacts and reflected, in part, the process of aggregation rather than showing 'real' school-level influences. Indeed, it was the socio-economic status of the school which largely constituted the general aptitude of a class in a coeducational school (γ*=.19). However, it should also be noted that independent coeducational schools showed a higher level of performance than government or catholic schools. Likewise, regional differences had a direct effect on Aptitude (γ*=.07) as well as a direct effect on science achievement (γ*=.07) which in both instances meant that schools located in certain areas, such as Queensland or Canberra (ACT), scored higher.

The two-level HLM model developed for coeducational schools in Australia highlights also the complexity of gender effects on science achievement. Gender had a noticeable direct student-level effect on achievement favouring the boys (γ*=-.14). But it became influential by operating through Aptitude (γ*=.03), SciValue (γ*=-.08), Effort (γ*=.15), SciAtt (γ*=-.09), and Practice (γ*=.09). The different signs of these effects should be noted since they are not only significant but also meaningful.

FIG. 8.3 *Two-level path model with science achievement as the outcome measure for coeducational schools in Australia, 14-year-old level, 1983/84; N=188*

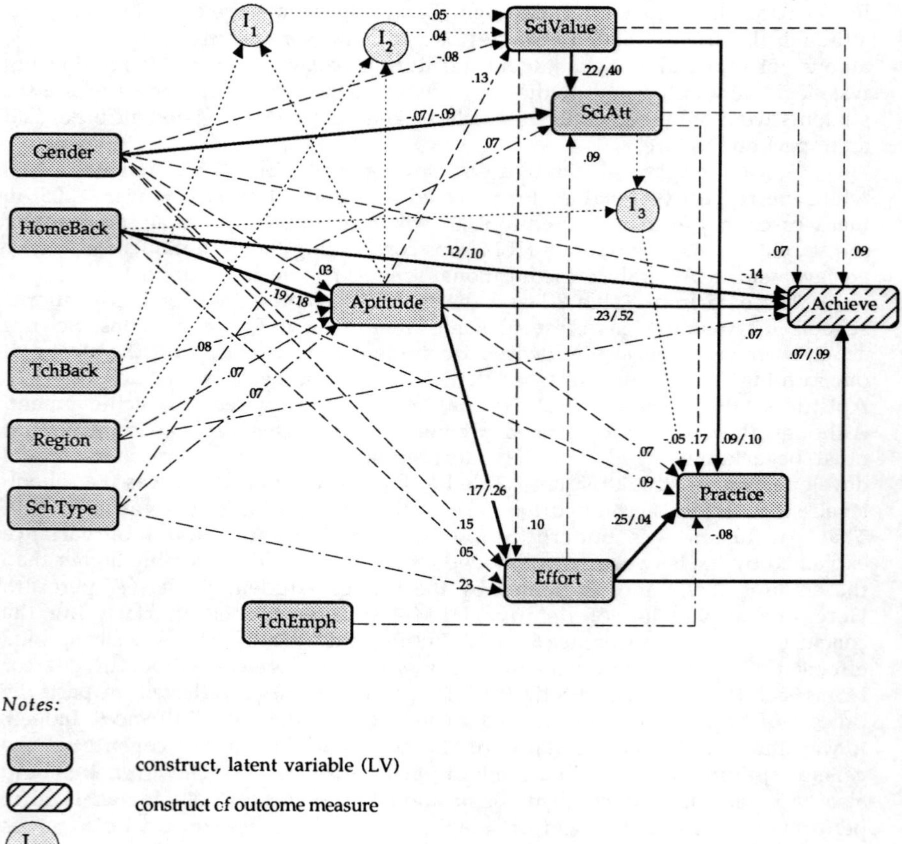

Notes:

construct, latent variable (LV)

construct of outcome measure

I_2 interaction effect

.17/.26 path, relevant on both student and school level, with γ^* coefficients; the first value (.17) shows the effect on the school level, the second (.26) on the student level

path, relevant on school level only

path, relevant on student level only

path for interaction effect

TABLE 8.1 *Selected statistics from HLM-2 runs for a standardized two-level path model for coeducational schools in Australia, 14-year-old level, 1983/84; N=188*

Fixed effects					Random effects		
school level construct	student level construct	γ*	standard error	p-value	slopes of...	estimated parameter variance	p-value
Outcome → Achieve							
HomeBack		.12	.02	.000			
Region		.07	.02	.000			
Aptitude		.23	.02	.000			
Effort		.07	.03	.008			
	HomeBack	.10	.01	.000	HomeBack	.00	.067
	Aptitude	.52	.01	.000	Aptitude	.00	.117
	Gender	-.14	.01	.000	Gender	.01	.310
	SciValue	.09	.01	.000	SciValue	.00	.265
	Effort	.09	.02	.000	Effort	.01	.318
	SciAtt	.07	.01	.000	SciAtt	.01	.261
Outcome → Practice							
TchEmph		-.08	.03	.004			
SciValue		.09	.03	.002			
Effort		.25	.04	.000			
	Aptitude	.07	.02	.000	Aptitude	.01	✎ .049
	Gender	.09	.02	.000	Gender	.01	✎ .006
	SciValue	.10	.02	.000	SciValue	.01	✎ .008
	Effort	.04	.02	.018	Effort	.01	.387
	SciAtt	.17	.02	.000	SciAtt	.01	✎ .041
	⇑ HomeBack	-.05	.02	.016			
Outcome → SciAtt							
Gender		-.07	.03	.005			
SchType		.07	.03	.005			
SciValue		.22	.02	.000			
	Gender	-.09	.01	.000	Gender	.00	✎ .046
	SciValue	.40	.02	.000	SciValue	.01	<.500
	Effort	.09	.02	.000	Effort	.00	.172

to be continued

Notes:

Region (boxed) — *school-level (between-unit) construct*

Gender (shaded) — student-level (within-unit) construct

⇑ HomeBack — interaction effect of school-level construct

✎ — only in these cases the variation of the slopes is significant, but only for SciAtt is this variation explained by Gender

TABLE 8.1 (ctd.) *Selected statistics from HLM-2 runs for a standardized two-level path model for coeducational schools in Australia, 14-year-old level, 1983/84; N=188*

Fixed effects					Random effects		
school level construct	**student level construct**	**γ***	**standard error**	**p-value**	**slopes of...**	**estimated parameter variance**	**p-value**
Outcome → Effort							
SchType		.23	.03	.000			
Aptitude		.17	.02	.000			
	HomeBack	.05	.02	.001	HomeBack	.01	<.500
	Aptitude	.26	.02	.000	Aptitude	.02	.052
	Gender	.15	.02	.000	Gender	.01	.003
	SciValue	.10	.02	.000	SciValue	.01	.447
Outcome → SciValue							
	Aptitude	.13	.02	.000	Aptitude	.02	.019
	⇑ TchBack	.05	.02	.022			
	⇑ Region	.04	.02	.028			
	Gender	-.08	.02	.000	Gender	.01	<.500
Outcome → Aptitude							
HomeBack		.19	.03	.000			
Region		.07	.03	.006			
SchType		.07	.04	.049			
TchBack		.08	.02	.002			
	HomeBack	.18	.02	.000	HomeBack	.01	.424
	Gender	.03	.02	.039	Gender	.01	<.500

Notes:

Region (boxed)	school-level (between-unit) construct
Gender (shaded)	student-level (within-unit) construct
⇑ TchBack	interaction effect of school-level construct

Another quite substantial difference between the findings of the between-school PLS model and the HLM model for science achievement was that Gender obviously did not operate on the school level at all. This implied that the direct gender effect found at the between-school level (see Figure 7.3) must have been an artefact caused by the aggregation of data. Alternatively, it could have been a consequence of the differences between single-sex boys and single-sex girls schools in the analysis that it included all three types of school. With the exception of an influence of Gender on SciAtt (γ*=-.07) at the school level, all other gender effects occurred at the between-student level only. In other words, except for a very minor effect of Gender on the expression of science-related attitudes, where male-dominated classes showed more favourable attitudes towards science, the gender-typical composition of the class, i.e. predominantly male or predominantly female science classes, did not have a noticeable effect on the achievement in science in coeducational schools in Australia.

A further unexpected result of the HLM runs was that practical work did not affect science achievement at either the student or school level. To some extent this contrasts with the findings of Postlethwaite and Wiley (1991, p. 129) who reported a weak positive effect of experimental work on science achievement at the student level.

Some comments should also be made about the relationship of science-related attitudes, values, and motivation (Effort) and their effect on achievement. The PLS results presented earlier suggested that neither SciValue, Effort, nor SciAtt had a sizeable influence on science achievement at either of the two measurement levels. Although this could be confirmed for the coeducational schools (all effects were $\gamma^* < .10$; the cut-off point for relevant effects for PLS was $p \geq .10$) it was, however, interesting to confirm that the interrelations among these three constructs, as hypothesized beforehand (see Section 6.2.1), seemed to exist only at the student level. Nevertheless, the rather tight relationship between SciValue and SciAtt could also be identified at the school level ($\gamma^* = .22$) although it was less pronounced than at the student level ($\gamma^* = .40$). Of the three constructs only Effort showed a small effect on science achievement at the school level ($\gamma^* = .07$). But, in general, science-related attitudes, values, and motivation played no important role in influencing achievement in science for Australia's 14-year-olds enrolled in coeducational schools.

It was mentioned above that the HLM analyses discovered three interaction effects, I_1, I_2, and I_3. The direct effect that Aptitude had on SciValue ($\gamma^* = .13$) signalled the likelihood of an additional random effect which could explain about 1 percent of variance (p-value .019). It was found that both TchBack (I_1; $\gamma^* = .05$) and Region (I_2; $\gamma^* = .04$) caused this variation in the slopes of SciValue regressed upon Aptitude. These effects show that students with a high aptitude were more likely to express positive science-related values if their teacher was a more experienced (in this context: older) person and/or if they lived in certain states such Queensland or ACT. The third interaction effect influenced the slopes of SciAtt when estimating Practice (I_3; $\gamma^* = -.05$). The positive impact science-related attitudes had on doing practical work and experiments at the student level were negatively affected by the socio-economic status of the school (HomeBack). It was of particular interest to realize that this interaction was the only measurable effect the social climate of the school exercised on science-related attitudes and values. Viewing it differently, it appeared that the socio-economic status of Australian coeducational schools did not determine the manifestation of students' attitudes and values towards science and their effect on other latent variables within the path model over and above the effect of HomeBack found at the student-level. As will be seen similar findings were made with respect to single-sex schools.

The data presented in Table 8.1 and its visual representation of the respective two-level path model for Australia's coeducational schools underlined that science achievement was both a function of several student-level effects (HomeBack, Gender, Aptitude, SciValue, Effort, and SciAtt) as well as school-level effects (HomeBack, Aptitude, Region, and Effort). New insights into the complexity of gender effects operating on science achievement were gained. Clarity could also be obtained with respect to the relevance of certain paths. Though a number of paths had $|\gamma^*|$-values of ≤10, which actually was the somewhat artificial cut-off criterion used for relevant paths in PLS analyses, the HLM data supported the assumption that a cut-off point of $|\beta|<.10$ was, in fact, very conservative. Thus, HLM brought forth a few extra paths not having been considered strong enough when running PLS at the between-student or between-school level. Due to a very low sampling error (see Section 3.4.3) it was possible to even consider effects as 'real' if the p-value of their γ^* was ≤.100 - an error probability of approximately 10 percent. In this respect, HLM performed at a much superior level over PLS, though PLS was definitely the more satisfactory exploratory technique.

In any case, the HLM results confirmed the PLS findings reported in Chapters 6 and 7 that, with respect to gender differences, at the 14-year-old level boys had considerable advantage over girls in science achievement in coeducational schools. It seemed that, with the exception of a school-level effect of Gender on science-related attitudes, gender effects operated only at the between-student level. The gender-typical composition of the class did not influence at all achievement in science. Here, the PLS results reported in Chapter 7 were obviously misleading. However, they applied to both single-sex and coeducational schools, and it is now necessary to examine whether the effects recorded were associated with type of school differences.

→*Single-Sex Boys Schools in Australia* A sample of 20 single-sex boys schools was examined using HLM. Figure 8.4 shows the two-level path model. The data necessary for interpreting the path model are summarized in Table 8.2.

At first sight, the path model revealed that only a comparatively limited number of paths appeared to be effective in influencing science achievement. In fact, SciAtt (γ^*=.22) and Aptitude (γ^*=.56) were the only two factors directly effecting achievement at the between-student level. Further to this, it was Aptitude alone that showed a strong impact on science achievement (γ^*=.47). Thus, the achievement of 14-year-old boys enrolled in single-sex schools was basically dependent on their individual Word Knowledge test scores as well as the general aptitude of the school they attended. In addition, the science-related attitudes of the students showed considerable influence on their achievement in science but only within the school. The general aptitude of the school, i.e. Aptitude measured at the between-school level, was strongly dependent on the type of school (γ^*=.21). Here, students from independent schools showed a higher aptitude level than students from catholic schools (the number of government single-sex schools was negligible).

But it was stunning to see that school-level effects other than SchType and Aptitude were of no relevance for the achievement of boys in science. This could possibly be attributable to the homogeneity of the students enrolled in single-sex boys schools with respect to their socio-economic background. However, it must be recalled that the number of single-sex boys schools in Australia was around 10 percent of all junior secondary schools and that boys schools were (nearly) all run by private institutions or religious denominations. As students had to pay fees for attending these schools, in general, only the financially better off families could afford to send their boys there.

FIG. 8.4 *Two-level path model with science achievement as the outcome measure for single-sex boys schools in Australia, 14-year-old level, 1983/84; N=20*

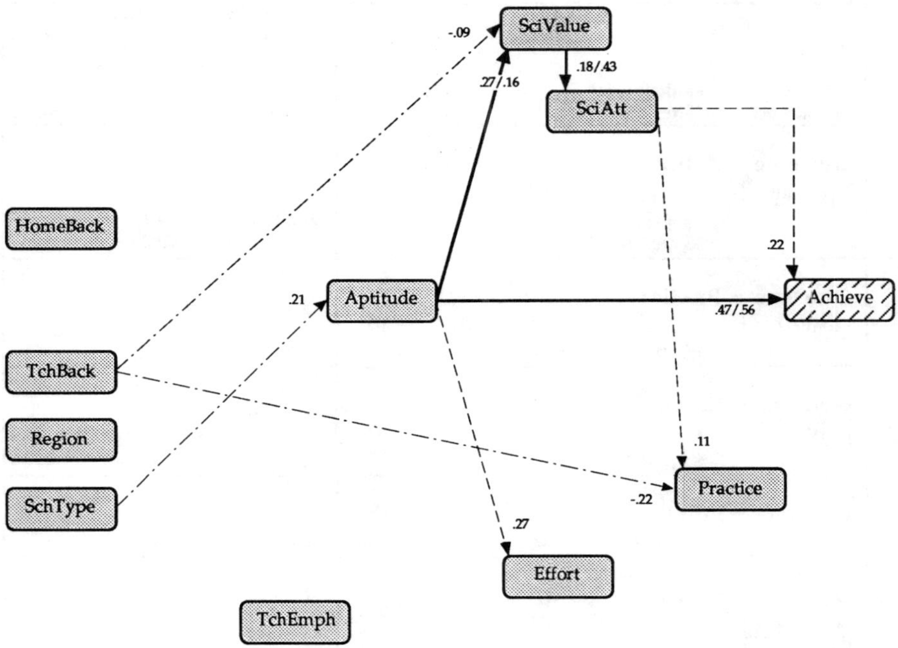

Notes:
for further details see legend for Figure 8.3

Thus, the socio-economic climate within a school was supposed to be rather homogeneous. In fact, at both the student and the school level, the effect of HomeBack on any other variable including achievement was nil. It must be acknowledged, however that statistical significance is dependent on the sample size, and that with a large sample of single-sex boys schools other variables might be found to be of consequence.

A further indication for the homogeneity of boys-only schools in Australia was the fact that hardly any variation seemed to occur between the schools with respect to geographical differences (i.e. Region) and the emphasis put on investigation when teaching science (i.e. TchEmph).

TABLE 8.2 Selected statistics from HLM-2 runs for a standardized two-level path model for single-sex boys schools in Australia, 14-year-old level, 1983/84; N=20

Fixed effects					Random effects		
school level construct	student level construct	γ*	standard error	p-value	slopes of...	estimated parameter variance	p-value
Outcome → Achieve							
Aptitude		.47	.07	.000			
	Aptitude	.56	.05	.000	Aptitude	.01	<.500
	SciAtt	.22	.04	.000	SciAtt	.00	<.500
Outcome → Practice							
TchBack		-.22	.10	.035			
	SciAtt	.11	.05	.033	SciAtt	.01	.365
Outcome → SciAtt							
SciValue		.18	.07	.017			
	SciValue	.43	.05	.000	SciValue	.01	.223
Outcome → Effort							
	Aptitude	.27	.06	.000	Aptitude	.02	.173
Outcome → SciValue							
TchBack		-.09	.06	.115▲			
Aptitude		.27	.07	.001			
	Aptitude	.16	.06	.016	Aptitude	.01	<.500
Outcome → Aptitude							
SchType		.21	.07	.007			

Notes:

▲ due to the strength of γ* and the overall stability of this HLM step the effect of TchBack on SciValue was considered non-trivial and the path was not omitted

Region (boxed): school-level (between-unit) construct

Gender (shaded): student-level (within-unit) construct

In contrast to this evident homogeneity between single-sex boys schools in Australia it was of note that more experienced teachers influenced negatively both the expression of science-related values (γ*=-.09) and the amount of practical work done (γ*=-.22) at the school level. Although this did not seem to have a direct effect on the achievement in science, the experience of the science teachers as measured by TchBack inhibited significantly the performance of experiments during science lessons in boys schools at the 14-year-old level.

Another major difference between single-sex boys schools and coeducational schools should be seen in the findings that neither SciValue nor Effort had a significant impact on science achievement at all. Hence, in Australia's boys schools achievement in science was largely dependent on the general aptitude of the school and to some extent on the type of school which affected the aptitude level. Further to this, the achievement level of an individual boy enrolled in a boys school was positively influenced by two major factors, his aptitude and his attitudes towards science.

It should be added that no random or interaction effects were found. As will be seen, some similarities but also differences could be detected when examining the achievement in science in Australia's single-sex girls schools.

→*Single-Sex Girls Schools in Australia* The sample consisted of 11 girls schools which was examined, again, using the HLM technique. Figure 8.5 presents the two-level path model. All data necessary for interpreting the path model has been included in Table 8.3.

Comparing the model for Australian girls schools with that of boys schools revealed that in both cases just a small number of paths turned out to be of significance which was a consequence of the limited number of schools. In both types of single-sex schools science achievement was influenced by Aptitude (on the student and school level) as well as SciAtt. In the girls schools student's aptitude ($\gamma^*=.53$) was the strongest predictor together with the girl's attitude towards science ($\gamma^*=.19$) and the general aptitude of the school ($\gamma^*=.35$). As with boys schools this could be taken as an indication of the homogeneity of the student body enrolled in the single-sex girls schools. Another similarity in single-sex science education seemed to be that SciAtt had a medium-sized positive effect on the amount of practical work a student did in school ($\gamma^*=.19$). Also, TchEmph did not show much variation between schools so that it did not influence any other construct in the model.

However, these were all the similar findings when comparing girls schools with boys schools. In contrast, several results were found to be unique for Australian single-sex girls schools. One of these gender-typical findings was that whether a girls school was run by the Catholic Church or an independent body was of relevance to the motivation of the school only (SchType→Effort $\gamma^*=.27$). In turn, the effort a school made in science had a positive effect on the number of experiments being done during science instruction (Effort→Practice $\gamma^*=.32$). As with coeducational schools Effort was influenced by Aptitude at both the student and school level. But while in boys schools the general aptitude was largely affected by the type of school the aptitude level of a girls-only school was widely determined by the socio-economic status of the school ($\gamma^*=.33$). These observations led to the conclusion that, unlike boys schools, considerable status differences existed between single-sex girls schools. These status differences, however, were neither related to the experience of the staff nor to the location of the school (TchBack and Region showed no effect at all).

FIG. 8.5 *Two-level path model with science achievement as the outcome measure for single-sex girls schools in Australia, 14-year-old level, 1983/84; N=25*

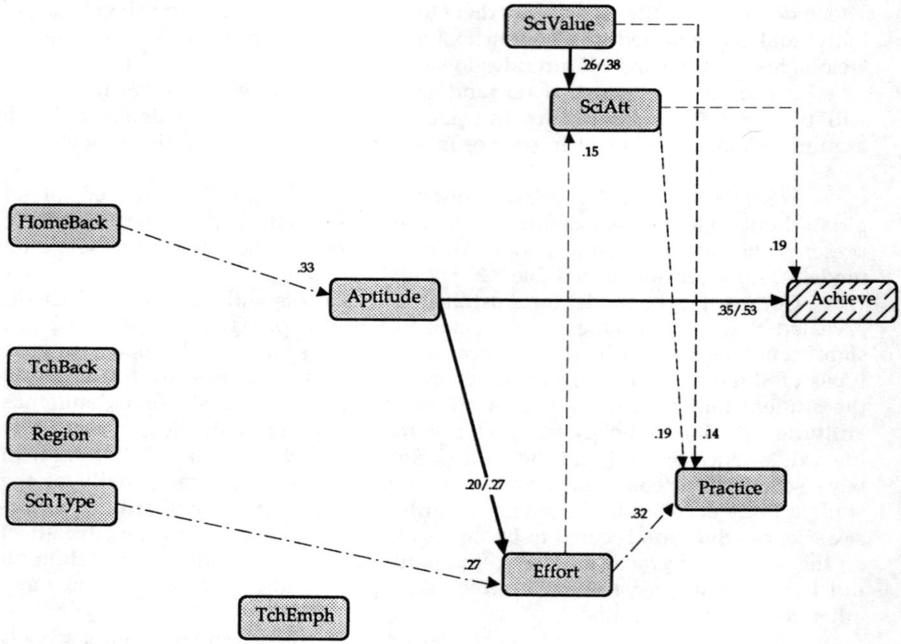

Notes:
for further details see legend for Figure 8.3

In some contrast to other Australian schools was that SciValue was not affected by Aptitude or any of the antecedents. This underlined that science-related values showed considerable variation among the girls in a school but it seemed that this was not the result of home, aptitude, teaching, or school factors. It could, however, be hypothesized that Australian 14-year-old girls, at least those attending single-sex girls school, held quite different values towards science which appeared to be unrelated to other factors including their cognitive level of aptitude. This was very different from the case of boys schools This showed how much values and, in effect, attitudes were determined by gender. Furthermore, if one takes into account the positive stimulation science-related values and attitudes had on science achievement more careful analyses are needed to investigate the question of how values and attitudes towards science are established and how they can be promoted by the parents, teacher, and school environment.

TABLE 8.3 *Selected statistics from HLM-2 runs for a standardized two-level path model for single-sex girls schools in Australia, 14-year-old level, 1983/84; N=25*

Fixed effects					Random effects		
school level construct	**student level construct**	**γ***	**standard error**	**p-value**	**slopes of...**	**estimated parameter variance**	**p-value**
Outcome → Achieve							
Aptitude		.35	.04	.000			
	Aptitude	.53	.04	.000	Aptitude	.00	<.500
	SciAtt	.19	.04	.000	SciAtt	.01	.220
Outcome → Practice							
Effort		.32	.07	.000			
	SciValue	.14	.05	.007	SciValue	.01	.110
	SciAtt	.19	.05	.001	SciAtt	.01	<.500
Outcome → SciAtt							
SciValue		.26	.06	.000			
	SciValue	.38	.04	.000	SciValue	.01	.342
	Effort	.15	.05	.005	Effort	.01	.106
Outcome → Effort							
SchType		.27	.07	.001			
Aptitude		.20	.06	.004			
	Aptitude	.27	.04	.000	Aptitude	.01	.246
Outcome → SciValue							
Aptitude		.03	.04	.315▲			
Outcome → Aptitude							
HomeBack		.33	.07	.000			

Notes:

▲ the final HLM-2 run showed this trivial effect so that SciValue could not be predicted by any other student or school level construct

[Region] school-level (between-unit) construct

Gender (shaded) student-level (within-unit) construct

An inspection of the path model (Figure 8.5) shows that science-related values, attitudes and motivation played a much stronger, positive role in science instruction. This was especially true for the amount of practical and experimental work being done by girls, although experimental work did not affect science achievement as measured by the tests in any Australian lower secondary school (SciValue→Practice $\gamma^*=.14$; Effort→Practice $\gamma^*=.32$; SciAtt→Practice $\gamma^*=.19$).

As with boys schools Australia's single-sex girls schools appeared to be rather homogeneous in terms of science teaching: neither TchBack, TchEmph, Effort, nor Practice had any effect on the achievement level of a school in science. The socio-economic status of girls schools, however, varied significantly as expressed by the strong path leading from HomeBack to Aptitude ($\gamma^*=.33$). This variation, in turn, was partly responsible for the motivation of the school ($\gamma^*=.20$) but basically the school's status decided about the achievement of the school in science. It was interesting to note that between Australia's girls schools the amount of practical work being done apparently varied a great deal.

→*Conclusions* The three different path models for Australia's coeducational schools (Figure 8.3), single-sex boys schools (Figure 8.4), and single-sex girls schools (Figure 8.5) showed remarkable differences. But it should be pointed out that some similarities between the three forms of schooling were also found.

First of all, the path model for coeducational schools exhibited a sizeable number of paths among the variables at both the student and school level. Here, influential differences could be identified in each antecedent considered in the path model.

Gender differences in values, attitudes, and achievement were always in favour of the boys. However, the gender-typical composition of a class did not contribute to better or worse achievement in science. In single-sex schools the achievement level of the school was either largely dependent on whether it was run by an independent body or by the Catholic Church (for boys-only schools) or it was dependent on the socio-economic climate of the school (for girls-only schools). Regional differences became important only for coeducational schools.

In coeducational schools the influence of teacher- and teaching-related variables was seemingly different from single-sex schools, but it should be recalled that the amount of practical work done had no consequences at the individual or school achievement level in science. Likewise, the effect Effort had on science achievement was found to be small in coeducational schools.

For all Australian 14-year-olds, wherever they were enrolled, science-related attitudes influenced science achievement positively. This was particularly true in single-sex schools. Science-related values, however, had a small positive effect only in coeducational schools.

The homogeneity of students within single-sex schools could be best illustrated by the absence of home background effects. Though marked differences were found between single-sex boys schools across Australia, the socio-economic variation within single-sex schools was minimal and affected neither the girls' nor boys' achievement in science. Alternatively, variation in the socio-economic status of the home contributed directly and indirectly to the achievement of students enrolled in coeducational schools. Students from higher-status homes performed much better. The results allow for the interpretation that parents with similar educational background and of similar occupational status enrolled their daughters or sons in a single-sex school. It should be recalled that the statistics produced by HLM-2 did not permit any inferences about the quality, or the educational or social suitability of single-sex education in Australia. The findings described merely highlighted significant differences among the factors influencing science achievement between coeducational schools, single-sex boys and single-sex girls schools.

In which way some of these findings could be generalized over and above the Australian setting remains to be seen when the three gender types of schooling are analyzed for the English school system.

8.3 Two-level HLM Models for England

Politically, in England educational goals and administrative guidelines are the responsibility of the government represented by the Secretary of State for Education and Science. However, much of the daily business and administrative work to be done in education is carried out by local administrative bodies known as *Local Education Authorities*, LEAs for short. The LEAs are, for example, in charge of the employment of teachers, the construction of schools, or the acquisition of textbooks or laboratory equipment (Booth, 1988, pp. 691-692).

In the 1983/84 sample from England there were four different types of schools:

Private or Public Schools	Commonly, these private schools, run by private individuals, companies, or charitable trusts form a relatively small proportion of all schools in England. About five percent of all students aged 5 to 16 are educated in private schools.
Voluntary or Maintained Schools	Another minority type of school which is often established by religious groups, such as the Protestant or Catholic Church, but, in general, subsidized by the LEAs.
Grammar Schools	These are selective, academically oriented secondary schools. This type of school was the common before educational reforms took place in 1965. Today, grammar schools are small in numbers as many of these schools had been converted into comprehensive schools since 1965.
Comprehensive Schools	Today the most common type of secondary school which offers a wide spectrum of courses to students and at the sixth form level offer specialized courses.

For the data analysis voluntary schools and grammar schools have been combined into the group of 'other maintained schools' (see Appendix 4.3) as they are basically financed by the government or LEAs but at times run by private institutions.

In addition to the classification by this formal or administrative type of schooling, in England schools could also be categorized by gender type. For the 1983/84 data collection Oon Chye (1991, p. 33) reported that about eight percent of all lower secondary schools were boys schools, seven percent girls schools, and the remaining were coeducational. This reflected a sharp decrease in single-sex schooling since the first data collection which took place in 1970/71 when approximately 36 percent of all 14-year-old students were enrolled in single-sex schools (Comber and Keeves, 1973, p. 151).

For each of the three gender types of schooling, coeducational, single-sex boys, and single-sex girls schools, separate HLM path models were developed and tested based on the model presented in Figure 8.1. Again, Gender was omitted from the models for single-sex schools on account of the lack of variance on this construct. In the following sections these results are discussed.

→*Coeducational Schools in England* The sample consisted of 145 schools which were examined with HLM-2. Figure 8.6 presents the two-level path model. The corresponding statistics of the path model have been included in Table 8.4.

One of the first impressions when analyzing the path structure was that similar to the coeducational schools in Australia a multitude of student and school level paths was operating among the constructs. With the exception of SciAtt and Practice all other student-related LVs showed effects both at the within-unit and the between-unit level. If it is taken into consideration that each of the school-level constructs (i.e. TchBack, Region, RegType, TchEmph) had some influence on, at least, one other construct, this will support the conclusion that both students in comprehensive schools and comprehensive schools as such apparently differ remarkably from one another in science achievement.

In general, the results of HLM-2 are similar to the results of the PLS analyses discussed in Chapter 6 (between-student level) and Chapter 7 (between-school level). However, Aptitude turned out to be the strongest predictor of achievement at both measurement levels (γ^*=.25/γ^*=.48 [the first γ^*-coefficient indicates the path at the school level, the second the strength of the path at the student level]). Although no distinction between direct and total effects is made in HLM, it can be inferred from the previous PLS analyses that further to the direct effects of the socio-economic status on science achievement (γ^*=.09/γ^*=.10) HomeBack also had considerable indirect influence by operating through Aptitude (γ^*=.19/γ^*=.21) and Effort (γ^*=.13/γ^*=.08). In addition, Aptitude had a sizeable effect on Effort (γ^*=.19/γ^*=.20). Thus, students and school with higher aptitude and higher socio-economic status performed better in the science tests.

While Aptitude was partly predictable by the socio-economic status of the student it was also a function of Region (γ^*=.06) and RegType (γ^*=.16). This, in fact, showed that it made a difference for the achievement in science (1) in which region a student was attending a school; and (2) in what kind of school he or she was enrolled. Though Region and RegType did not exercise any direct influence on achievement, they operated indirectly through Aptitude (both) and Effort (RegType only). It would appear safe to conclude that regional differences between coeducational schools accounted for less variation in achievement than did the derived variable of region and school type. It might be worth noting that the product-moment correlation of Region with RegType was r=-.15. Thus, it was even more interesting to find that both constructs had a positive effect on Aptitude and, therefore, indirectly on science achievement.

FIG. 8.6 *Two-level path model with science achievement as the outcome measure for coeducational schools in England, 14-year-old level, 1983/84; N=145*

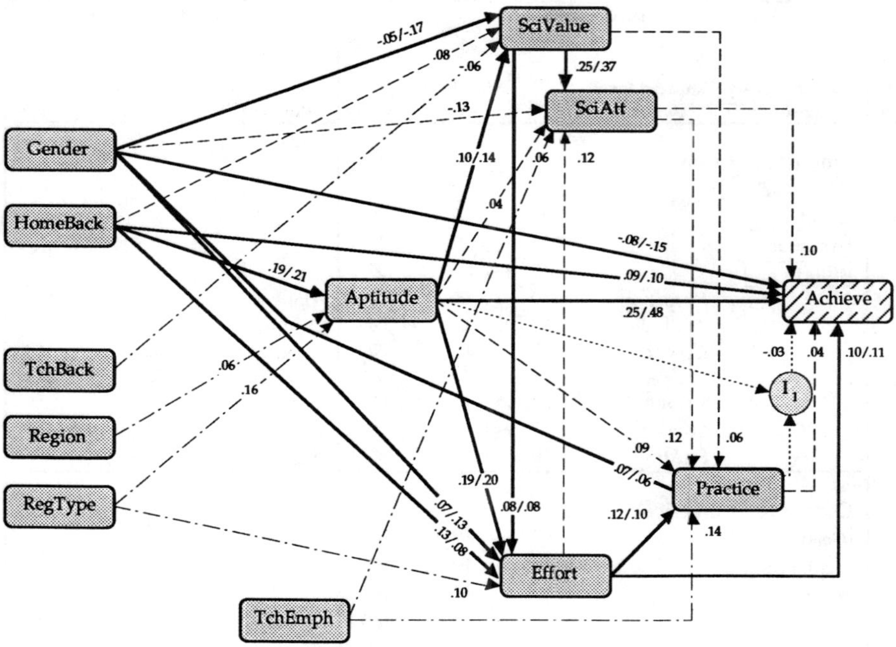

Notes:
for further details see legend for Figure 8.3

These results suggest that it was more favourable for students to be enrolled in an independent or public school in the southern counties of England. In contrast, students attending one of the comprehensive schools in the Midlands or northern counties were disadvantaged simply because of the type of school and its location. These findings were underlined by the very close relations found between RegType and Aptitude (r=.55), Effort (r=.54), and science achievement (r=.51). Clearly, marked status differences existed on the one hand between the governmental-run coeducational schools and private ones and on the other hand between different regions within England. It could not be rejected that students of financially well-off parents were attending the more 'exclusive' independent or private coeducational schools, mainly located in the south of England. This was also underscored by the extremely high positive correlation between HomeBack and RegType (r=.50).

TABLE 8.4 *Selected statistics from HLM-2 runs for a standardized two-level path model for coeducational schools in England, 14-year-old level, 1983/84; N=145*

Fixed effects					Random effects		
school level construct	student level construct	γ*	standard error	p-value	slopes of...	estimated parameter variance	p-value
Outcome → Achieve							
HomeBack		.09	.03	.001			
Gender		-.08	.02	.000			
Aptitude		.25	.03	.000			
Effort		.10	.03	.001			
	HomeBack	.10	.02	.000	HomeBack	.01	.017
	Aptitude	.48	.02	.000	Aptitude	.03	.003
	Gender	-.15	.02	.000	Gender	.02	.163
	Practice	.04	.02	.023	Practice	.02	.019
	⥣ Aptitude	-.03	.02	.077			
	Effort	.11	.02	.000	Effort	.01	.151
	SciAtt	.10	.02	.000	SciAtt	.02	.254
Outcome → Practice							
Gender		.07	.03	.038			
TchEmph		.14	.03	.000			
Effort		.12	.03	.000			
	Aptitude	.09	.03	.001	Aptitude	.03	.031
	Gender	.06	.02	.005	Gender	.00	<.500
	SciValue	.06	.02	.007	SciValue	.01	.429
	Effort	.10	.03	.000	Effort	.02	<.500
	SciAtt	.12	.02	.000	SciAtt	.02	.379
Outcome → SciAtt							
TchEmph		.06	.02	.006			
SciValue		.25	.02	.000			
	Aptitude	.04	.02	.027	Aptitude	.01	<.500
	Gender	-.13	.02	.000	Gender	.01	.010
	SciValue	.37	.02	.000	SciValue	.01	.137
	Effort	.12	.02	.000	Effort	.01	<.500

to be continued

Notes:

Region (boxed) school-level (between-unit) construct

Gender (shaded) student-level (within-unit) construct

⥣ Aptitude interaction effect of school-level construct

TABLE 8.4 (ctd.) Selected statistics from HLM-2 runs for a standardized two-level path model for coeducational schools in England, 14-year-old level, 1983/84; N=145

Fixed effects					Random effects		
school level construct	student level construct	γ*	standard error	p-value	slopes of...	estimated parameter variance	p-value
Outcome → Effort							
HomeBack		.13	.04	.001			
Gender		.07	.03	.029			
RegType		.10	.04	.007			
Aptitude		.19	.04	.000			
SciValue		.08	.03	.006			
	HomeBack	.08	.02	.000	HomeBack	.01	.417
	Aptitude	.20	.02	.000	Aptitude	.01	.276
	Gender	.13	.02	.000	Gender	.00	.095
	SciValue	.08	.02	.000	SciValue	.01	<.500
Outcome → SciValue							
Gender		-.05	.02	.043			
TchBack		-.06	.02	.015			
Aptitude		.10	.02	.000			
	HomeBack	.08	.02	.000	HomeBack	.01	.315
	Aptitude	.14	.02	.000	Aptitude	.01	.437
	Gender	-.17	.02	.000	Gender	.01	.025
Outcome → Aptitude							
HomeBack		.19	.03	.000			
Region		.06	.03	.030			
RegType		.16	.03	.000			
	HomeBack	.21	.02	.000	HomeBack	.01	.361

Notes:

Region (boxed)	school-level (between-unit) construct
Gender (shaded)	student-level (within-unit) construct
⇑ TchBack	interaction effect of school-level construct

As Region and, especially, RegType showed some influence on Aptitude, and as Aptitude proved to be the strongest predictor of science achievement at both measurement levels, it must be accepted that students attending the high status independent or private schools scored, indeed, higher on the achievement tests. This finding was also supported by the strong relationship between RegType and science achievement expressed by a correlation of r=.51.

But while achievement in science in coeducational schools was strongly dependent on the status of the home of the students as well as on the status and location of the school it should not be overlooked that several other significant factors contributed to the variability of the outcome measure. As was to be expected, the effect of Gender seemed to favour the achievement of boys at the student level (γ^*=-.15) and, to some extent, male-dominated classes at the school level (γ^*=-.08). But apart from these direct effects Gender also influenced science achievement indirectly. While girls and female-dominated groups tended to spend slightly more time doing homework (Gender→Effort γ^*=.07/γ^*=.13), boys showed more favourable values (γ^*=-.17) and attitudes towards science (γ^*=-.13). A weaker effect was the amount of practical or experimental work done in science classes which was higher among boys and in male-dominated groups. These HLM results were in accordance with the findings reported for England for the between-student level (Chapter 6) and the between-school level (Chapter 7) where, in both cases, the total effect of Gender on science achievement amounted to t=-.14. In other words, in England coeducational schooling had, at least in 1983/84, not yet succeeded in removing all gender-typical effects in science achievement at the 14-year-old level.

Moreover, the results of the HLM analyses for coeducational schools in England suggested that considerable heterogeneity existed among lower secondary schools. This can be seen from the numerous school-level paths found to be operating within the path model. But over and above school- or school-type-related differences it must be acknowledged that very pronounced differences seemed to prevail between students enrolled in coeducational schools per se. Rich and poor, comprehensive or public school, interested or disinterested in science, boy or girl were the most obvious pairs of opposites which were identified as key factors influencing achievement in science at the 14-year-old level in coeducational schools in England. It remains to be seen whether a similar variability was found among single-sex schools.

→*Single-Sex Boys Schools in England* Using HLM a sample of 11 schools was analyzed. Figure 8.7 shows the two-level path model with the relevant data recorded in Table 8.5. Though the γ^*-coefficients obtained were found to be significant, it could not be overlooked that the size of the sample somewhat limited the adequacy and generalizability of the results. This should be borne in mind for the following discussion.

In comparison to coeducational schools, the number of significant paths detected was lower. However, it came as a great surprise that within the whole model (see Figure 8.6) only two student-level effects could be identified as significant in terms of science achievement.

The larger of these student-level effects came from Aptitude (γ^*=.45) and the smaller one from HomeBack (γ^*=.19). Thus, the more able boys, i.e. those with higher scores on the Word Knowledge subtest, as well as boys from higher status families could expect a higher achievement in science. It was interesting to see that the strong positive effect of Aptitude detected at the student level was even more pronounced at the school level (γ^*=.54).

FIG. 8.7 *Two-level path model with science achievement as the outcome measure for single-sex boys schools in England, 14-year-old level, 1983/84; N=11*

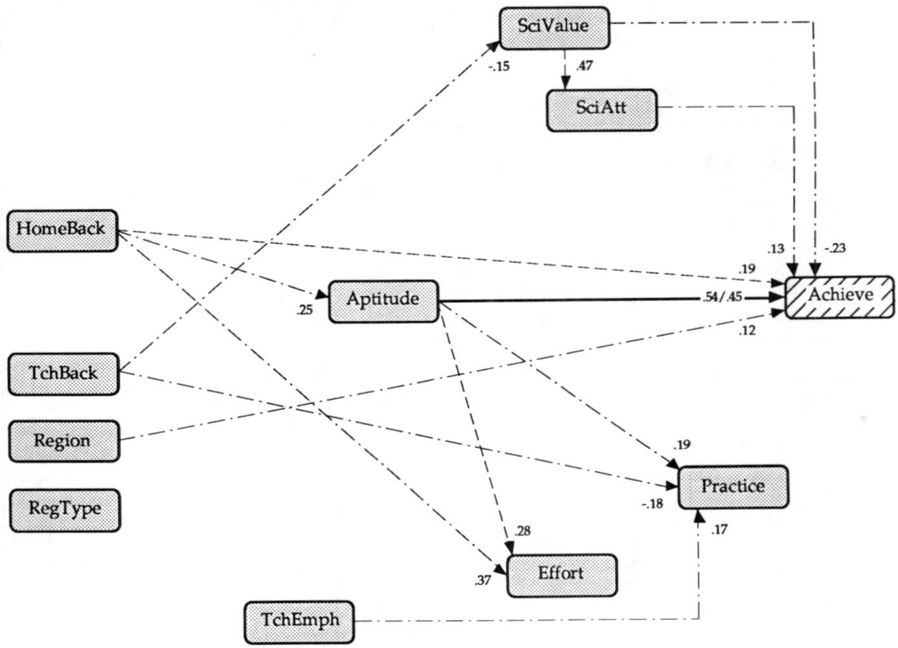

Notes:
for further details see legend for Figure 8.3

This observation and the fact that all other paths were operating only at the school level clearly indicated the cardinal differences which apparently existed between single-sex boys schools. School- and teacher-related factors did, to a much greater extent than with coeducational schools, give rise to heterogeneity among single-sex boys schools.

However, while considerable differences between the schools seemed to exist in the amount of time spent on homework and experimental work carried out in science lessons, neither Effort not Practice showed a significant effect on science achievement. In sharp contrast to coeducational schools, there was the further fact that Effort failed to predict Practice. What this meant was that single-sex boys schools seemed to differ significantly in their way of teaching science.

TABLE 8.5 Selected statistics from HLM-2 runs for a standardized two-level path model for single-sex boys schools in England, 14-year-old level, 1983/84; N=11

Fixed effects					Random effects		
school level construct	**student level construct**	**γ***	**standard error**	**p-value**	**slopes of...**	**estimated parameter variance**	**p-value**
Outcome → Achieve							
Region		.12	.06	.045			
Aptitude		.54	.05	.000			
SciValue		-.23	.05	.006			
SciAtt		.13	.05	.034			
	HomeBack	.19	.06	.012	HomeBack	.01	.035
	Aptitude	.45	.08	.002	Aptitude	.03	.057
Outcome → Practice							
TchBack		-.18	.08	.047			
TchEmph		.17	.08	.053			
Aptitude		.19	.09	.058			
Outcome → SciAtt							
	SciValue	.47	.06	.000	SciValue	.01	<.500
Outcome → Effort							
HomeBack		.37	.06	.000			
	Aptitude	.28	.08	.005	Aptitude	.01	.165
Outcome → SciValue							
TchBack		-.15	.07	.041			
Outcome → Aptitude							
HomeBack		.25	.07	.007			

Notes:

Region — school-level (between-unit) construct

Gender — student-level (within-unit) construct

While some schools put more emphasis on homework, in particular those of higher socio-economic status (γ*=.37) and/or those with more able students (γ*=.28), others stressed experimental work, especially those with younger, less experienced (or newly qualified) teachers (γ*=-.18). But, as can be seen, neither form of teaching or instructional organization helped to boost the achievement in science.

In addition to this instructional heterogeneity, there was also considerable variability in the science-related values and attitudes. In stark contrast to a close link, frequently noticed between SciValue and SciAtt at both the student and the school level (see Chapters 6 and 7), no such relationship seemed to exist for single-sex schools in England: value and attitude statements were somewhat related at the school level but no connection was identified at the student level. Even more interesting was the finding that SciValue had a negative influence on science achievement ($\gamma^*=-.23$) while the effect of SciAtt was positive ($\gamma^*=.13$). Yet the correlation between SciAtt and science achievement was negative (r=-.13) which might have been caused by collinearity effects between SciValue and SciAtt. This reason could help to explain the influence of SciValue on SciAtt detected at the school level which was found to be strong and positive ($\gamma^*=.47$).

How can these rather divergent findings be interpreted? At first sight it did not make much sense that although a very high relationship existed between SciValue and SciAtt ($\gamma^*=.47$ and r=.56) both constructs showed opposite effects on achievement at the school level. In terms of simple correlations for single-sex boys schools, SciValue had small but positive correlations with Effort (r=.30) and Practice (r=.24) but the correlations of SciAtt with Effort (r=-.16) and Practice (r=-.21) were negative. This would seem to suggest that although science-related values played a key role in the expression of students' attitudes towards science, this process led to different effects in different schools in science achievement. Unfortunately, no data was available to work out exactly how these different attitudinal mechanisms might be operating.

All in all, in England single-sex boys schools showed high homogeneity within the schools but at the same time the HLM results made the differences between schools become very obvious. Status and ability still dominated the individual boy's achievement in science but it must be recognized that major status and, thus, performance differences existed between the single-sex schools. There was some indication that schools located in the south of England showed higher achievement (Region→Achieve $\gamma^*=.12$) but as RegType did not have any effect whatsoever this could not be related to a particular type of school.

The factors which were important in influencing science achievement for single-sex boys schools at the between-student and the between-school level differed considerably from those found to be effective for coeducational schools with the exception of status and aptitude which appeared to be pervasive in explaining differences among all English secondary schools.

→*Single-Sex Girls Schools in England* A sample of 13 girls schools was examined with HLM. Figure 8.8 presents the two-level path model. All data necessary for interpreting the path model have been included in Table 8.6. As with boys schools some restrictions apply to the generalizability and the significance of the results because of the relatively small sample size.

In comparison with the two path models discussed for coeducational and single-sex boys schools, the HLM model for single-sex girls schools illustrated, again, the importance of the socio-economic status of the home for success in science achievement.

FIG. 8.8 *Two-level path model with science achievement as the outcome measure for single-sex girls schools in England, 14-year-old level, 1983/84; N=13*

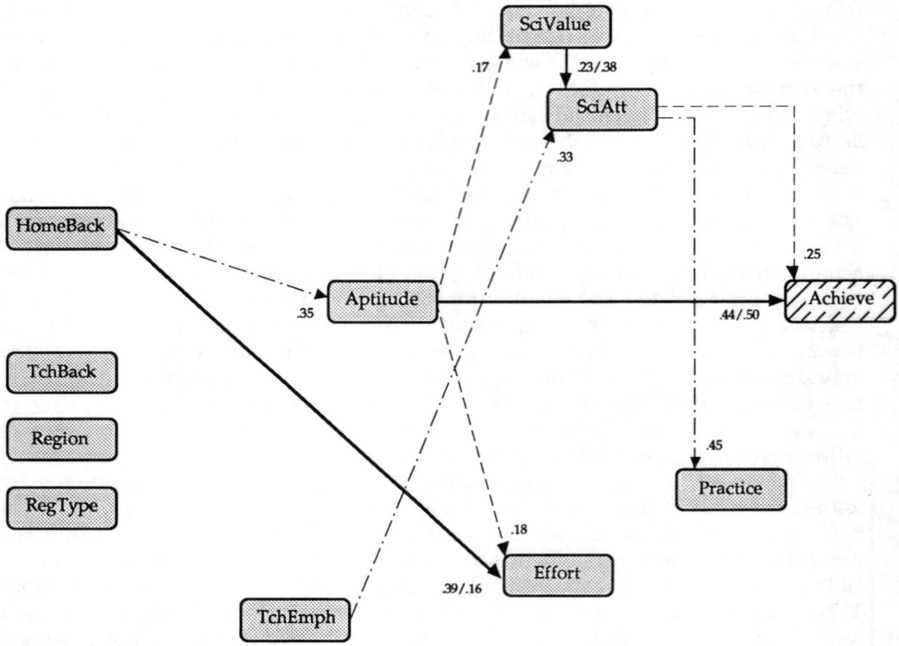

Notes:
for further details see legend for Figure 8.3

While HomeBack predicted Aptitude at the between-school level only (γ^*=.35), the general ability of the girl and the school she was attending turned out to be nearly equally strong key factors in influencing achievement in science (γ^*=.44/γ^*=.50). Except for SciAtt, which positively affected science achievement at the student-level (γ^*=.25) no other student- or school-level effect was identified as significant in the single-sex girls schools using HLM analyses.

It can be seen that neither TchBack, Region, nor RegType had any effect whatsoever. Nevertheless, girls schools differed greatly in terms of the amount of homework given which was largely dependent of the socio-economic climate (γ^*=.39). At the student level, Effort was positively influenced by both HomeBack (γ^*=.16) and Aptitude (γ^*=.18). These findings indicated that higher status schools as well as girls from higher status homes often did more homework but this did not increase to the achievement level of girls in science when other factors were taken into account.

TABLE 8.6 *Selected statistics from HLM-2 runs for a standardized two-level path model for single-sex girls schools in England, 14-year-old level, 1983/84; N=13*

Fixed effects					Random effects		
school level construct	**student level construct**	**γ***	**standard error**	**p-value**	**slopes of...**	**estimated parameter variance**	**p-value**
Outcome → Achieve							
Aptitude		.44	.04	.000			
	Aptitude	.50	.06	.000	Aptitude	.01	.176
	SciAtt	.25	.07	.003	SciAtt	.02	.217
Outcome → Practice							
SciAtt		.45	.18	.025			
Outcome → SciAtt							
TchEmph		.33	.11	.010			
SciValue		.23	.09	.022			
	SciValue	.38	.06	.000	SciValue	.00	.380
Outcome → Effort							
HomeBack		.39	.06	.000			
	HomeBack	.16	.07	.045	HomeBack	.01	<.500
	Aptitude	.18	.07	.028	Aptitude	.01	.231
Outcome → SciValue							
	Aptitude	.17	.07	.035	Aptitude	.01	.355
Outcome → Aptitude							
HomeBack		.35	.06	.000			

Notes:

Region (boxed): school-level (between-unit) construct

Gender (shaded): student-level (within-unit) construct

As could be expected values and attitudes towards science operated differently in girls schools than in boys schools. The most astonishing finding in this respect was that TchEmph showed a very marked positive effect on SciAtt at the between-school level (γ*=.33). Thus, groups in which the teacher put more emphasis on scientific investigations apparently had more favourable attitudes towards science which massively contributed to a higher amount of experimental work in science (γ*=.45). Further to this, it was found that the girls' achievement in science profited a great deal if the girls held more positive attitudes towards science (γ*=.25). However, science-related attitudes were positively influenced by

SciValue (γ^*=.23) and indirectly also by Aptitude (Aptitude→SciValue γ^*=.17). This lent support to the hypothesis that science-related attitudes were operating differently at different levels of measurement.

In all, the path model for girls schools in England showed only a very small number of paths which were found to be effective at the student and school levels. This could be interpreted that science instruction at the 14-year-old level in single-sex girls school did not vary much from school to school and certainly much less so, than in coeducational schools. However, there were major discrepancies between girls schools in the emphasis put on the teaching of scientific investigation and, subsequently, in science-related attitudes and practical and experimental work being done.

The path model also highlighted that similar to single-sex boys schools the socio-economic status of the girls' homes and, hence, of the girls schools they were attending influenced science achievement substantially but it was operating indirectly through aptitude.

→*Conclusions* The three path models presented for England's coeducational schools (Figure 8.6), single-sex boys schools (Figure 8.7), and single-sex girls schools (Figure 8.8) identified considerable differences but, at the same time, showed some similar patterns.

As in the case of Australia the greatest complexity in the model for science achievement was found in coeducational schools and students enrolled in such schools. The network of effects detected within the path model for coeducational schools varied greatly both between schools and also between students. While a certain amount of homogeneity was reported for the single-sex schools in England, it seemed that success in science was almost entirely a function of the socio-economic status of the students and schools. Since gender differences, of course, could not be examined for single-sex schools, it became obvious that, in particular with respect to science-related values and attitudes, the achievement of boys and girls in science was affected differently. This had special significance in the case of single-sex girls schools where science achievement was a function of status (HomeBack, Aptitude) and interest in science (SciAtt) only.

While the amount of time spent on homework and experimental work made no difference for the success in science in single-sex schools it had a limited but significant effect on achievement for students attending coeducational schools.

As with Australia, regional and school-type differences were also detected.

In general, the HLM analyses conducted for Australia and England enriched the information gathered so far by PLS. Quite clearly, it could be seen how complex the mechanisms were which, finally, led to higher or lower achievement in science.

Though separate HLM models were tested and estimated for coeducational and single-sex schools marked differences between the genders in science education and science achievement could also be identified.

8.4 Coeducational versus Single-Sex Schooling

The results from the separate path models discussed in the previous sections for coeducational, single-sex boys, and single-sex girls schools in Australia and England illustrated major differences as well as some similarities of the factors influencing science achievement. Though some of these factors appeared to operating in Australia and England it would be too premature to make inferences for coeducational and single-sex schooling in general. Above all, it should not be neglected that between the first and second data collection, 1970/71 and 1983/84, more and more single-sex schools were abandoned or converted into mixed schools a general trend observed in many countries world-wide (Keeves and Kotte, 1991, Table 6.2). Historically similar educational policies left Australia and England to remain the only two countries in the sample to offer the possibility of single-sex schooling at the lower secondary school level which was taken by a sizeable number of students. As was described, the proportions of single-sex boys and girls schools were about equal in both countries.

However, the types of factors believed to affect science achievement in coeducational schools were different in each of the two countries and also different again in boys and girls schools. Nevertheless, some identical observations made in Australia and England are worth mentioning.

→*Coeducational Schools* Status, gender, aptitude, school-type, and regional differences accounted for much of the variation at both the school and the student level. HomeBack and Aptitude turned out to be the factors which had a strong effect, both directly and indirectly, on science achievement. While science-related values and attitudes had only a minor impact at the between-student level, home and class efforts helped to increase the achievement both for the individual student as well as for the school. A few interaction effects, previously undetected, could be isolated but their overall contribution to explaining variance on the outcome variable was minor. Nevertheless, the interaction effects did provide evidence of the processes operating in schools to influence science achievement.

→*Boys Schools* In Australia and England the student's aptitude was the best predictor of science achievement in single-sex boys schools both at the student and at the school level. While the status of the home positively and directly influenced achievement, in England no such effect was found for Australia. This and other findings led to the conclusion that single-sex boys schools tended to be more heterogeneous in England than in Australia. This could also be illustrated by the very small numbers of paths found to be effective within the path model for single-sex schools in Australia. In both countries teaching-related effects were reported to be non-significant in influencing science achievement.

→*Girls Schools* In both countries science achievement turned out to be solely a function of Aptitude, which operated at both levels of measurement, and SciAtt, which was effective at the student level only. Both constructs had a marked positive effect on achievement. Thus, girls, who were generally more able and who held more favourable attitudes towards science showed higher performance. Though no 'masculine dominance' believed to hamper the achievement in science of girls was present in science instruction in girls schools (Kelly, 1978; Keeves, 1988; Duncan, 1989), still considerable variation existed within schools and between schools with respect to Aptitude, SciValue, Effort, SciAtt, and Practice. Despite these differences, with the exception of SciAtt no effect was found to be influential on science achievement, either directly or indirectly. This held true at both the student and the school level.

→*Coeducational versus Single-Sex Schools* The hierarchical, two-level analyses undertaken demonstrated remarkable differences within the inner models for coeducational and single-sex schools. It became evident that in coeducational schools a multitude of significant effects determined the achievement in science. Gender itself was an important aspect for success in science, but gender differences were also influential in a network which operated both at the between-student and at the between-school level. However, it might seem to be justified to conclude that the within-school effects had a stronger influence on the achievement in science than school-level effects. In contrast, the network of factors affecting science achievement in single-sex schools tended to be much more limited with status, aptitude, and science-related values and attitudes being the significant factors. In single-sex schools, the effects of TchBack, TchEmph, Effort, or Practice could be ruled out. However, due to the very small sample sizes of single-sex schools the findings should be treated with caution.

Kelly, in her research on the effects of coeducational versus single-sex education using data from the FISS study, summarized that "the proposition that girls achieve better in science in girls-only schools than in coeducational schools must be rejected" (1978, p. 70). Though her level of argumentation differed from the analytical approach pursued in this report, the results presented here were in support of her findings. While the path models for single-sex schools identified basically only two key factors which influenced science achievement, namely the aptitude and the attitude towards science, this limited the possibilities of girls enrolled in single-sex schools to compensate a possible 'inability', low interest in science, or perception of science as difficult by spending more time on homework or doing more experiments, strategies which were found to be effective in coeducational schools. It must also be recognized that aptitude was, in general, strongly dependent of the socio-economic environment.

However, the detailed analyses of coeducational, single-sex boys, and single-sex girls schools have cast more light on the complexity of the problem. They show, on the one hand, how to avoid the creation and manifestation of gender differences, and how, on the other hand, a maximization of equitable opportunities could be achieved in terms of science instruction.

In this study, the analysis of gender-typical forms of schooling was limited to schools in Australia and England. And although the trend favouring coeducational schooling has probably continued since the time of the data collection in 1983/84, recent observations of 'male dominance' in the classrooms and school laboratories have received even more prominence due to the introduction of computers and the instructional problems related to it.

8.5 Summary and Conclusions

When for the first time international comparative results investigating the factors which influenced science achievement were compiled by Comber and Keeves, the authors in a somewhat mindful foresight realized the complexity of the analysis of achievement differences between the sexes:

> "Only by regression analysis can this complex pattern of relationships be unravelled. Unfortunately, however, in the regression analyses reported..., it is not possible to examine fully the contribution of single-sex and coeducational school types, because this distinction has been incorporated into a more complex type of school variable based on the way in which the sample was stratified. The possible relationships indicated above must await more detailed analyses in the future." (1973, p. 153)

One form of complex multivariate regression techniques, partial least squares, was used for between-student and between-school analyses to bring some more clarity to the problem of gender differences in science achievement. It was seen that the aggregation of student data in conjunction with a set of newly introduced latent variables (TchBack, TchGend, SchFacil, and TchEmph) succeeded in boosting the amount of explained variance in all 10 countries. It was evident that in most of the countries a network of factors became effective to increase or decrease the achievement of a school in science. It was, however, impossible to partial out adequately purely student-related effects from effects which were caused by school-level variables.

To analyze simultaneously student, teacher, and school effects of the data available a multivariate, multi-level statistical approach such as HLM was required. It was decided that it would be advantageous to employ the HLM-2 program to develop and test separate path models for coeducational, single-sex boys, and single-sex girls schools for two selected countries, Australia and England. The aim was to gain more insight into the question of different forms of gender-typed schooling. This was considered conceptually more rewarding than to arrive at a more precise estimate for the bias expected to occur when aggregating student-level data to the school level. Besides, aspects of the feasibility had also to be taken into consideration. Furthermore, it was envisaged that a more selective use of HLM would allow some inferences about the possible amount of aggregation bias which may have led to an over-interpretation of the results from the between-school analyses described in Chapter 7.

The HLM analyses presented in this chapter did show the existence of aggregation bias. However, as could be shown the major findings concerning the path models for Australia and England in previous chapters were, in general, not called into question. The results obtained by means of HLM simply allowed a more detailed examination of the relationships of the processes inherent in the path models for the different gender-types of schooling in the two countries. As such significant differences were identified both between coeducational and single-sex schools as well as between single-sex boys and single-sex girls schools. But, as was pointed out earlier, the findings did not allow a ranking of the various gender-typical forms of schooling in terms of effective science teaching.

Moreover, the results presented for Australia and England should have made obvious that gender differences, though they did exist, were related to, in many instances, alterable variables. Further carefully conducted multivariate analyses would be desirable to determine how best to identify and carry out the necessary changes. This chapter illustrated how complex the network of school-type effects could be, especially in the setting of different gender-typed forms of schooling.

The next and final chapter summarizes the major results of this study and relates them to the hypotheses put forward in Chapter 2.

9. Summary and Conclusions

After an introduction to the issue of gender differences in science education and science achievement (Chapter 1), the current state of research was described from which the general and specific research questions were derived and postulated (Chapter 2). Following a description of the outline of the IEA FISS and SISS studies conducted in 1970/71 and 1983/84 respectively (Chapter 3), which served as a data base for this dissertation, the methodological difficulties and statistical properties of the two relevant statistical techniques used in the analysis of cross-national data from 10 countries (Australia, England, Finland, Hungary, Italy, Japan, the Netherlands, Sweden Thailand, and the United States), PLS and HLM, have been discussed in greater detail (Chapter 4).

It was conceived advantageous to break down the results of the data analyses into four separate chapters. At first, an account of gender differences found in the achievement in science and the science subjects, biology, chemistry, and physics was given for the different levels of schooling across the sample of 10 countries (Chapter 5).

In the next chapter (Chapter 6), an analysis of gender differences in science-related attitudes and values and their relationship to achievement in science was presented before other student-level effects were examined. Based on the results from the path models of each country using data from SISS, an average path model incorporating student variables with science achievement as outcome measure was developed highlighting common gender effects on science achievement across the set of 10 countries. In addition, the influence of student factors on future enrolment of students in science courses at the upper secondary school level was investigated as previous research had indicated important relationships between science achievement and participation in science.

Carrying the student level analyses further, the simultaneous effects of student, teacher, and school variables on the achievement in science were examined using aggregated data in between-school PLS analyses for each country (Chapter 7). Likewise, these factors were inspected for path models with future participation as the outcome measure. Resulting from these between-school analyses brief summary statements were given for each country specifying major recommendations helping to reduce possible gender differences and to increase achievement in science at the school level.

Completing the analyses of gender differences in science achievement, the issue of gender-typed single-sex schooling was examined in Australia and England employing a two-level HLM approach separately for single-sex boys schools, single-sex girls schools, and coeducational schools (Chapter 8). These analyses allowed, even if restricted to two countries, a parallel comparison of student, teacher, and school factors within one path model for every gender-typed form of schooling without running the risk of obtaining aggregation or disaggregation bias.

At this stage, it is considered appropriate to summarize the most important findings from each of the analyses chapters (Chapters 5 through 8) of the dissertation concerning the general research questions posed in Section 2.2.1. In addition, the major outcomes of this thesis will be evaluated against results from other research.

This chapter is closed with a critical review of the shortcomings of this study before some thought is spent on open questions about gender differences in science achievement and about the important issues future research in this field may have to deal with.

9.1 Answers to the General Research Questions

In planning the type of analyses to be performed for this thesis, eight general research questions considered important to be addressed were developed based on the current state of research. These general research questions have been presented in Section 2.2.1. The more specific research questions put up in Section 2.2.2 have already been dealt with in the relevant chapters and need not be picked up again in this summary. However, it is of importance to give a summary statement to each of the eight general research questions which could help as a basis for discussion and future research necessary when faced with the issue of gender differences in science achievement in school.

Following the numbering of the research questions set out in Section 2.2.1, each major research question is presented below with a brief summary statement that comprises the results discussed in the previous chapters. The relevant chapter(s) in which more detailed information can be obtained is given in brackets (e.g. ➡Chapter 5).

Research Question 1
What differences, if any, are found between boys and girls at the 10-year-old level, 14-year-old level, and terminal secondary school level in terms of achievement in science (Biology, Chemistry, Physics) in the 10 countries? (➡Chapter 5)

Gender differences were identified in all 10 countries at all three levels of schooling for all subjects. While gender differences in biology achievement were small or non-existent when compared with chemistry or physics, it was in physics that these differences were most pronounced in almost all 10 countries. Though biology, in comparison to the other science subjects, was perceived as more attractive by girls in most countries, gender differences could still be identified.

Gender differences were apparent, even if to a small degree, already at the 10-year-old level. It was found that gender differences seemed to 'grow' during formal schooling in the countries in all science subjects with achievement differences between boys and girls showing the greatest imbalance at the terminal secondary school level. Thus, while science achievement varies considerably between the boys and girls, in nearly all instances female students are clearly disadvantaged.

Research Question 2
What changes occurred in the differences between boys and girls at the three grade levels tested in terms of achievement in science between the two data collections in 1970/71 and 1983/84 in the 10 countries? (➡Chapter 5)

Noticeable changes had occurred between the two data collections in 1970/71 and 1983/84 across the 10 countries. As a general trend, gender differences in science achievement were found to have receded during the 14-year period, but the extent to which this happened varied considerably between the countries and was also very different between the three science subjects.

It should be acknowledged, though, that despite the finding that gender differences in science achievement were, on average, less in 1983/84 than in 1970/71 this 'improvement' only illustrates a tendency. Still, at the 14-year-old level, even in 1983/84 gender differences were quite pronounced in the Netherlands and the United States while they were smallest in Hungary.

While gender differences in science achievement decreased between the two data collections in all other countries, girls were even more disadvantaged at the 14-year-old level in England and Thailand in 1983/84. This reverse trend continued at the terminal secondary level only for Thailand.

Research Question 3
What differences, if any, are found in the attitudes towards science of boys and girls in the 10 countries? (➡Chapter 6)

Consistent and sizeable gender differences were detected across all six attitude scales examined. There was evidence that, in general, no major differences were found between the countries. This suggests a rather congruent or parallel development and manifestation of attitudes and values towards science across the countries. However, it became necessary to distinguish between attitudes and values towards science when investigating gender differences.

It can also be concluded that gender differences in values (career interest in science, beneficial aspects of science) as distinct from attitudes (interest in science, ease of learning science) were less susceptible to teaching- and school-related factors. Consequently, it can be argued that the student's primary socialization would account for most of the gender differences exhibited with respect to values towards science.

Except for the scale that measured students' attitude towards school and school learning, it was the boys who voiced a stronger acceptance of science. Boys were more interested in science, found learning science-related tasks easier, showed a more pronounced interest in starting a career in science, and rated science to be more beneficial than their female classmates.

It would therefore appear justified to summarize that, on average, boys had a stronger affinity and interest towards science.

Research Question 4
What factors contribute to gender differences in science achievement in the 10 countries and how do they operate? (➡Chapters 6 and 7)

A general path model with science achievement as the outcome measure was developed for the between-student analyses and tested for each of the 10 countries. Despite country-specific differences in the direct and total effects of the latent variables in the model, it was possible to make some important generalizations across the countries.

Though the socio-economic status and the students' general aptitude turned out to be the strongest overall predictors of science achievement at the 14-year-old level, gender had a remarkably intense effect on science achievement throughout all countries ($\bar{t}$=-.18). As the average direct effect of gender was only minimally lower ($\bar{p}$=-.16), this allowed the conclusion that, even when considering the effects of other latent variables in the model, gender differences favouring the boys were still substantial. Obviously, gender was strongly associated with science achievement.

The student-level path model employed was found appropriate in most countries. However, the comparatively low amount of variance accounted for by this path model in Finland, Japan, Sweden, and Thailand suggested to take into consideration further teacher and school variables to help explaining gender differences in science achievement more concisely.

For the understanding of the magnitude of school-level effects it would be desirable to relate the observed between-school variance to the between-student overall variance in each country. Analyses of the SISS data showed large differences between the intra-class correlations coefficients (roh) for the countries at the 14-year-old level which varied between roh=.04 for Japan and roh=.50 for the Netherlands (see Postlethwaite and Wiley, 1991, Table 3.3).

The examination of the between-school path models showed that the socio-economic status of the school and general ability level of the school group tested were most influential on science achievement. In addition, several other factors were important in each of the countries. However, these effects varied quite remarkably between the 10 countries. It seems safe to conclude, though, that science-related values and attitudes as well as the school- and homework-related efforts also accounted for much of the variance of the outcome measure. Though not applicable in each of the 10 countries, the notion that more homework would improve the achievement in science, would seem true.

In contrast, teacher and school factors (e.g., teaching experience and qualification, sex of teacher, availability of science laboratories) were of little effect on science achievement except, perhaps, in Thailand. The special situation in Thailand, the only developing country in the sample, would seem to suggest that school- and teacher-related variables became less influential the higher the general educational standard (or economic status) of a country is. Once certain standards were achieved regarding school science equipment and science teacher preservice and inservice training, other factors, some of them not yet identified, stimulated or inhibited students' achievement in science at the lower secondary school level.

The between-school analyses presented some evidence that what was taught in the classroom or laboratory (science curriculum) and how this was done (teaching strategies) had considerable influence, positively or negatively, on science achievement as well as on gender differences in science achievement.

It was, however, the socio-economic status which accounted for most of the differences in science achievement at the school level. In comparison with the other countries this became most evident in Australia, England, the Netherlands, and the United States.

On average, the between-school path models were able to explain 55 percent of the total variance. The results demonstrated that, additional to the gender differences identified at the between-student level, emphasis has to be put also on teacher- and school-related variables in the attempt to understand the complex network of effects promoting gender differences in science achievement at the 14-year-old level. However, it also became clear that a sizeable proportion of the variance explained was the result of aggregation bias.

It should not be overlooked that further to the variables incorporated and tested in the path models for the student and school level, other factors for which no or incomplete data were available across the 10 countries might also play an important role in fostering gender differences in science achievement. In the cases of the Netherlands and the United States science achievement was believed to be determined, to a large degree, by the various school types or science programs offered. In other countries differences between rural and urban schools as well as the ethnicity of the school group appeared to gain significant influence on the gender differences in achievement.

However, it should be reiterated that the complex network of factors contributing, positively or negatively, to gender differences for 14-year-olds at the school level were very much country-specific.

Research Question 5
What factors contribute to gender differences in participation rates in science in the 10 countries and how do they operate? (➡Chapters 6 and 7)

The research question was analyzed by asking 14-year-old students about their interest to enrol in future science courses at the terminal secondary school level. Similarly to the investigation of factors influencing gender differences in science achievement, this question has been dealt with at both the student level and the school level as it became obvious that teacher and school factors cannot be cast aside when identifying the major factors affecting participation of students in future science courses.

The decision to enrol in science courses in the upper secondary grades was very much correlated with the students' general educational expectations and attitudes. Future enrolment depended only indirectly on key factors such as socio-economic status, general aptitude, or gender. It appeared to be a function of values, attitudes, and aspirations.

There was strong evidence that educational expectations of students, and not their occupational expectations, were conditional for future enrolment in science courses. In the 10 countries examined, students from higher status homes commonly expressed higher educational expectations than students with a less favourable home background. Under certain circumstances, though, science-related attitudes and values apparently countered the influence of the home background on educational expectations.

The interpretation of the country-specific path models led to the conclusion that student attitudes were apparently both independent of the general student background as well as interdependently connected with interest to participate in future science courses. However, the balance between independence and interdependence varied considerably from country to country.

As participation in future science courses was not linked to occupational expectations, this suggested that it was probably premature to ask 14-year-olds about their future careers. It is understandable that students of that age may not be able to develop concise ideas about their future careers.

Another point of interest in the analyses was the relationship between science achievement and future participation in science courses. It emerged that the close positive link between achievement in science and the students' interest to enrol in upper secondary science courses could not be verified in most countries. It remained unclear why science achievement had a very pronounced negative effect on future enrolment in Hungary, the country in which students scored top marks in the science tests.

Small gender differences in the participation in science courses at the upper secondary school level were found in six of the 10 countries. Though male-dominated classes were more likely to enrol in future science courses, gender-typing of the class was only marginally influential.

Overall, the most dominant factors in promoting future participation in science courses at the between-school level for the 14-year-olds were the socio-economic status of the class, the science-related values, and the educational expectations. However, much ambiguity and country-specific differences made it difficult to trace any further similarities between the countries.

Research Question 6
How can the participation rate of students in science courses be increased?
(➡Chapters 6 and 7)

From the analyses conducted it would seem advisable to make students aware of all options and advantages that exist when proceeding to further study science in courses given at the upper secondary school level. Raising students' educational expectation appears to make future enrolment in science courses more likely.

Though students' home background is widely regarded as a non-malleable variable it would still make sense to try also for a greater awareness among parents which educational and occupational opportunities there are for their children. Parents should understand that students who continue studying science at the terminal secondary school level increase their occupational opportunities.

Further to this, it would certainly be helpful if students' attitudes and values towards science could be positively stimulated. How this could be done, however, is a question which is far beyond the scope of this study and cannot be discussed here.

Examining the path models of 10 countries, gender showed only a minor effect on participation in science at the school level. Here, in six out of 10 countries students in male-dominated classes or single-sex boy schools were more likely to express an interest in further studying science. This finding would seem to advocate educational measures being taken to avoid the existence of gender-typed classrooms wherever possible. Having classes with equal proportions of boys and girls at the lower secondary school level should be seen as a precondition to avert male dominance in science courses at the terminal secondary school level found in many countries.

It is, of course, acknowledged that major country-specific differences existed which suggests the undertaking of additional investigations in order to arrive at more specific recommendations of how to increase the participation of students in science courses in each of the countries.

Research Question 7
Is coeducational schooling more effective than single-sex schooling for the achievement level of boys and girls in science? (➡Chapter 8)

Using two-level, hierarchical linear modelling techniques it was possible to identify remarkable differences in the network of factors operating on science achievement in coeducational and single-sex schools of Australia and England. These were the only two countries which still had a sizeable proportion of single-sex schools in operation in 1983/84.

In coeducational schools a multitude of significant effects determined the achievement in science. Male students achieved better than females in science, but gender differences were also operating indirectly in a complex network both at the between-student and at the between-school level. There was evidence that the within-school effects had a stronger influence on the achievement in science than school-level effects in coeducational schools. Status, gender, aptitude, school-type, and regional differences accounted for much of the variation at both the student and the school level. Science-related values and attitudes played only a minor role influencing achievement in science at the between-student level, homework and classwork efforts helped to increase the achievement both for the individual student as well as for the school group in coeducational schools in Australia and England.

In contrast, the network of factors affecting science achievement in single-sex schools was much simpler with status, aptitude, and attitudes being the only significant effects. In single-sex schools, teacher and teaching characteristics as well as homework, classwork, and experimental work did not influence science achievement significantly. In both countries the general aptitude, which operated at both levels of measurement, and science-related attitudes, which were effective at the student level only, had a marked direct positive effect on science achievement in single-sex schools. Thus, those boys and girls enrolled in single-sex schools, who were generally more able and held more favourable attitudes towards science, achieved better.

Only two major factors which influenced science achievement could be identified in single-sex schools: the aptitude level and the attitudes towards science. This somewhat limited the chances of girls enrolled in single-sex schools to compensate a possible 'inability' or low interest in science by spending more time on homework or by doing more experiments, strategies which were found to be effective to increase the level of achievement in science in coeducational schools. It should also not be overlooked that aptitude was, in general, strongly dependent on the socio-economic environment.

It became apparent that single-sex schools in Australia and England were highly status-oriented: students coming from a more favourable home background were more likely to achieve better. Unfortunately, the path models for single-sex schools indicated that apart from status and aptitude the effects of malleable variables to reduce achievement differences in science were marginal. This finding was in strong contrast to the situation in coeducational schools where several malleable variables, such as homework, classwork, and to a minor degree, attitudes towards science, could be identified.

Though the analyses shed more light on the network of factors affecting science achievement at coeducational and single-sex schools, a statement on which form of gender-typed schooling proved more effective would be premature and unjustified. 'Effectiveness' cannot be judged on grounds of a limited number of abstracted latent variables. But for students to compensate a lower ability or disinterest in science certainly more possibilities were at hand in coeducational schools.

Research Question 8

What measures can be recommended to help balance gender differences in science achievement? (➡Chapter 7)

It proved difficult to arrive at general recommendations which could be applied to all 10 countries. This was the reason why country-specific recommendations to improve science achievement at large and to reduce gender differences in science achievement in particular have already been given in Chapter 7. These need not be repeated here.

The investigation of the various path models demonstrated very clearly that the socio-economic status (of the school, class and student), as in most other studies, was **the** key factor influencing science achievement throughout all countries. Teacher and teaching characteristics as well as the effort undertaken inside and outside the classroom were also of considerable importance in influencing science achievement for boys and girls in many countries. School organization, school status and school setting played a relatively important role in fostering gender differences in only a few countries. The effects of science-related attitudes and values were more pronounced at the student than at the school level.

Given these findings, it would seem appropriate to suggest that of all malleable variables found to have an effect on the achievement level in science, homework- and classroom-related efforts still make the difference: increasing the amount of homework to be done would be advantageous to achieve better in science both for the student and for the school.

However, as it turned out that in many countries general ability was another main predictor of success in science, steps have to be taken to prevent achievement differences in science as well as gender differences from being established at an early stage within the process of formal schooling. As was shown, gender differences in science achievement apparently 'increase' during schooling, meaning that measures need to be taken as early as possible, probably already in primary education, which aim at the suppression of factors known to lead to the establishment of gender differences in science achievement. As results indicated, it would be desirable to implement strategies in the curriculum as well as in the pre- and inservice training which would help to reduce gender differences. The scope of this dissertation has not resulted in more specific propositions. But further research on these types of strategies is certainly needed.

Although the results presented suggested that coeducational schooling does not guarantee the elimination of gender differences in science achievement at the student or school level, it is not justified to conclude that single-sex schooling would be the organizational alternative. No evidence was obtained that segregating boys and girls during science instruction would remove gender differences in science achievement.

Educational planners, however, should note that it would seem legitimate to recommend that students, parents, and teachers be made more aware that although boys and girls apparently held gender-typical attitudes and values towards science, there was no evidence for the assumption that boys and girls could not achieve equally well in science. Carrying this statement further, it would be unwise to apply different educational measures separately to boys and girls in school hoping to reduce gender differences. Even if boys were found to hold more positive attitudes towards science, this does not necessarily mean that it would be wise to develop female-specific strategies aimed at stimulating the girls' attitudes towards science or teaching girls separately from boys.

The results presented in this thesis indicated clearly that gender effects on science achievement and participation in future science courses were different between the countries examined. The school level path models were too heterogeneous in each of the countries to allow for generalizations or recommendations applicable to other countries not examined here. However, even if the network of factors found to influence gender differences in science achievement, the general achievement level of students and schools as well as the participation in future science courses showed a great deal of variation between the countries, the identification of influential factors can surely be of help in the analysis of gender differences in science achievement that other countries wish to undertake.

9.2 Evaluation of Major Findings

As could be noted from the answers to the general research questions given in the previous section, the comparison of data gathered during the data collections of the IEA FISS and SISS studies in 1970/71 and 1983/84 from the 10 countries offered several new and interesting findings. In addition, some of the results presented in earlier chapters also lent support to research results from related studies.

In this thesis, multivariate analyses were carried out for the first time, using PLS and HLM techniques, to study a network of effects determining gender differences in science achievement at the 14-year-old level on the basis of data from 10 countries. For the understanding of these factors, which were found to operate at both the student and the school level, it was imperative to take into consideration also gender differences in science achievement at the 10-year-old and terminal secondary school level.

It became apparent that gender differences in science achievement could be identified in all countries. However, gender differences varied between...

- ☞ the levels of analyses (student or school level);
- ☞ the three grade-levels tested;
- ☞ the three science subjects biology, chemistry, and physics;
- ☞ the different types of school; and
- ☞ the 10 countries examined.

The finding that gender differences in science achievement were, in general, always in favour of boys or male-dominated classes underlined similar results from the FISS study made by Comber and Keeves (1973).

Another important aspect pointed out by Comber and Keeves was that differences in science achievement between boys and girls increased during secondary education from the 10-year-old level to the terminal secondary school level. Results from this study verified that observation quite clearly.

Further to this, the analyses presented in this thesis lent support to the statement, also made by the Comber and Keeves, that gender differences in science achievement were larger in physics than in chemistry, and lowest in biology. However, it became apparent that numerous country-specific changes had occurred since the 1970/71 data collection on which the results of FISS were based which indicated that, on the whole, differences in science achievement between boys and girls decreased until 1983/84. This decrease was found to have occurred for all three science subjects.

From the analyses of the student and school level path models it was possible to see the importance of science-related attitudes and values in their effect on science achievement in several of the countries investigated. The findings suggest that gender-typed attitudes and values were, indeed, conditional for gender differences observed in science achievement at the three age groups tested. Here, similar findings from the analysis of the educational system in Botswana made by Duncan (1989) were found to hold true also in several developed countries such as Finland, the Netherlands, Sweden, or the United States.

Despite the fact that in most countries several latent variables were identified to have a significant influence on the achievement level of students and schools in science, it was the effects of the home background, the aptitude level, and gender - in exactly this sequence - which proved to be the strongest predictors of science achievement at the 14-year-old level. Though it was nothing new that status and aptitude were important for the students' performance level in science (see Comber and Keeves, 1973; Kelly, 1978; Humrich, 1988 [1]; Keeves, 1991; Postlethwaite and Wiley, 1991) it was highly interesting to note the relative effect gender had, given all other latent variables in the path models held constant. It would, however, be ignorant to conclude that boys or male-dominated classes achieved better because of their sex. Expressed differently, boys probably did **not** achieve better in science **because** of their sex but because boys underwent different enculturation, socialization, and elementary education from the girls (Maccoby and Jacklin; 1974; Kelly, 1978; Marini and Greenberger, 1978; Keeves, 1991; Keeves and Kotte, 1991). It was these underlying latent variables not represented in the path models which made gender an important factor in promoting gender differences in science achievement.

Although the results showed large variations in the network of factors influencing science achievement and, thus, promoting gender differences in science achievement between the 10 countries, it became apparent how important the issue of gender differences in science still was in 1983/84.

Therefore, particular emphasis was put on the analyses of gender-typed schooling in the cases of Australia and England, the only countries with a sizeable proportion of sex-segregated schools still in operation in 1983/84. Separate multivariate, two-level analyses, using hierarchical linear modelling techniques showed major differences among the factors determining science achievement in coeducational and single-sex schools. However, generalizations applicable to other countries would require further research.

9.3 Limitations and Shortcomings

The analyses presented in this thesis were based on data from the FISS and SISS study conducted under the auspices of IEA. As pointed out in Chapter 3, the interpretability of the results should be evaluated against several statistical criteria specified in the design and conceptualization of the two major science studies. Among others, these were restrictions based on the sampling (weighting of data, sampling error, sampling bias, design effect), the selection of variables and scales to be included for multivariate purposes, and problems related to the levels of analyses (aggregation and disaggregation bias). These problems have already been discussed in the relevant chapters and need not be repeated here.

However, when making statements on gender differences in science achievement of a more general kind, it must be acknowledged that only a limited number of manifest variables were included in tests and questionnaires during the data collection of the FISS study in 1970/71 and the SISS study in 1983/84. Finally, only an even more reduced number of student, teacher, and school variables could be used for examination of the student and school level path models. Even when discarding the possibility of measurement errors, it is evident that other variables, not yet considered, would have certainly contributed to a higher amount of variance explained by the different path models. The path analyses suggested that for several countries it would have been desirable to have added variables such as those measuring ethnicity, regionality, type of school, science program, instructional method employed for science teaching, or students' motivation to name just a few important ones.

Though the major findings of this study are, in principle, only applicable to the set of 10 countries examined, it appears likely that similar observations would have been made in other countries. Although several aspects underlined the usefulness and validity of cross-country comparisons, it should not be forgotten that gender differences in science achievement were basically very much country-specific.

Another key problem related to the interpretability of path models was the nature of 'causal gender effects'. Can sex of student, sex of teacher, and sex of school really have a causal effect on science achievement? Or are gender effects only associated with other variables in the model? Even if evidence of association has been found how is it to be interpreted? Evidence of association is certainly prerequisite if causal relations are to be inferred, but it is insufficient. When we find an association between attitudes towards science and achievement in science it is not automatically evident whether attitudes promote achievement or whether it is achievement that promotes attitudes.

At several occasion evidence of associations has been presented in this thesis and inferences have been made which seemed most likely. It is, however, recognized that "in the last resort the interpretation must depend upon memory, introspection, and testimony and these may differ from one interpreter to another. These are grounds for caution in interpretation. They are not grounds for refraining from the attempt to interpret." (Postlethwaite, 1967, p. 131)

As was pointed out beforehand, *causal gender effects* should be understood as effects of variables, not measured in the different path models, which are represented by the sex of the student, teacher, or school. These 'latent' variables brought into the path models examined could have derived from the different enculturation, socialization, or gender-specific experiences of boys and girls during pre-school or primary education. Keeping this in mind, it seems legitimate to ascribe causality to the gender effects detected.

9.4 Implications for Future Research

Despite several theoretical and practical limitations, it is hoped that new insights have been gained into the issue of gender differences in science achievement at the secondary school level. The cross-country comparison of descriptive statistics as well as the analyses of student and school level path models gave further clarity about the network of effects determining achievement in science.

It has already been discussed that it would have been desirable to employ additional variables such as the science program a student is enrolled or the type of instruction given to help understanding the gender differences observed in science achievement at the various levels of schooling.

Apart from this, further country-specific analyses need to be conducted in order to have access to a larger base of data to be used for cross-country comparisons.

Likewise, it should be acknowledged that the results presented in this thesis described the situation in 1983/84 and it is likely that numerous changes in terms of science education and science achievement have occurred in the 10 countries investigated.

Though more research is necessary to obtain a better understanding of the various networks of factors found to influence gender differences in science achievement at the secondary school level in different countries, it may be a sign of optimism to finalize this thesis with the conclusion that during the 14-year-period between 1970 and 1984 gender differences in science achievement did actually decrease in most countries. And that, indeed, is promising.

References

Adams, F. (ed.) (1985) Science & Computers in Primary Education. A Report of the Educational Workshop Held in Edinburgh, 3-6 September, 1984. Scottish Council for Research in Education, Edinburgh.

Adams, R. S. and Biddle, B. J. (1970) Realities of Teaching: Explorations with Video Tape. Holt, Rinehart, and Winston, New York.

Aitkin, M., and Longford, N. (1986) Statistical modelling issues in school effectiveness studies. Journal of The Royal Statistical Society, Series A, Vol. 149 (Part 1), pp. 1-26.

Amthauer, R. (1970) Intelligenz Struktur Test I-S-T 70. Hogrefe, Göttingen, Germany.

Anderson, L. W., Ryan, D. W., and Shapiro, B. J. (eds.) (1989) The Classroom Environment Study. International Studies in Educational Achievement. Pergamon Press, Oxford.

Baker, S. W. (1980) Biological influences on human sex and gender. Journal of Women in Culture and Society, Vol. 6, pp. 80-96.

Bank, B. J. (1985) Student Sex and Classroom Behaviour. Husén, T. and Postlethwaite, T. N. (eds.) International Encyclopedia of Education, pp. 4878-4881. Pergamon Press, Oxford.

Báthory, Z. (1988) Hungary. Postlethwaite, T. N. (ed.) The Encyclopedia of Comparative Education and National Systems of Education, pp. 338-345. Pergamon Press, Oxford.

Báthory, Z. (1989) How Two Educational Systems Learned from Comparative Studies. The Hungarian Experience. Purves, A. C. (ed.) International Comparisons and Educational Reform, pp. 45-50. Association for Supervision and Curriculum Development, Alexandria, Virginia.

Beaton, A. E. (1969) Some Mathematical and Empirical Properties of Criterion Scaled Variables. Mayeske, G. W. et al. (eds.) A Study of Our Nation's Schools, Appendix I, pp. 339-343. United States Department of Health, Education, and Welfare, Washington, DC.

Becker, B. J. and Chang, L. (1986) Measurement of Science Achievement and Its Role in Gender Differences. Paper presented at the Annual Meeting of the American Educational Research Association, San Francisco, CA.

Bem, S. L. (1974) The measurement of physical androgyny. Journal of Consulting and Clinical Psychology, Vol. 42, No. 2, pp. 162-185.

Block, J. H. (1973) Conceptions of sex role: Some cross-cultural and longitudinal perspectives. American Psychologist, Vol. 28, pp. 512-526.

Bock, R. D. (ed.) (1989) Multilevel Analysis of Educational Data. Academic Press, San Diego, CA.

Bonora, D. (1974) Educational Aims and Curriculum in France: An IEA Survey. Comparative Education Review, Vol. 18, No. 2, pp. 217-227.

Booth, C. (1988) United Kingdom. Postlethwaite, T. N. (ed.) The Encyclopedia of Comparative Education and National Systems of Education, pp. 691-698. Pergamon Press, Oxford.

Bornstein, R. (1982) Sexism in education. Sadker, M. P. and Sadker, D. M. (eds.) Sex equity handbook for schools. Longman, New York.

Bradley, J. and Hutchings, D. (1973) Concepts of Science and Scientists as Factors Affecting Subject Choice in Secondary Schools. School Science Review, Vol. 55, No. 8.

Brehmer, I., Küllchen, H., and Sommer, L. (1989) Mädchen, Macht (und) Mathe. Geschlechtsspezifische Leistungskurswahl in der reformierten Oberstufe. Dokumente und Berichte 10 der Parlamentarischen Staatssekretärin für die Gleichstellung von Mann und Frau. Landesregierung Nordrhein-Westfalen, Düsseldorf.

Bridgham, R. (1972) Ease of Grading and Enrollments in Secondary School Science. Journal of Research and Science Teaching, Vol. 9, pp. 323-329.

Bryk, A. S. and Raudenbush, S. W. (1988) Methodology for cross-level organizational research. Research in Organizational Behavior. JAI Press, Greenwich, CT.

Bryk, A. S., Raudenbush, S. W., Seltzer, M., and Congdon, R. T. (1988) An introduction to HLM: Computer program and users' guide version 2.0. University of Chicago, Chicago, IL.

Bryk, A. S. and Raudenbush, S. W. (1989) Toward a more appropriate conceptualization of research on school effects: A three-level hierarchical linear model. Bock, D. (ed.) Multilevel analysis of educational data, pp. 159-204. Academic Press, New York.

Burstein, L. (1980) The Analysis of Multi-level Data in Educational Research and Evaluation. Review of Research in Education, No. 8, pp. 158-233.

Burstein, L., Miller, M. D., and Linn, R. L. (1981) The use of within-group slopes as indices of group outcomes. Report No. 171. Center for the Study of Evaluation, Graduate School of Education, Los Angeles.

Cahill, S. E. (1983) From Babies into Boys and Girls: The Acquisition of Gender Identity. Paper presented at the Annual Meeting of the American Sociological Association, Detroit, MI.

Chantavanich, S. and Fry, G. W. (1988) Thailand. Postlethwaite, T. N. (ed.) The Encyclopedia of Comparative Education and National Systems of Education, pp. 657-662. Pergamon Press, Oxford.

Cheung, K. C. (1988) Outcomes of schooling: Mathematics achievement and attitudes towards mathematics learning in Hong Kong. Educational Studies in Mathematics, Vol. 19 (Part 2), pp. 209-220.

Cheung, K. C. and Keeves, J. P (1990) The Problems with Current Analytical Procedures. Cheung, K. C. et al. The Analysis of Multivariate Data in Educational Research: Studies of Problems and Their Solutions. International Journal of Educational Research, Vol. 14, No. 3, pp. 233-244. Pergamon Press, Oxford.

Cheung, K. C. and Tsoi, S. C. (1990) Model Specification and Related Issues. Cheung, K. C. et al. The Analysis of Multivariate Data in Educational Research: Studies of Problems and Their Solutions. International Journal of Educational Research, Vol. 14, No. 3, pp. 245-256. Pergamon Press, Oxford.

Cheung, K. C., Keeves, J. P., Sellin, N., and Tsoi, S. C. (1990) The Analysis of Multivariate Data in Educational Research: Studies of Problems and Their Solutions. International Journal of Educational Research, Vol. 14, No. 3. Pergamon Press, Oxford.

Chipman, S. F. (1988) Word Problems: Where Test Bias Creeps In. Paper presented at the Annual Meeting of the American Educational Research Association, New Orleans, LA.

Choppin, B. H. (1974) The Introduction of New Science Curricula in England and Wales. Comparative Education Review, Vol. 18, pp. 196-208.

Cobb, P. (1987) Information-processing psychology and mathematics education - a constructivist perspective. Journal of Mathematical Behavior, Vol. 6, pp. 3-40.

Cohen, J. (1969) Statistical power analysis for the behavioural sciences. Academic Press, New York.

Collis, B. (1985) Sex-Related Differences in Attitudes toward Computers: Implications for Counselors. School Counselor, Vol. 33, No. 2, pp. 120-130.

Comber, L. C. and Keeves, J. P. (1973) Science Education in Nineteen Countries. International Studies in Evaluation I. Almqvist and Wiksell, Stockholm.

Cooksey, D. (1981) Social class and academic performance: a Cameroun case study. Comparative Education Review, Vol. 25, No. 3, pp. 403-418.

Cooley, W. W. and Lohnes, P. R. (1971) Multivariate Data Analysis. Wiley, New York.

Cronbach, L. J. and Furby, L. (1970) How should we measure 'change' - or should we? Psychological Bulletin, Vol. 74, No. 1, pp. 68-80.

Cronbach, L. J. (with assistance of Deken, J. E. and Webb, N.) (1976) Research on classrooms and schools: Formulation of questions, design and analysis. Occasional Paper of the Stanford Evaluation Consortium. Stanford University, Stanford Evaluation Consortium, Stanford, CA.

De Corte, W. and Brusselmans-Dehairs, C. (1985) Gender and mathematics: Some results from the Belgian (Flemish) SIMS participation. Laboratory for Applied Psychology, Seminar and Laboratory for Didactics, Ghent State University, Ghent, Belgium.

de Leeuw, A. and Raat, J. H. (1986) Interest of girls and new teaching material. Women Challenge Technology, Vol. 1, pp. 179-188. European Conference on Women, Natural Sciences and Technology, Elsinore, Denmark, 15-19 November.

de Leeuw, J. and Kreft, I. (1986) Random coefficients models for multi-level models. Journal of Educational Statistics, Vol. 11 (Part 1), pp. 57-85.

Degenhart, R. E. (ed.) (1990) Thirty Years of International Research. An annotated bibliography of IEA publications (1960-1990). IEA Headquarters, The Hague.
Dempster, A. P., Laird, N. M., and Rubin, D. B. (1977) Maximum Likelihood from Incomplete Data via the EM Algorithm. Journal of the Royal Statistical Society, Series B, Vol. 39, pp. 1-38.
Dickson, D. (1986) Britain's Royal Society condemns sex bias in mathematics teaching. Science, Vol. 233, pp. 618-619.
Digranes, J. L. (1985) A Look at High School Students and Computers: An Observational Study. Paper presented at the Rocky Mountain American Educational Research Association, Las Cruces, NM.
Duncan, W. (1989) Engendering School Learning. Science, Attitudes and Achievement Among Girls and Boys in Botswana. Studies in Comparative and International Education, No. 16. Institute of International Education, University of Stockholm, Stockholm.
Dunkin, M. J. (1985) Teacher Sex and Instruction. Husén, T. and Postlethwaite, T. N. (eds.) International Encyclopedia of Education, pp. 5032-5035. Pergamon Press, Oxford.
Dupont, B. (1981) Unequal Education. A Study of Sex Differences in Secondary School Curricula. UNESCO, Paris.
Eastman, S. T. and Krendl, K. A. (1984) Computers and Gender: Differential Effects of Electronic Search on Students' Achievement and Attitudes. Paper presented at the Annual Meeting of the Communication, Language and Gender Conference, Oxford, England.
Eccles, J. S. (1983) Expectancies, values and academic behaviors. Educational Researcher, Vol. 15, No. 6, pp. 15-19. Freeman, San Francisco, CA.
Eccles, J. S. (1984) Sex differences in mathematics participation. Spence, J. T. (ed.) Achievement and Achievement Motives. Psychological and Sociological Approaches. JAI Press, Greenwich, CT.
Eccles, J. S. (1986) Gender-roles and women's achievement. Steinkamp, M. W. and Maehr, M. L. (eds.) Women in Science.
Eckstein, M. A. (1988) United States. Postlethwaite, T. N. (ed.) The Encyclopedia of Comparative Education and National Systems of Education, pp. 698-705. Pergamon Press, Oxford.
Enders-Dragesser, U. and Fuchs, C. (eds.) (1990) Frauensache Schule. Fischer, Frankfurt.
Epler, D. M. (1982) Teachers' Perceptions of Predicted Career Success of Students Who Choose Nontraditional Occupational Careers. ERIC database, ref. ED226113.
Erickson, G. L. and Erickson, L. J. (1984) Females and Science Achievement: Evidence, Explanations, and Implications. Journal of Science Education, Vol. 68, No. 2, pp 63-89.
Eysenck, H. J. (1967) The Biological Basis of Personality. Thomas, Springfield.
Eysenck, H. J. (1970) The Structure of Human Personality. Methuen, London.
Eysenck, H. J. (1979) The Structure and Measurement of Intelligence. Springer, Berlin.
Fägerlind, I. (1988) Status Attainment Models. Keeves, J. P. (ed.) Educational Research, Methodology, and Measurement: An International Handbook, pp. 555-559. Pergamon Press, Oxford.
Fennema, E. and Sherman, J. (1977) Sex-Related Differences in Mathematics Achievement, Spatial Visualization and Affective Factors. American Educational Research Journal, Vol. 14, pp. 51-71.
Fennema, E. and Sherman, J. (1978) Sex-Related Differences in Mathematics Achievement and Related Factors: A Further Study. Journal for Research in Mathematics Education, Vol. 9, pp. 189-203.
Finn, J. D. and Kotlewski, L. D. (1975) Components of Achievement in Science. Department of Educational Psychology, University of New York, Buffalo, NY.
Finn, J. D., Dulberg, L., and Reis, J. (1979) Sex differences in educational attainment: A cross-national perspective. Harvard Educational Review, Vol. 49, No. 11, pp. 477-503.
Finn, J. D., Reis, J., and Dulberg, L. (1980) Sex Differences in Educational Attainment: The Process. Comparative Education Review, Vol. 24, No. 6, pp. 33-52.
Fleming, M. L. and Malone, M. R. (1983) The relationship of student characteristics and student performance in science as viewed by meta-analysis research. Journal of Research in Science Teaching, Vol. 20, pp. 481-495.

Foshay, A. W., Thorndike, R. L., Hotyat, F., Pidgeon, D. A., and Walker, D. A. (1962) Educational Achievement of Thirteen-Year-Olds in Twelve Countries. International Studies in Education. UNESCO Institute for Education, Hamburg.
Fuller, B. and Heynemann, S. P. (1989) Third World school quality: Current collapse, future potential. Educational Researcher, Vol. 18, No. 2, pp. 12-19.
Gallagher, J. J. (1987) A summary of research in science education. Science Education, Vol. 71, pp. 277-284.
Golbeck, S. L. (1986) The Role of Physical Content in Piagetian Spatial Tasks: Sex Differences in Spatial Knowledge? Journal of Research in Science Teaching, Vol. 23, No. 4, pp. 365-376.
Goldstein, H. (1986) Multilevel mixed linear model analysis using iterative generalized least squares. Paper presented at the Annual Meeting of the American Educational Research Association, San Francisco, CA.
Goldstein, H. (1989) Multilevel Modelling in Large Scale Achievement Surveys. Biometrika, Vol. 73, No. 1, pp. 43-56.
Good, T. L., Sikes, J. N., and Brophy, J. E. (1972) Effects of teacher sex and student sex and student achievement on classroom interaction. Technical Report No. 61. Center for Research in Social Behaviour, University of Missouri, Columbia, MI.
Good, T. L. and Findley, M. J. (1985) Sex role expectations and achievement. Dusek, J. (ed.) Teacher Expectancies. Lawrence Earlbaum, Hilldale, NY.
Gorman, T. P., Purves, A. C., and Degenhart, R. E. (eds.) (1989) The IEA Study of Written Composition I: The International Writing Tasks and Scoring Scales. Pergamon Press, Oxford.
Grayson, D. (1987) Evaluating the Impact of the Gender Expectations and Student Achievement (GESA) Program. Paper presented at the Annual Meeting of the American Educational Research Association, Washington, DC.
Grossman, P. L., Wilson, S. M., and Shulman, L. S. (1989) Teachers of substance: Subject matter knowledge for teaching. Reynolds, M. C. (ed.) Knowledge base for the beginning teacher, pp. 37-68. Pergamon Press, Oxford.
Guilford, J. P. (1965) Fundamental Statistics in Psychology and Education. McGraw-Hill, New York.
Guilford, J. P. (1967) The Nature of Human Intelligence. McGraw-Hill, New York.
Guilford, J. P. and Fruchter, B. (1978) Fundamental Statistics in Psychology and Education. McGraw-Hill, New York.
Gullahorn, J. E. (1979 [1]) Sex-related behaviors: Historical and psychobiological perspectives. Gullahorn, J. E. (ed.) Psychology and Women: In Transition, pp. 1-8. Winston and Sons, Washington, DC.
Gullahorn, J. E. (1979 [2]) Sex-related factors in cognition and in brain lateralization. Gullahorn, J. E. (ed.) Psychology and Women: In Transition, pp. 9-35. Winston and Sons, Washington, DC.
Haertel, G. D., Walberg, H. J., Junker, L., and Pascarella, E. T. (1981) Early adolescent sex differences in science learning: Evidence from the National Assessment of Educational Progress (NAEP). American Educational Research Journal, Vol. 18, No. 10, pp. 329-341.
Handley, H. M. and Morse, L. W. (1984) Two-year study relating adolescents' self-concept and gender role perceptions to achievement and attitudes towards science. Journal of Research in Science Teaching, Vol. 21, No. 6, pp. 599-607. ERIC database, ref. ED238902.
Hanna, G. and Kuendiger, E. (1986) Differences in Mathematical Achievement Levels and in Attitudes for Girls and Boys in Twenty Countries. Paper presented at the Annual Meeting of the American Educational Research Association, San Francisco, CA.
Hanushek, E. A. (1986) The economics of schooling: Production and efficiency in public schools. Journal of Economic Literature, Vol. 14, pp. 351-388.
Harman, H. H. (1976) Modern Factor Analysis. University of Chicago Press, Chicago, IL.
Hegarty, E. (1987) Science laboratory teaching. Dunkin, M. J. (ed.) International Encyclopedia of Teaching and Teacher Education, pp. 304-305. Pergamon Press, Oxford.
Heinrichs, U. and Schulz, T. (1989) Mädchen und Naturwissenschaft. Pädagogik, Vol. 9.

Heynemann, S. P. (1976) A brief note on the relationship between socio-economic status and test performance among Ugandan primary school children. Comparative Education Review, Vol. 20, No. 1, pp. 42-47.
Heynemann, S. P. and Loxley, W. A. (1982) Influences on academic achievement across high and low income countries: a reanalysis of IEA data. Sociology of Education, Vol. 55, No. 1, pp. 13-21.
Hildebrand, G. M. (1987) A causal model and an intervention model. Contributions to the Fourth GASAT Conference. Vol. 3, pp. 208-216. University of Michigan, Ann Arbor.
Hilgard, E. R. and Bower, G. H. (1966) Theories of Learning. Meredith Publishing Company, New York.
Hoffmann, L. (1987) The significance of school and out-of-school conditions in the development of male and female students' interest: The example of physics. Contributions to the Fourth GASAT Conference. Vol. 3, pp. 95-102. University of Michigan, Ann Arbor.
Humrich, E. (1988 [1]) Sex Differences in Science Attitude and Achievement. Ph.D. dissertation, Teachers College, Columbia University, New York.
Humrich, E. (1988 [2]) Sex Differences in the Second IEA Science Study: U.S. Results in an International Context. Paper presented at the Annual Meeting of the National Association for Research in Science Teaching.
Husén, T. (ed.) (1967) International Study of Achievement in Mathematics. A Comparison of Twelve Countries. Volume II. Almqvist and Wiksell, Stockholm.
Husén, T. (1989) Talent, Opportunity and Career. Volume II. Almqvist and Wiksell, Stockholm.
Husén, T. and Keeves, J. P. (eds.) (1991) Issues in Science Education. Science Competence in a Social and Ecological Context. Pergamon Press, Oxford.
Hyde, J. S. (1981) How large are cognitive gender differences? American Psychology, Vol. 36, pp. 892-901.
IEA (International Association for the Evaluation of Educational Achievement) (1988) Science Achievement in Seventeen Countries. A Preliminary Report. Pergamon Press, Oxford.
IEA (International Association for the Evaluation of Educational Achievement) (1990) Material for National Centers. Manuals. International Coordinating Center of the Reading Literacy Study, Department of Comparative Education, University of Hamburg.
Imsen, G. (1987) Gender, values, and technology: A cultural perspective. Contributions to the Fourth GASAT Conference. Vol. 3, pp. 103-111. University of Michigan, Ann Arbor.
Jayaratne, T. (1983) Sex Differences in Children's Math Achievement: Parental Attitudes. Paper presented at the Biennial Meeting of the Society for Research in Child Development, Detroit, MI.
Jensen, A. R. (1970) Hierarchical theories of mental ability. Dockrell, W. B. (ed.) On Intelligence. Methuen, London.
Jensen, A. R. (1973) Educability and Group Differences. Methuen, London.
Jimenez, E., Lockheed, M., and Wattanawaha, N. (1988) The Relative Efficiency of Private and Public Schools: The Case of Thailand. World Bank Economic Review, Vol. 2, No. 2. The World Bank, Washington, DC.
Johnson, R. T., Johnson, D. W., Scott, L. E., and Ramolae, B. A. (1985) Effects ot single-sex and mixed-sex cooperative interaction on science achievement and attitudes and cross-handicap and cross-sex relationships. Journal of Research in Science Teaching, Vol. 22, pp. 207-220.
Jones, G. M. (1989) Gender Issues in Teacher Education. Journal of Teacher-Education, Vol. 40, No. 1, pp. 33-38.
Jöreskog, K. G. and Sörbom, D. (1983) LISREL IV: Analysis of Linear Structural Relationships by Maximum Likelihood and Least Squares Methods. National Educational Resources, Chicago, IL.
Jöreskog, K. G. and Sörbom, D. (1986) LISREL VI. Analysis of Linear Structural Relationships by Maximum Likelihood, Instrumental Variables and Least Squares Methods. Scientific Software, Mooresville, IN.

Jöreskog, K. G. and Sörbom, D. (1989) LISREL 7. A guide to the program and applications (2nd ed.) SPSS Publications, Chicago, IL.
Kahle, J. B. and Lakes, M. K. (1983) The myth of equality in science classrooms. Journal of Research in Science Teaching, Vol. 20, pp. 131-140.
Kahle, J. B. (1989) Images of Scientists: Gender Issues in Science Classrooms. What Research Says to the Science and Mathematics Teacher, No. 4, pp. 1-7. The Key Centre for School and Mathematics, Curtin University of Technology, Perth, Western Australia.
Kanaya, T. (1988) Japan. Postlethwaite, T. N. (ed.) The Encyclopedia of Comparative Education and National Systems of Education, pp. 403-408. Pergamon Press, Oxford.
Kann, U. (1981) Career Development in a Changing Society. The Case of Botswana. Institute of International Education, Stockholm.
Keeves, J. P. (1972) Educational Environment and Student Achievement. Almqvist and Wiksell, Stockholm.
Keeves, J. P. (ed.) (1986) Aspiration, motivation and achievement: different methods of analysis and different results. International Journal of Educational Research, Vol. 10, No. 2, pp. 115-243.
Keeves, J. P. (1988) Sex Differences in Ability and Achievement. Keeves, J. P. (ed.) Educational Research, Methodology, and Measurement: An International Handbook, pp. 482-487. Pergamon Press, Oxford.
Keeves, J. P. and Sellin, N. (1988) Multilevel Data. Keeves, J. P. (ed.) Educational Research, Methodology, and Measurement: An International Handbook, pp. 689-700. Pergamon Press, Oxford.
Keeves, J. P. (1991) Changes in Science Education and Achievement, 1970-1984. Pergamon Press, Oxford.
Keeves, J. P. and Cheung, K. C. (1990) Testing Multi-Level Product-Process Networks. Cheung, K. C. et al. The Analysis of Multivariate Data in Educational Research: Studies of Problems and Their Solutions. International Journal of Educational Research, Vol. 14, No. 3, pp. 307-315. Pergamon Press, Oxford.
Keeves, J. P. and Kotte, D. (1991) Sex Differences in Science Education: 1970 to 1984. Keeves, J. P. (ed.) Changes in Science Education and Achievement, 1970-1984. Pergamon Press, Oxford.
Keeves, J. P. and Muller, E. W. (1991) Changes in the Opportunity to Learn Science: 1970-84. Keeves, J. P. (ed.) Changes in Science Education and Achievement, 1970-1984. Pergamon Press, Oxford.
Keeves, J. P. and Saha, L. J. (1991) Home Background Factors and Educational Outcomes. Keeves, J. P. (ed.) Changes in Science Education and Achievement, 1970-1984. Pergamon Press, Oxford.
Kelly, A. (1978) Girls and Science. An International Study of Sex Differences in School Science Achievement. International Association for the Evaluation of Educational Achievement, IEA Monograph Studies No. 9. Almqvist and Wiksell, Stockholm.
Kelly, A. (1981) The Missing Half. Girls and Science Education. Manchester University Press, Manchester.
Kelly, A. and Smail, B. (1986) Sex stereotyping and attitudes to science among 11-year-old schoolchildren. British Journal of Educational Psychology, Vol. 56, pp. 158-168.
Keys, W. (1986) The Second IEA Science Study: Aspects of science education in English schools. National Foundation of Educational Research, Slough, England.
Keys, W. and Matsubara, S. (1986) The Constitution of Attitude and Descriptive Scales. Second International Science Study: internal document IEA/SISS/192. International Association for the Evaluation of Educational Achievement, Stockholm.
Kida, H. (1991) Issues in the Scientific Education of the Citizen in the High Technology Era. Comments. Husén, T. and Keeves, J. P. (eds.) Issues in Science Education, pp. 192-197 Pergamon Press, Oxford.
Kish, L. (1965) Survey sampling. John Wiley, New York.
Kish, L. and Frankel, M. R. (1970) Unbalanced repeated replications for standard errors. Journal of the American Statistical Association, Vol. 65, pp. 1071-1094.
Kojima, S. (1974) IEA Science Study in Japan with Special Reference to the Practical Test. Comparative Education Review, Vol. 18, No. 2, pp. 262-267.

Kolata, G. B. (1980) Math and sex: Are girls born with less ability? Science, Vol. 210, pp. 1234-1235.

Krathwohl, D. R., Bloom, B. S., and Masia, B. B. (1964) Taxonomy of Educational Objectives: The Classification of Educational Goals. Handbook 2: Affective Domain. McKay, New York.

Krendl, K. A. et al. (1989) Children and Computers: Do Sex-Related Differences Persist? Journal of Communication, Vol. 39, No. 3, pp. 85-93.

Kroeger, J. E. (1990) An Investigation of the Effects of a Non-Sexist Curriculum on One Preschool Group's Sex-Role Stereotypical Behavior. ERIC database, ref. ED320688.

Kulik, J. A., Bangert, R. L., and Williams, G. W. (1983) Effects of computer-based teaching on secondary school students. Journal of Educational Psychology, Vol. 75, pp. 19-26.

Languis, M. and Naour, P. (1985) Sex Differences and Neurodevelopmental Variables: A Vector Model Paper presented at the Annual Meeting of the American Educational Research Association, Chicago, IL.

Lau, M. Y. (1986) The relationship of parental attitudes and of the socio-economic status of students to students' attitudes towards mathematics. Unpublished M.Ed. Dissertation, University of Hong Kong, Hong Kong.

Lee, P. C. and Gropper, N. B. (1974) Sex-role culture and educational practice. Harvard Educational Review, Vol. 44, pp. 369-410.

Lee, V. R. and Bryk, A. S. (1986) Effects of Single-Sex Secondary Schools on Student Achievement and Attitudes. Journal of Educational Psychology, No. 78, pp. 381-395.

Leimu, K. (1988) Finland. Postlethwaite, T. N. (ed.) The Encyclopedia of Comparative Education and National Systems of Education, pp. 260-267. Pergamon Press, Oxford.

Leinhardt, G., Seewald, A. M., and Engel, M. (1979) Learning what's taught: Sex differences in instruction. Journal of Educational Psychology, Vol. 71, pp. 432-439.

Levy, B. (1972) "The School" 's in the sex-role stereotyping of girls: A feminist review of the literature. Feminist Studies, Vol. 1, pp. 5-23.

Levy, G. D. (1989) Gender Schematization and Gender Scripts: Sometimes a Little Knowledge Can Be Dangerous. Paper presented at the Biennial Meeting of the Society for Research in Child Development, Kansas City, MO.

Levy, J. and Levy, J. M. (1978) Human lateralization from head to foot: Sex-related factors. Science, Vol. 200, pp. 1291-1292.

Levy-Agresti, J. and Sperry, R. W. (1968) Differential perceptual capacities in major and minor hemispheres. Proceedings of the National Academy of Sciences, Vol. 61, p. 1151.

Lewis, G. L. and Massad, C. E. (eds.) (1975) The Teaching of English as a Foreign Language in Ten Countries. International Studies in Evaluation IV. Almqvist and Wiksell, Stockholm.

Li, C. C. (1986) The explication of a model of mathematics learning in the context of the IEA mathematics study. Unpublished M.Ed. Dissertation, University of Hong Kong, Hong Kong.

Linn, M. C., De Benedictis, T., Delucchi, K., Harris, A., and Stage. E (1987) Gender differences in National Assessment of Educational Progress science items: What does "I don't know" really mean? Journal of Research in Science Teaching, Vol. 24, No. 3, pp. 267-278.

Linn, M. C. and Hyde, J. S. (1989) Gender, Mathematics, and Science. Educational Researcher, Vol. 18, No. 8, pp. 17-27.

Lockheed, M. E. (1975) The modification of female leadership behavior in the presence of males. ERIC database, ref. ED106742.

Lockheed, M. E., Hunter, B., Anderson, R. E., Beazley, R. M., and Esty, E. T. (1983) Computer Literacy: Definition and Survey Items for Assessment in School. U.S. Department of Education/National Center for Educational Statistics, Washington, DC.

Loewen, J. W. et al. (1988) Gender Bias in SAT Items. Paper presented at the Annual Meeting of the American Educational Research Association, New Orleans, LA. ERIC database, ref. ED294915.

Lohmoeller, J.-B. (1983) LVPLS 1.6 Program Manual. University of the Armed Forces, Munich.

Lohmoeller, J.-B. (1988) Latent Variables Path Analysis with Partial Least Squares. Version 1.8 Instant Manual. Department of Political Science, Free University Berlin, Berlin.

Lohnes, P. R. (1988) Discriminant Analysis. Keeves, J. P. (ed.) Educational Research, Methodology, and Measurement: An International Handbook, pp. 628-633. Pergamon Press, Oxford.

Long, J. E. et al. (1986) The Effects of Teacher Sex Equity and Effectiveness Training on Classroom Interaction at the University Level. Paper presented at the Annual Meeting of the American Educational Research Association, San Francisco, CA.

Longford, N. T. (1987) A fast scoring algorithm for maximum likelihood estimation in unbalanced mixed models with nested effects. Biometrika, Vol. 74, pp. 817-827.

Maccoby, E. E. and Jacklin, C. N. (1974) The Psychology of Sex Differences. Stanford University Press, Stanford, CA.

Mappa, S. (1982) Radio Broadcasting Program for Out Of School Education in South Sulawesi (Indonesia). Evaluation in Education, Vol. 6, pp. 31-51. Pergamon Press, Oxford.

Marini, M. M. and Greenberger, E. (1978) Sex differences in educational aspirations and expectations. American Educational Research Journal, Vol. 15., pp. 67-79.

Marklund, S. (1988) Sweden. Postlethwaite, T. N. (ed.) The Encyclopedia of Comparative Education and National Systems of Education, pp. 630-637. Pergamon Press, Oxford.

Marks, E. S. (1947) Sampling in the revision of the Stanford-Binet scale. Psychological Bulletin No. 44, pp. 413-434.

Marsh, H. W. (1989) Sex Differences in the Development of Verbal and Mathematics Constructs: The High School and Beyond Study. American Educational Research Journal, Vol. 26, No. 2, pp. 191-225.

Matlin, M. W. and Matkoski, K. M. (1985) Gender-Stereotyping of Cognitive Abilities. Paper presented at the Annual Meeting of the Eastern Psychological Association, Boston, MA.

McCullagh, P. and Nelder, J. A. (1983) Generalized Linear Models. Chapman and Hall, London.

McGaw, B. (1985) Meta-Analysis. Husén, T. and Postlethwaite, T. N. (eds.) International Encyclopedia of Education, pp. 3322-3230. Pergamon Press, Oxford.

McKenzie, P. (1988) Australia. Postlethwaite, T. N. (ed.) The Encyclopedia of Comparative Education and National Systems of Education, pp. 106-113. Pergamon Press, Oxford.

Merton, R. K. (1962) Social Theory and Social Structure. Free Press, Glencoe, IL.

Meyer, G. R. and Penfold, D. M. (1961) Factors Associated with Interest in Science. British Journal of Educational Psychology, Vol. 31, 33.

Meyer, J. (1988) Hair-Pulling & Name Calling: A Study of Female/Female Competition. Paper presented at the Annual Meeting of the National Women's Studies Association, Minneapolis, MN.

Miller, A. and Asher, W. (1983) Gender Differences as a Factor in Teachers' Perceptions of Students: A Comment. ERIC database, ref. EJ284864.

Mitchell, G. E. (1981) Human sex differences. Van Nostrand Reinhold, New York.

Montgomery, M. J. (1986) Promote Sex Equity in Vocational Education. ERIC database, ref. ED268690.

Morgenstern, C. (1990) Determinanten naturwissenschaftlicher Leistung in allgemeinbildenden Schulen im internationalen Vergleich. Faculty of Sociology, University of Hamburg, Hamburg.

Morse, L. W. and Handley, H. M. (1982) Relationship of Significant Others, Parental and Teacher Influences to the Development of Self Concept, Science Attitudes and Achievement among Adolescent Girls. Final Report. ERIC database, ref. ED238902.

Moss, J. D. (1982) Towards Equality: Progress by Girls in Mathematics in Australian Secondary Schools. Occasional Paper No. 16. The Australian Council for Educational Research, Hawthorn, Victoria, Australia.

Mosteller, F. and Tukey, J. W. (1968) Data Analysis, Including Statistics. Lindsay, G. and Aronson, E. (eds.) The Handbook of Social Psychology, Vol. II, pp. 80-203. Addison-Wesley, Reading, MA.

Mottier, I. (1986) The debate on single sex education in the Netherlands. Women Challenge Technology. Vol. 1, 337-346. European Conference on Women, Natural Sciences and Technology, Elsinore, Denmark, 15-19 November.

NCES (National Center for Education Statistics) (1990) The Conditions of Education. Volume 1: Elementary and Secondary Education. U.S. Department of Education, Office of Educational Research and Improvement, Washington, DC.

Nie, N. H., Hull, C. H., Jenkins, J. G., Steinbrenner, K., Bent, D. H. (1975) SPSS Statistical Package for the Social Sciences. McGraw-Hill, New York.
Nielsen, J. and Roepstorff, L. (1985) Girls and Computers - A World of Difference? Paper presented at the Girls and Science and Technology Conference, Kingston, England.
Nielsen, L. (1986) The role of physics teachers. Women Challenge Technology. Vol. 4, pp. 823-828. European Conference on Women, Natural Sciences and Technology, Elsinore, Denmark, 15-19 November.
NIER (National Institute for Educational Research) (1988) Basic Facts and Figures about the Educational System in Japan. National Institute for Educational Research, Tokyo.
Nijhof, W. J. and Streumer, J. N. (1988) Netherlands. Postlethwaite, T. N. (ed.) The Encyclopedia of Comparative Education and National Systems of Education, pp. 498-506. Pergamon Press, Oxford.
Noonan, R. D. (1982) An Evaluation of the IEA Science Attitude and Descriptive Scales. Working Paper Series, No. 34. Institute for International Education, University of Stockholm, Stockholm.
Noonan, R. D. and Wold, H. (1988) Partial Least Squares Analysis. Keeves, J. P. (ed.) Educational Research, Methodology, and Measurement: An International Handbook, pp. 710-716. Pergamon Press, Oxford.
Norisah, Naimah, H., Abu Bakar Nordin, and Solehan bin Rehot (1982) Remedial Reading Support Program for Children in Grade 2 in Malaysia. Evaluation in Education, Vol. 6, pp. 137-160. Pergamon Press, Oxford.
Norusis, M. J. (1985) SPSS-X Advanced Statistics Guide. SPSS Inc., Chicago, IL.
Nurmi, J. E. (1988) Parent-Child Interaction and Adolescents' Future Orientation: The Effects of Age and Sex. Paper presented at the European Conference on Developmental Psychology, Budapest.
Oon Chye, Y. (1991) The Contexts of Science Education. Postlethwaite, T. N. and Wiley, D. E. (eds.) Science Achievement in Twenty-Three Countries, Chapter 2, pp. 13-48. Pergamon Press, Oxford.
Ormerod, M. B. (1975) Single-Sex and Co-Education: An Analysis of Pupils' Science Preferences and Choices and Their Attitudes to Other Aspects of Science under These Two Systems. Paper presented at the Girls and Science Education Conference, Chelsea College, London.
Ormerod, M. B. and Duckworth, D. (1975) Pupils' Attitudes to Science: A Review of Research. National Foundation of Educational Research, Slough, England.
Parsons, T. and Bales, R. F. (1955) Family, Socialization, and Interaction Process. Free Press, Glencoe, IL.
Peaker, G. F. (1975) An Empirical Study of Education in Twenty-One Countries: A Technical Report. IEA - International Studies in Evaluation VIII. John Wiley, New York.
Pedhazur, E. J. (1982) Multiple Regression in Behavioral Research: Explanation and Prediction. Holt, Rinehart, and Winston, New York.
Peng, D. R. and Jaffe, S. (1979) Gender-typical factors determining the selection of occupational careers. ERIC database, ref. ED179712.
Petersen, A. C. et al. (1982) Sex Differences. Mitzel, H. E. (ed.) Encyclopedia of Educational Research, pp. 1696-1712. Academic Press, New York.
Petersen, A. C. (1983) The Development of Sex-Related Differences in Achievement. Paper presented at the Annual Convention of the American Psychological Association, Anaheim, CA.
Postlethwaite, T. N. (ed.) (1986) International Educational Research. Papers in Honor of Torsten Husén. Pergamon Press, Oxford.
Postlethwaite, T. N. (1967) School Organization and Student Achievement. A Study Based on Achievement in Mathematics in Twelve Countries. Stockholm Studies in Educational Psychology 15. Almqvist and Wiksell, Stockholm.
Postlethwaite, T. N. (1991) Achievement in Science Education in 1984 in 23 Countries. Husén, T. and Keeves, J. P. (eds.) Issues in Science Education. Science Competence in a Social and Ecological Context, pp. 35-59. Pergamon Press, Oxford.
Postlethwaite, T. N. and Wiley, D. E. (1991) Science Achievement in Twenty-Three Countries. Pergamon Press, Oxford.

Raudenbush, S. W. and Bryk, A. S. (1986) A Hierarchical Model for Studying School Effects. Sociology of Education, Vol. 59, No. 1, pp. 1-17.

Raudenbush, S. W. (1988) Educational applications of hierarchical linear models: a review. Journal of Educational Statistics, Vol. 13, pp. 85-116.

Raudenbush, S. W. (1989) An Introduction to the Three-Level HLM: Computer Program and Users' Guide (for HLM Version 1.0). Michigan State University, East Lansing, MI.

Reynolds, A. J. and Walberg, H. J. (1991) A Structural Model of Science Achievement. Journal of Educational Psychology, Vol. 83, No. 1, pp. 97-107.

Riis, U. et al. (1988) Naturvetenskaplig undervisning i svensk skola - huvudresultat från en IEA-undersökning. Skolöverstyrelsen, Stockholm.

Robitaille, D. F. and Garden, R. A. (eds.) (1989) The IEA Study of Mathematics II: Contexts and Outcomes of School Mathematics. Pergamon Press, Oxford.

Roloff, C., Metz-Göckel, Koch, C., and Holzrichter, E. (1987) Nicht nur ein gutes Examen. Forschungsergebnisse aus dem Projekt: Studienverlauf und Berufseinstieg von Frauen in Naturwissenschaft und Technologie - Die Chemikerinnen und Informatikerinnen. Dortmunder Diskussionsbeiträge zur Hochschuldidaktik, Vol. 11. Hochschuldidaktisches Zentrum der Universität Dortmund, Dortmund, Germany.

Rosenthal, R. and Jacobson, L. (1968) Pygmalion in the Classroom: Teachers' Expectations and Pupils' Intellectual Development. Holt, Rinehart, and Winston, New York.

Rosier, M. J. (1989) Administration and Sampling Report of the Second International Science Study. ERIC database.

Rosier, M. J. and Keeves, J. P. (1991) Science Education and Curricula in Twenty-Three Countries. Pergamon Press, Oxford.

Ross, K. (1978) Sample design for educational survey research. Keeves, J. P. (ed.) Educational Research, Methodology, and Measurement: An International Handbook, pp. 537-541.

Ross, K. (1988) Sampling Errors. Evaluation in Education, Vol. 2, pp. 105-195. Pergamon Press, Oxford.

Rust, K. F. (1984) Techniques for estimating variances for sampling surveys Unpublished Ph. D. dissertation, University of Michigan, Ann Arbor.

Saario, T. N., Jacklin, C. N., and Tittle, C. K. (1973) Sex role stereotyping in the public schools. Harvard Education Review, Vol. 43, pp. 386-416.

Sadker, D. et al. (1984) Teacher Reactions to Classroom Responses of Male and Female Students. Paper presented at the Annual Meeting of the American Educational Research Association, New Orleans, LA.

Sanguinetty, J. A. (1983) Academic Achievement, School Quality and Family Background: Study in Seven Latin American Countries. Paper presented at the Annual Meeting of the Comparative and International Education Society, Atlanta, GA.

Scandura, J. M. (1973) Structural Learning I. Theory and Research. Gordon and Breach, New York.

Schreiber, D. A. (1984) Factors Affecting Female Attitude Formation toward Science. Specific Reference to 12-14 Year Old Female Adolescents and Their Affective Orientation toward Middle School Science. ERIC database, ref. ED256617.

Sellin, N. (1989) PLSPATH Version 3.01. Application Manual., Hamburg.

Sellin, N. (1990 [1]) PLSPATH Version 3.01. Program Manual., Hamburg.

Sellin, N. (1990 [2]) On Aggregation Bias. Cheung, K. C. et al. (eds.) The Analysis of Multivariate Data in Educational Research: Studies of Problems and Their Solutions. International Journal of Educational Research, Vol. 14, No. 3, pp. 257-268. Pergamon Press, Oxford.

Shukla, S. (1974) Achievement of Indian Children in Mother Tongue (Hindi) and Science. Comparative Education Review, Vol. 18, No. 2, pp. 237-247.

Shymanski, J. A. and Kyle, W. C. (1988) A summary of research in science education - 1986. Science Education, Vol. 72, pp. 245-373.

Silver, E. A. (1987) Foundations of cognitive theory and research for mathematics problem solving. Schoenfeld, A. H. (ed.) Cognitive science and mathematics education, pp. 33-60. Erlbaum, Hillsdale, NJ.

Silver, P. (1985) Sex Equity in Educational Employment. Husén, T. and Postlethwaite, T. N. (eds.) International Encyclopedia of Education, pp. 4550-4555. Pergamon Press, Oxford.

Smail, B. (1984 [1]) Spatial visualization skills and technical crafts education. National Foundation for Educational Research, Slough, England.
Smail, B. (1984 [2]) Girl-friendly science: Avoiding sex bias in the curriculum. Longman, London.
Smith, W. S. and Litman, C. I. (1979) Early adolescent girls' and boys' learning of a spatial visualization skill. Science Education, Vol. 63, pp. 671-676.
Smith, W. S. and Schroeder, C. K. (1979) Pre-adolescents' learning and retention of a spatial visualization skill. School Science and Mathematics, Vol. 81, No. 8, pp. 705-709.
Spearritt, D. (1988) Factor Analysis. Keeves, J. P. (ed.) Educational Research, Methodology, and Measurement: An International Handbook, pp. 644-654. Pergamon Press, Oxford.
Sperry, R. W. (1961) Cerebral organization and behaviour. Science, Vol. 133, pp. 1749-1757.
Sperry, R. W. (1972) Mental Unity Following Surgical Disnections of the Cerebral Hemispheres. Washburn, S. L. and Dolphinow, P. (eds.) Perspectives on Human Evolution, pp. 395-420.
Srocke, B. (1989) Mädchen und Mathematik. Deutscher Universitäts-Verlag, Wiesbaden, Germany.
Staver, J. R. and Walberg, H. J. (1986) An analysis of factors that affect public and private school science achievement. Journal of Research in Science Teaching, Vol. 23, pp. 97-112.
Steinkamp, M. W. (1982) Sex-Related Differences in Attitude toward Science: A Quantitative Synthesis of Research. Paper presented at the Annual Meeting of the American Educational Research Association, New York.
Steinkamp, M. W. and Maehr, M. L. (1984) Gender differences in motivational orientations toward achievement in school science: A quantitative synthesis. American Educational Research Journal, Vol. 21, pp. 39-59.
Stockard, J., Schmuck, P. A., Kempner, K., Williams, P., Edson, S. K., and Smith, M. A. (1980) Sex Equity in Education. Academic Press, New York.
Stone, M. (1974) Cross-validatory choice and assessment of statistical predictions. Journal of the Royal Statistical Society, Series B, No. 36, pp. 111-147.
Taylor, H. E. (1982) British Columbia Science Assessment 1982. Summary Report. ERIC database, ref. ED225873.
Thorndike, R. L. (1973) Reading Comprehension Education in Fifteen Countries. International Studies in Evaluation III. Almqvist and Wiksell, Stockholm.
Thurstone, L. L. (1938) Primary Mental Ability. University of Chicago, Chicago, IL.
Tobin, K. (1987) It can't happen here. Fraser, B. J. and Giddings, G. J. (eds.) Gender issues in science education. Faculty of Education, Curtin University of Technology, Perth, Western Australia.
Tracy, D. M. (1987) Toys, Spatial Ability, and Science and Mathematics Achievement - Are They Related? Sex Roles, Vol. 17, No. 3-4, pp. 115-138.
Travers, K. J. and Westbury, I. (1989) The IEA Study of Mathematics I: International analysis of mathematics curricula. Pergamon Press, Oxford.
Tukey, J. W. (1977) Exploratory Data Analysis. Addison-Wesley, Reading, MA.
Tyler, L. E. (1965) The Psychology of Human Differences. Appleton-Century-Crofts, New York.
Visalberghi, A. (1988) Italy. Postlethwaite, T. N. (ed.) The Encyclopedia of Comparative Education and National Systems of Education, pp. 385-391. Pergamon Press, Oxford.
Walberg, H. J. (1969) Physics, Feminity, and Creativity. Review of Educational Research, Vol. 61, No. 1, pp. 25-69.
Walberg, H. J. and Fowler, W. J. (1987) Expenditure and size efficiencies of public school districts. Educational Researcher, Vol. 16, No. 7, pp. 5-13.
Walberg, H. J. (1989) Science, Mathematics, and National Welfare: Retrospective and Prospective Achievements. Purves, A. C. (ed.) International Comparisons and Educational Reform, pp. 99-111. Association for Supervision and Curriculum Development, Alexandria, VA.
Walberg, H. J. (1990) Productive Teaching and Instruction: Assessing The Knowledge Base. Phi Delta Kappan, Vol. 71, No. 6, pp. 470-478.
Walberg, H. J. (1991) Improving School Science in Advanced and Developing Countries. Developmental Psychology, Vol. 1, pp. 47-54.

Walker, D. A. (ed.) (1976) The IEA Six-Subject Survey: An Empirical Study of Education in Twenty-One Countries. International Studies in Evaluation, Vol. IX. Almqvist and Wiksell, Stockholm.
Weinbach, I. (1988) Das Verhältnis von Mädchen zu naturwissenschaftlichen Fächern - Abriß der Forschungssituation. Empirische Pädagogik, Vol. 2, No. 2, pp. 105-125.
Weinberg, S. and Rabinowitz, J. (1970) A sex difference in the Wechsler IQ vocabulary scores as a predictor of strategy in a probability-learning task performed by adolescents. Developmental Psychology, Vol. 3, pp. 218-224.
Whyte, J. (1986) Girls into science and technology. Routledge and Kegan Paul, London.
Wigfield, A. (1983) Students' Perceptions of Their Parents' Beliefs Concerning Their Academic Competence. Paper presented at the Biennial Meeting of the Society for Research in Child Development, Detroit, MI.
Wiley, D. E. (1976) Another hour, another day: Quantity of schooling, a potent path for policy. Sewell, W. H., Hauser, R. M., and Featherman, D. L. (eds.) Schooling and achievement in American society. Academic Press, New York.
Wiley, D. E. and Wolfe, R. G. (1991) Opportunity and Achievement: What they tell us about curriculum. Postlethwaite, T. N. and Wiley, D. E. (eds.) Science Achievement in Twenty-Three Countries, Chapter 5, pp. 91-107. Pergamon Press, Oxford.
Witelson, S. F. (1977) Early hemisphere specialization and interhemisphere plasticity: An empirical and theoretical review. Segalowitz, S. and Gruber, F. (eds.) Language development and neurological theory. Academic Press, New York.
Wittig, M. A. and Petersen, A. C. (1979) Sex-Related Differences in Cognitive Functioning: Development Issues. Academic Press, New York.
Wold, H. (1982) Soft modeling: The basic design and some extensions. Jöreskog, K. G. and Wold, H. (eds.) Systems under indirect observation, Part II, pp. 1-54. North Holland, Amsterdam.
Wolfe, L. R. and Tucker, J. (1987) Beyond the Star Trek syndrome to an egalitarian future: Where no one has gone before. Contributions to the Fourth GASAT Conference. Vol. 3, pp. 192-207. University of Michigan, Ann Arbor.
Yee, D. K. et al. (1986) Sex Equity in the Home: Parents' Influence on Their Children's Attitudes about Math. Paper presented at the Annual Meeting of the American Educational Research Association, San Francisco, CA.
Zerega, M. E., Haertel, G. D., Tsai, S., and Walberg, H. J. (1986) Late adolescent sex differences in science learning. Science Education, Vol. 70, pp. 447-460.

Appendices

Note:
Further information regarding the data sets and variables used as well as the FISS and SISS studies may be obtained from the

IEA Headquarter Sweelinckplein 14 **2517 GK THE HAGUE** THE NETHERLANDS

Particular print-outs for the PLSPATH and HLM-2 runs reported in this book may be obtained as disk files on 3.5" HD floppy disks in either MS-DOS or Macintosh format by sending in a blank disk to the author

Dr. Dieter Kotte International Consultant Ueckerstr. 38 **2000 HAMBURG 53** FEDERAL REPUBLIC OF GERMANY

Appendix 3.1

Achieved Sample Sizes in Target Populations, 1983/84

☞ This appendix presents some information about the number of students, teachers, and schools in the achieved samples of the target populations of the 1983/84 data collection. The data were compiled from Volume 2 (Postlethwaite and Wiley, 1991) and Volume 3 (Keeves, 1991) of the final report for the SISS study.

Target Population / Sample	Australia	England	Finland	Hungary	Italy	Japan	Netherlands	Sweden	Thailand	United States
Population 1										▲
Students	4259	3698	1600	2590	5152	7924	-	1270	-	2909
Teachers	819	390	121	97	309	543	-	137	-	120
Schools	220	181	106	100	119	221	-	134	-	121
Population 2										
Students	4917	3069	2546	2497	4622	7610	5025	1181	3780	1958
Teachers	1630	1077	244	354	1425	314	393	190	96	88
Schools	233	146	90	99	224	199	224	137	96	88
Population 3										
Biology Students	1631	884	1652	301	147	1212	-	619	1171	659
Chemistry Students	1177	892	971	143	217	1468	-	1172	1168	537
Physics Students	1073	917	810	398	1766	1187	-	1156	1168	485
Non-Specialist Students	995	1004	-	1036	2455	2230	-	1281	3685	2055
Teachers	700	790	84	173	535	123	-	nc	431	114
Schools	165	127	86	77	317	193	-	214	98	164

Notes:

▲ The data for the United States samples was taken from the 1983/84 testing (Phase A); for details see Keeves, 1991, Chapter 2 and especially Table 2.2

nc no information collected

\- not tested

Appendix 3.2

Marker Variable Checks and Standard Errors: Percent Male Students in Target Population and Sample

☞ This appendix presents some information about marker variable checks and the sampling errors of sample estimates for 10 selected countries. The data were extracted from Appendix C8 of Volume 2 of the final report for the SISS study (Postlethwaite and Wiley, 1991).

	Australia	England	Finland	Hungary	Italy	Japan	Netherlands	Sweden	Thailand	United States
Population 1										
% Male Students in Population	51.0	51.0	52.0	51.0	52.0	51.0	-	51.0	-	49.0
% Male Students in Sample	49.0	51.5	51.7	50.1	52.7	51.5	-	50.9	-	49.2
Std. Error of Sample Estimate	1.37	1.18	1.13	1.15	0.79	0.38	-	1.12	-	0.91
Population 2										
% Male Students in Population	51.0	51.0	na	51.0	na	51.0	52.0	51.0	50.0	52.0
% Male Students in Sample	54.4	52.2	51.6	49.2	52.0	51.6	50.3	51.9	47.6	50.9
Std. Error of Sample Estimate	2.40	2.28	1.02	1.00	1.05	0.31	2.00	1.46	2.45	1.43
Population 3 Biology										
% Male Students in Population	33.0	40.0	39.0	na	na	na	-	55.0	47.0	51.0
% Male Students in Sample	33.3	40.0	47.1	35.6	81.8	27.2	-	55.6	45.9	41.0
Std. Error of Sample Estimate	2.67	3.84	3.19	4.20	nc	nc	-	4.54	2.59	8.44
Population 3 Chemistry										
% Male Students in Population	61.0	65.0	46.0	na	na	na	-	75.0	47.0	51.0
% Male Students in Sample	56.9	66.5	47.6	36.2	69.4	73.6	-	70.1	45.9	58.8
Std. Error of Sample Estimate	2.96	3.33	5.71	6.87	nc	nc	-	2.82	2.68	11.07
Population 3 Physics										
% Male Students in Population	73.0	79.0	61.0	na	71.0	na	-	75.0	47.0	51.0
% Male Students in Sample	68.9	78.0	63.3	71.2	80.9	87.1	-	70.4	48.3	78.0
Std. Error of Sample Estimate	2.81	2.89	5.20	2.79	5.99	12.22	-	2.48	2.62	7.47

Notes:

na no information supplied by the National Center

nc not computed because of small Ns, but, in general, standard errors were high

\- not tested

Appendix 4.1

Information about the Variables Used in Analyses

☞ This appendix presents some information about the variables and composite scales used in analyses of gender differences in science achievement. For further details see the related publications of Comber and Keeves (1973), Rosier and Keeves (1991), Postlethwaite and Wiley (1991), and Keeves (1991).

Variable, Scale	Contents; Coding
Student Questionnaire	
Father's Occupation	occupations were classified into a four-category scale; low values indicated low occupational status and high values indicate high occupational status
Mother's Occupation	occupations were classified into a four-category scale; low values indicate low occupational status and high values indicated high occupational status; missing data had been coded differently for different countries
Father's Education	this variable was formed from information about the student's father's secondary education and post-secondary education; expressed in years; missing data had been assigned to national mean
Mother's Education	this variable was formed from information about the student's mother's secondary education and post-secondary education; expressed in years; missing data had been assigned to national mean
Family Size	family size in numbers (not numbers of brothers and sisters)
Books in Home	number of books in the home
Use of Dictionary	use of a dictionary at home; three-category variable
Year Level	year level the student was enrolled at; applicable only in Australia
Age	age of the student recorded in months; for the 14-year-olds the maximum value had been set to 200 months; values in excess of 200 were set to missing
Language	language spoken at home; the values were recoded to '1=language of the country' and '0=another language'; missing data were recoded to the modal category '1'
Sex of Student	the variable was coded '1=male' and '2=female'
Expected Occupation	expected occupation of student; occupations had been classified into a four-category scale; low values indicated low occupational status and high values indicate high occupational status
Expected Education	expected education of student; this variable was formed from information about the student's expected secondary and post-secondary education; expressed in years
Post-Science Education	student's expectation to study science as part of further education; dichotomous variable recoded '0=no' and '1=yes'
Hours Homework	hours spent on homework in all subjects

Variable, Scale	Contents; Coding
Attitude Questionnaire	
Beneficial Aspects of Science	scale value for seven attitudinal items measuring the construct of 'beneficial aspects of science'; Likert-type scale; see Appendix 4.2 for more details
Interest in Science	scale value for four attitudinal items measuring the student's interest science; Likert-type scale; see Appendix 4.2 for more details
School and School Learning	scale value for seven attitudinal items measuring the student's interest in school and school learning; Likert-type scale; see Appendix 4.2 for more details
Non-Harmful Aspects	scale value for six attitudinal items measuring the student's perception of science being non-harmful; Likert-type scale; see Appendix 4.2 for more details
Career Interest in Science	scale value for seven attitudinal items measuring the student's interest in a future science-related career; Likert-type scale; see Appendix 4.2 for more details
Ease of Learning Science	scale value for four attitudinal items measuring the student's perception of school science being an easy-to-learn subject; Likert-type scale; see Appendix 4.2 for more details
Descriptive Scale Questionnaire	
Homework Effort	scale value for six descriptive items assessing the student's effort undertaken with respect to homework; Likert-type scale; see Appendix 4.2 for more details
Classroom Effort	scale value for five descriptive items assessing the student's effort undertaken in the classroom; Likert-type scale; see Appendix 4.2 for more details
Practical Work	scale value for five descriptive items assessing the amount of practical and experimental work being done in science classes; Likert-type scale; see Appendix 4.2 for more details
Aptitude Tests	
Word Knowledge	corrected Word Knowledge test score; the test was based on 40 items
Mathematics	corrected Mathematics test score; the test was based on 20 items assessing computational skills
Achievement Tests	
Biology Score	corrected Biology score based on 10 items from the core and rotated tests
Chemistry Score	corrected Chemistry score based on 10 items from the core and rotated tests
Physics Score	corrected Physics score based on 10 items from the core and rotated tests
Opportunity to Learn Questionnaire *(variables from this questionnaire were only employed in preliminary PLS analyses)*	
OTL Score	opportunity to learn items in the core and rotated tests
OTL Biology Score	opportunity to learn biology items in the core and rotated tests
OTL Chemistry Score	opportunity to learn chemistry items in the core and rotated tests
OTL Physics Score	opportunity to learn physics items in the core and rotated tests

Variable, Scale	Contents; Coding
Teacher Questionnaire	
Age of Teacher	the variable was recoded to a rank scale score with values of '1' to 5', the higher value indicating higher age of teacher
Teacher Post-Second. Education	number of years of post-secondary education
Teacher Post-Second. Science Education	proportion of time spent on the study of science during post-secondary education
Years of Teaching Experience	number of years of teaching experience; the variable was recoded to a rank scale score with values of '1' to 5', the higher value indicating a higher amount of experience
Time Spent Inservice Training	amount of time spent on inservice training; the variable was recoded to a rank scale score with values of '1' to 5', the higher value indicating more time spent on inservice training
Sex of Teacher	the variable was coded '1=male' and '2=female'
Proportion Teaching Science	this derived variable was formed as the proportion of the time a teacher spent teaching science to the time spent on teaching science and other subjects
Emphasis on Knowledge	this derived variable described the emphasis which a teacher put on knowledge; it was formed by combining three variables, namely: - importance of official curriculum or syllabus, - importance of prescribed textbook, - importance of external examinations
Emphasis on Student Needs	this derived variable described the emphasis which a teacher put on student needs; it was formed by combining four variables, namely: - class divided into small groups, - students follow individualized programs, - importance of student needs when they leave school, - importance of student needs in the next grade
Emphasis on Scientific Investigation	this derived variable described the emphasis which a teacher put on knowledge; it was formed by combining five variables, namely: - time spent on practical work, - teaching in a laboratory equipped for science teaching, - importance to develop students' ability to think scientifically, - importance of knowledge of scientific concepts, - assessment of performance on project work
School Questionnaire	
No. of Science Laboratories	number of laboratories equipped for science teaching
Sex of School	*(used for HLM analyses in Australia and England only)* actually percent male students in school this was a derived, categorical variable which indicated single-sex or coeducational schooling
Region	*(used for HLM analyses in Australia and England only)* regional location of school; coding varied for Australia and England (see Section 8.1)
SchType	*(used for HLM analyses in Australia only)* type of school; for coding details see Section 8.1
RegType	*(used for HLM analyses in England only)* derived variable combining type of school and region in which the school was located; for coding details see Section 8.1

Appendix 4.2

Information about Attitudinal and Descriptive Scales Used in Analyses

☞ This appendix was prepared using information originally compiled for the SISS Vol. 2 released by Postlethwaite and Wiley (1991, pp. 201-206). The information is restricted to the 14-year-old level (Population 2).

Introduction

The first part of this appendix presents for the titles of the scales and a note in parentheses on the countries for which scale scores were not calculated for that country either because a country had not administered the scale or because an item had been dropped.

The second part of the appendix presents the means, standard deviations, and reliability coefficients (α) at the between-student level, as well as the N used for each scale in each country.

Where a country had not administered the items or had too few items (e.g. United States) then the country was not included in the analyses.

The following steps were undertaken:

1. A principal components analysis was undertaken for each set of items said to belong to each scale for each country.
2. Where a factor loading for an item was below .30 then the item was examined in terms of whether it could belong to the scale or not. In general, the item was dropped from the scale for all countries where a factor loading fell below .30 for more than two countries.
3. Even though the reliability of a scale was low at the between-student level, the scale was nevertheless retained on the assumption that at the aggregated level of between-schools or classrooms the reliability would be higher.

Items and Scales

Notes:

ATT	attitudinal item
DES	descriptive item
(r)	the direction of the scoring of the item was reversed
Interest in Science	composite scale

The scale score range in the international version of the measure is given in brackets.

SCALE ☞ Interest in Science (Range: 4-12)

(Italy not included)

direction: *A higher scale score indicates that a student has a more positive interest in science.*

LIKESCI	I like science.
P2ATT34	Science is an enjoyable school subject.
P2ATT35	The science taught at school is interesting.
P2DES09	The teacher makes science lessons interesting.

SCALE ☞ Ease of Learning Science (Range: 4-12)

direction: *A higher scale score indicates that a student perceives learning science as easy.*
(r)P2ATT36 Science is a difficult subject.
(r)P2ATT37 Science is difficult when it involves calculations.
(r)P2ATT38 Science is difficult when it involves handling apparatus.
(r)P2ATT39 There are too many facts to learn in science.

SCALE ☞ School and School Learning (Range: 7-21)

direction: *A higher scale score indicates that a student has a positive attitude towards school and school learning.*
(r)P2ATT06 School is not very enjoyable.
P2ATT09 I enjoy everything about school.
(r)P2ATT14 I am bored most of the time at school.
(r)P2ATT19 There are many school subjects I don't like.
P2ATT23 The most enjoyable part of my life is the time I spend at school.
P2ATT26 I want as much education as I can get.
(r)P2ATT28 I generally dislike my schoolwork.

SCALE ☞ Career Interest in Science (Range: 7-21)

direction: *A higher scale score indicates that a student shows an increased interest to resume a career in science.*
P2ATT04 Working in a science laboratory would be an interesting way to earn a living.
P2ATT08 In the future most jobs will require a knowledge of science.
P2ATT11 People who understand science are better off in our society.
P2ATT13 It is important to know science in order to get a good job.
P2ATT18 Science is a very good field for creative people to enter.
P2ATT21 In my future career I would like to use the science I learned at school.
P2ATT29 I want to learn more about world we live in.

SCALE ☞ Beneficial Aspects of Science (Range: 7-21)

direction: *A higher scale score indicates that a student holds more positive values about the beneficial aspects of science.*
P2ATT01 Science is useful for solving the problems of everyday life.
P2ATT05 Science is very important for a country's development.
P2ATT07 Money spent on science is well worth spending.
P2ATT12 Public money spent on science in the last few years has been used wisely.
P2ATT15 Scientific inventions improve our standard of living.
P2ATT16 The government should spend more money on scientific research.
P2ATT22 Science will help to make the world a better place in the future.

SCALE ☞ Non-Harmful Aspects of Science (Range: 6-18)

direction: *A higher scale score indicates that a student holds less negative values about the harmful aspects of science.*
(r)P2ATT03 Science has ruined the environment.
(r)P2ATT10 Much of the anxiety in modern society is due to science.
(r)P2ATT17 Scientific inventions have made the world too complex.
(r)P2ATT20 Scientific inventions have increased tensions between people.
(r)P2ATT24 Scientific discoveries do more harm than good.
(r)P2ATT27 Science and technology are the cause of many of the world's problems.

SCALE ☞ Homework Effort (Range: 0-18)

(Australia, England, Finland, and Japan not included)

direction: *A higher scale score indicates that a student undertakes more efforts to do homework.*

P2DES32 When you have finished an assignment, do you check it?
P2DES33 When you have an assignment, do you do your best?
P2DES35 When you sit down to do your homework, how long does it take you to get started?
P2DES36 When you do your homework, are you disturbed by things going on around you?
P2DES37 When you have homework, do you do all of it?
P2DES38 When you have homework, do you hand it in on time?

SCALE ☞ Classroom Effort (Range: 0-15)

(Australia, England, and Finland not included)

direction: *A higher scale score indicates that a student undertakes more efforts to take part in classroom activities.*

P2DES25 What do you do when the teacher asks you a question in class?
P2DES26 What do you do when the teacher asks someone else a question?
P2DES27 When the teacher is presenting something new, do you pay attention?
P2DES39 When you are absent and miss a lesson, do you try to make up what you have missed?
P2DES40 Do you ever go to school but deliberately miss a lesson?

SCALE ☞ Practical Work (Range: 5-15)

(For HLM analyses of Australia and England only)

direction: *A higher scale score indicates that a student does more practical work or experiments in science.*

P2DES11 For science homework we write up reports of our laboratory and practical work.
P2DES18 We do practical work (experiments) as part of our science lessons.
P2DES19 The science class breaks up into small groups of students to do practical work (experiments).
P2DES20 When we do experiments the teacher gives us instructions about what to do.
P2DES21 When we do an experiment we use a book or other written instructions to show us how to do it.

Computation of Scores

Three response categories were, in general, used: 'agree,' 'disagree' and 'uncertain.' The scoring scheme employed involved the scoring of a favourable response to an item as '3'; an uncertain response as '2'; and an unfavourable response as '1'. At the Population 3 level some countries used five response categories which have been reduced to three categories to maintain consistency with other countries. The scoring of an agree or disagree response was reversed for items marked (r). High values indicate a positive/favourable attitude whereas low values represent a negative/unfavourable attitude. Special recoding was applied for selected descriptive items in Populations 2 and 3, namely: P3DES25 through P3DES40. Starting with item P3DES25 the scale range was collapsed to four points (0,1,2,3) from the original five points. As a result of such collapsing, an equal item and scale range could be produced. Following the data preparation an extensive data check was performed to check the valid item range after recoding and on item availability across countries. Where a country failed to administer a particular item in a particular scale, scale scores for that scale were not calculated for that country. After these steps had been accomplished, standardized scale values were computed by dividing each national scale by the number of items in it.

Statistics for Attitude and Descriptive Scales

Attitude Scale	Statistic	Australia	England	Finland	Hungary	Italy	Japan	Netherlands	Sweden	Thailand
Interest in Science										
	mean	2.13	2.19	2.06	2.50	nc	1.88	1.94	2.06	2.38
	sd	.59	.57	.57	.41	nc	.57	.53	.55	.40
	N	4862	2970	2492	2401	nc	6430	4799	658	3779
	α	.78	.77	.76	.50	nc	.74	.69	.74	.51
Ease of Learning Science										
	mean	1.94	1.94	2.13	1.88	2.00	1.81	1.88	1.94	1.75
	sd	.54	.52	.48	.55	.53	.56	.55	.53	.50
	N	4868	2959	2486	2400	4492	6424	4805	656	3779
	α	.51	.49	.45	.44	.48	.52	.58	.54	.42
School and School Learning										
	mean	1.94	2.00	1.63	1.88	2.06	2.06	1.81	1.81	2.50
	sd	.52	.52	.51	.44	.50	.46	.48	.48	.36
	N	4867	2974	2498	2381	4487	6421	4803	657	3780
	α	.74	.77	.78	.65	.71	.67	.72	.72	.49
Career Interest in Science										
	mean	2.13	2.25	2.13	2.44	2.44	1.94	2.06	2.19	2.63
	sd	.45	.43	.39	.30	.35	.45	.45	.37	.29
	N	4874	2977	2499	2381	4484	6429	4807	659	3779
	α	.63	.59	.57	.46	.51	.63	.63	.49	.39
Beneficial Aspects of Science										
	mean	2.31	2.38	2.31	2.63	2.50	2.07	2.18	2.25	2.56
	sd	.40	.39	.37	.26	.36	.39	.38	.34	.29
	N	4878	2984	2500	2380	4484	6427	4801	656	3780
	α	.62	.64	.62	.44	.62	.60	.61	.52	.42
Non-Harmful Aspects of Science										
	mean	2.13	2.13	2.19	2.25	2.13	2.00	2.13	1.88	2.13
	sd	.43	.44	.42	.41	.50	.46	.46	.43	.44
	N	4867	2969	2503	2380	4474	6418	4796	657	3779
	α	.59	.62	.64	.57	.71	.64	.67	.58	.50

Notes:

α Cronbach's α, reliability coefficient

\- population not tested

nc not calculated because of missing data

Descriptive Scale / Statistic	Australia	England	Finland	Hungary	Italy	Japan	Netherlands	Sweden	Thailand
Homework Effort									
mean	-	-	-	2.44	2.38	n c	2.06	2.25	2.31
sd	-	-	-	.35	.46	n c	.54	.58	.35
N	-	-	-	2392	4501	n c	4592	648	3774
α	-	-	-	.65	.76	n c	.77	.78	.56
Classroom Effort									
mean	-	-	-	2.44	2.38	2.13	2.13	2.25	2.56
sd	-	-	-	.33	.43	.49	.57	.53	.34
N	-	-	-	2395	4305	6401	4543	647	3773
α	-	-	-	.59	.62	.61	.67	.69	.42
Practical Work									
mean	2.38	2.44	2.06	2.13	1.81	2.38	2.00	n c	2.56
sd	.36	.37	.45	.34	.51	.30	.59	n c	.30
N	4853	2972	2450	2398	4451	4622	4564	n c	3780
α	.54	.60	.68	.58	.61	.45	.79	n c	.48

Notes:

α Cronbach's α, reliability coefficient
\- population not tested
nc not calculated because of missing data

Appendix 4.3

Information about the Variables Used in HLM Analyses

☞ This appendix lists the MVs constituting the constructs used in HLM analyses for Australia and England. The MVs have been selected on the basis of the PLS analyses performed for Australia and England. For further information regarding the MVs see Appendix 4.1.

Construct (LV)	Manifest Variable (MV)
School Files	
HomeBack	Father's Occupation
Gender	Sex of School
TchBack	Years of Teaching Experience
Region	regional location of school; original variable in Australia: SUPSTR; original variable in England: LOCATION
SchType	Australia only; type of school; original variable SCHTYPE
RegType	England only; derived variable; original variable REGTYPE
TchEmph	Emphasis on Knowledge
Aptitude	Word Knowledge test score
SciValue	Beneficial Aspects of Science
Effort	Hours Homework
SciAtt	Interest in Science
Practice	Practical Work
Achieve	total science score
Student Files	
HomeBack	Father's Occupation
Gender	Sex of Student
Aptitude	Word Knowledge test score
SciValue	Beneficial Aspects of Science
Effort	Hours Homework
SciAtt	Interest in Science
Practice	Practical Work
Achieve	total science score

'Region' which was thought to reflect regional or geographical differences between schools had a different meaning in Australia and England. For Australia 'Region' differentiated between the relative wealth of the states which were recoded as:

Northern Territories	0
Victoria	0.5
New South Wales	1.5
Tasmania	1.5
South Australia	2
West Australia	3.5
Queensland	4
Australian Capital Territory	6

In the case of England 'Region' reflected a more geographical stratification of the location of the school which was coded:

North England	1
Midlands	2
South England	3

Since it was hypothesized that the type of school played some importance in determining gender differences in science achievement, a school-type variable was also taken into account for the design of the two-level path model. In the case of Australia this variable was labelled 'SchType' and was:

Government	1
Catholic	4
Independent	8

For England the equivalent construct was named 'RegType' and was intended to differentiate between regional types of schools such as:

Comprehensive Schools/North England	1
Comprehensive Schools/Midlands	1
Comprehensive Schools/South England	2
Independent Private and Public Schools	9
Other Maintained Schools	2

'Other Maintained Schools' were, in most cases, the so-called 'Grammar Schools', a typical academically-oriented type of schooling common in England before the introduction of comprehensive schools in 1965 (Booth, 1988, pp. 692-693).

Appendix 5.1

Science Achievement Test Scores for 10 Countries, 1970/71 and 1983/84

☞ This appendix presents information about the total test scores (in percent) of the 10 countries in 1970/71 and 1983/84. The data were compiled from the final report for the FISS study (Comber and Keeves, 1973) and Volume 2 for the SISS study (Postlethwaite and Wiley, 1991).

		Australia	England	Finland	Hungary	Italy	Japan	Netherlands	Sweden	Thailand	United States
FISS Study, 1970/71											
10-year-old level	mean	-	39.2	43.8	41.8	41.2	54.2	38.2	45.8	24.8	44.2
	sd	-	21.2	20.5	20.0	21.5	19.2	19.0	18.2	16.2	23.2
14-year-old level	mean	30.8	26.6	25.6	36.4	23.1	39.0	22.2	27.1	19.5	27.0
	sd	16.6	18.0	13.2	15.9	12.8	18.5	12.5	14.6	10.1	14.5
terminal secondary level	mean	41.2	38.5	33.0	38.3	26.5	-	38.8	32.0	20.7	22.8
	sd	17.8	19.2	16.3	15.0	14.7	-	18.5	17.0	10.2	15.8
SISS Study, 1983/84											
10-year-old level	mean	56.9	53.8	66.0	61.7	59.2	66.4	-	◆ 62.8	-	▲
	sd	17.2	17.0	14.9	16.3	17.6	14.7	-	14.6	-	▲
14-year-old level	mean	58.8	55.9	60.3	70.7	■ 52.4	66.8	63.7	✚ 60.3	56.7	▲
	sd	16.0	15.7	13.9	15.1	15.3	16.1	16.1	16.0	12.6	▲
terminal secondary level											
Biology	mean	53.4	68.1	51.0	65.6	46.7	51.3	-	60.5	47.2	▼
	sd	13.3	11.9	11.5	12.3	14.1	15.4	-	14.0	12.7	▼
Chemistry	mean	53.5	73.1	40.5	53.4	42.9	57.8	-	47.3	38.8	▼
	sd	16.7	15	13.3	16.7	22.4	20.2	-	15	15.7	▼
Physics	mean	53.6	64.2	44.7	62.7	34.2	61.8	-	51.4	35.9	▼
	sd	14.6	13.3	13.5	15.3	13.4	16.1	-	14.5	16.9	▼

Notes:

- the total score is based on the core and two rotated test forms
- ■ Italy (Grade 8)
- ◆ Sweden (Grade 4)
- ✚ Sweden (Grade 8)
- ▲ the United States did not administer the rotated tests at Population 1 and 2 levels, so no total score could not be calculated
- ▼ subscores were not calculated for the United States because they were judged to be of insufficient accuracy due to the small number of items
- \- not tested

Appendix 6.1 PLSPATH Output

Country:	Australia (Population 2)
Input correlation matrix:	students overall
Outcome variable:	Achieve

LV-Correlations

	HomeBack	Aptitude	Gender	SciValue	Effort	SciAtt	Achieve
HomeBack	1.000						
Aptitude	.422	1.000					
Gender	.006	.052	1.000				
SciValue	.122	.162	-.110	1.000			
Homework	.295	.366	.151	.137	1.000		
SciAtt	.110	.132	-.128	.544	.145	1.000	
Achieve	.389	.673	-.133	.214	.302	.178	1.000

Outer Model Summary

Variable	Mode	Status	N of MVs	AvgRed	AvgCom	MinCom	MaxCom
HomeBack	Inward	Exogen	4	-	.489	.299	.764
Aptitude	Inward	Endogen	2	.136	.760	.759	.761
Gender	Unity	Exogen	1	-	1.000	1.000	1.000
SciValue	Inward	Endogen	2	.030	.735	.680	.789
Effort	Inward	Endogen	1	.176	1.000	1.000	1.000
SciAtt	Inward	Endogen	2	.176	.594	.194	.994
Achieve	Outward	Endogen	3	.343	.694	.679	.713

Outer Model Diagnostics

Variable	Mode	N of MVs	Status	Det(Block)	Min R**2	Max R**2
HomeBack	Inward	4	Exogen	.5173	.2032	.3099
Aptitude	Inward	2	Endogen	.7294	.2706	.2706
Gender	Unity	1	Exogen	1.0000	.0000	.0000
SciValue	Inward	2	Endogen	.7763	.2237	.2237
Effort	Inward	1	Endogen	1.0000	.0000	.0000
SciAtt	Inward	2	Endogen	.8644	.1356	.1356
Achieve	Outward	3	Endogen	.4375	.3730	.3920

List of Manifest Variables (MVs)

MV3	Father's Occupation	MV23	Interest in Science
MV7	Father's Education	MV26	Career Interest in Science
MV8	Mother's Education	MV27	Ease of Learning Science
MV13	Books in Home	MV33	Hours Homework
MV16	Sex of Student	MV37	Biology test score
MV20	Word Knowledge test score	MV38	Chemistry test score
MV21	Mathematics test score	MV39	Physics test score
MV22	Beneficial Aspects of Science		

Outer Model Equations

```
HomeBack= MV3   MV7   MV8   MV13
Aptitude= MV20  MV21
Gender  = MV16
SciValue= MV22  MV26
Effort  = MV33
SciAtt  = MV23  MV27
Achieve = MV37  MV38  MV39
```

Outer Model Results

Variable	Weight	Loading	Communality	Redundancy	Tolerance
HomeBack	Inward	Exogen	4 MVs		
MV3	.3048	.6744	.4549	.0000	.2301
MV7	.2524	.6622	.4385	.0000	.3099
MV8	.1368	.5464	.2986	.0000	.2391
MV13	.6323	.8739	.7636	.0000	.2032
Aptitude	Inward	Endogen	2 MVs		
MV20	.5726	.8714	.7594	.1355	.2706
MV21	.5744	.8723	.7609	.1357	.2706
Gender	Unity	Exogen	1 MVs		
MV16	1.0000	1.0000	1.0000	.0000	.0000
SciValue	Inward	Endogen	2 MVs		
MV22	.5207	.8245	.6798	.0274	.2237
MV26	.6423	.8885	.7895	.0319	.2237
Effort	Inward	Endogen	1 MVs		
MV33	1.0000	1.0000	1.0000	.1764	.0000
SciAtt	Inward	Endogen	2 MVs		
MV23	.9659	.9969	.9939	.2945	.1356
MV27	.0843	.4400	.1936	.0574	.1356
Achieve	Outward	Endogen	3 MVs		
MV37	.3831	.8240	.6790	.3356	.3774
MV38	.4149	.8446	.7134	.3526	.3920
MV39	.4017	.8309	.6904	.3412	.3730

Inner Model Diagnostics and Summary

Variable	N Pred	R-square	Det(Pred)	MinBeta	MaxBeta	MinDlta	MaxDlta	MaxTol
Aptitude	1	.178	1.0000	.422	.422	.178	.178	.000
SciValue	2	.040	.9973	-.119	.168	.014	.028	.003
Effort	3	.176	.8192	.135	.286	.018	.067	.181
SciAtt	1	.296	1.0000	.544	.544	.296	.296	.000
Achieve	3	.494	.8192	.124	.630	.013	.325	.181

Inner Model Equations

```
Aptitude= HomeBack
SciValue= Aptitude Gender
Effort  = HomeBack Aptitude Gender
SciAtt  = SciValue
Achieve = HomeBack Aptitude Gender
```

Direct Inner Model Effects

	HomeBack	Aptitude	Gender	SciValue	Effort	SciAtt	Achieve
HomeBack	.000	.000	.000	.000	.000	.000	.000
Aptitude	.422	.000	.000	.000	.000	.000	.000
Gender	.000	.000	.000	.000	.000	.000	.000
SciValue	.000	.168	-.119	.000	.000	.000	.000
Effort	.174	.286	.135	.000	.000	.000	.000
SciAtt	.000	.000	.000	.544	.000	.000	.000
Achieve	.124	.630	-.166	.000	.000	.000	.000

Total Inner Model Effects

	HomeBack	Aptitude	Gender	SciValue	Effort	SciAtt	Achieve
HomeBack	1.000	.000	.000	.000	.000	.000	.000
Aptitude	.422	1.000	.000	.000	.000	.000	.000
Gender	.000	.000	1.000	.000	.000	.000	.000
SciValue	.071	.168	-.119	1.000	.000	.000	.000
Effort	.295	.286	.135	.000	1.000	.000	.000
SciAtt	.039	.092	-.065	.544	.000	1.000	.000
Achieve	.390	.630	-.166	.000	.000	.000	1.000

Appendix 6.2 PLSPATH Output

Country: Australia (Population 2)
Input correlation matrix: students overall
Outcome variable: FutEnrol

LV-Correlations

	HomeBack	Aptitude	Gender	SciValue	Effort	SciAtt	Achieve	ExpEd	ExpOcc	FutEnrol
HomeBack	1.000									
Aptitude	.421	1.000								
Gender	.002	.052	1.000							
SciValue	.124	.158	-.112	1.000						
Homework	.293	.369	.151	.136	1.000					
SciAtt	.107	.118	-.121	.543	.141	1.000				
Achieve	.387	.673	-.133	.212	.302	.164	1.000			
ExpEd	.380	.414	.015	.257	.361	.211	.370	1.000		
ExpOcc	.256	.309	.026	.181	.247	.168	.283	.482	1.000	
FutEnrol	.147	.198	-.072	.236	.119	.208	.200	.220	.141	1.000

Outer Model Summary

Variable	Mode	Status	N of MVs	AvgRed	AvgCom	MinCom	MaxCom
HomeBack	Inward	Exogen	4	-	.503	.344	.704
Aptitude	Inward	Endogen	2	.135	.759	.730	.789
Gender	Unity	Exogen	1	-	1.000	1.000	1.000
SciValue	Inward	Endogen	2	.029	.733	.652	.813
Effort	Inward	Endogen	1	.177	1.000	1.000	1.000
SciAtt	Inward	Endogen	1	.294	1.000	1.000	1.000
Achieve	Outward	Endogen	3	.342	.694	.678	.714
ExpEd	Unity	Endogen	1	.286	1.000	1.000	1.000
ExpOcc	Unity	Endogen	1	.247	1.000	1.000	1.000
FutEnrol	Unity	Endogen	1	.100	1.000	1.000	1.000

Outer Model Diagnostics

Variable	Mode	N of MVs	Status	Det(Block)	Min R**2	Max R**2
HomeBack	Inward	4	Exogen	.5173	.2032	.3099
Aptitude	Inward	2	Endogen	.7294	.2706	.2706
Gender	Unity	1	Exogen	1.0000	.0000	.0000
SciValue	Inward	2	Endogen	.7763	.2237	.2237
Effort	Inward	1	Endogen	1.0000	.0000	.0000
SciAtt	Inward	1	Endogen	1.0000	.0000	.0000
Achieve	Outward	3	Endogen	.4375	.3730	.3920
ExpEd	Unity	1	Endogen	1.0000	.0000	.0000
ExpOcc	Unity	1	Endogen	1.0000	.0000	.0000
FutEnrol	Unity	1	Endogen	1.0000	.0000	.0000

List of Manifest Variables (MVs)

MV3	Father's Occupation	MV21	Mathematics test score
MV6	Expected Occupation	MV22	Beneficial Aspects of Science
MV7	Father's Education	MV23	Interest in Science
MV8	Mother's Education	MV26	Career Interest in Science
MV9	Expected Education	MV33	Hours Homework
MV13	Books in Home	MV37	Biology test score
MV16	Sex of Student	MV38	Chemistry test score
MV19	Post-Science Education	MV39	Physics test score
MV20	Word Knowledge test score		

Outer Model Equations

```
HomeBack= MV3   MV7   MV8   MV13
Aptitude= MV20  MV21
Gender  = MV16
SciValue= MV22  MV26
Effort  = MV33
SciAtt  = MV23
Achieve = MV37  MV38  MV39
ExpEd   = MV9
ExpOcc  = MV6
FutEnrol= MV19
```

Outer Model Results

Variable	Weight	Loading	Communality	Redundancy	Tolerance
HomeBack	Inward	Exogen	4 MVs		
MV3	.2830	.6701	.4490	.0000	.2301
MV7	.3239	.7181	.5157	.0000	.3099
MV8	.1723	.5864	.3439	.0000	.2391
MV13	.5683	.8389	.7037	.0000	.2032
Aptitude	Inward	Endogen	2 MVs		
MV20	.5375	.8542	.7296	.1295	.2706
MV21	.6088	.8884	.7893	.1401	.2706
Gender	Unity	Exogen	1 MVs		
MV16	1.0000	1.0000	1.0000	.0000	.0000
SciValue	Inward	Endogen	2 MVs		
MV22	.4911	.8076	.6522	.0258	.2237
MV26	.6693	.9016	.8128	.0321	.2237
Effort	Inward	Endogen	1 MVs		
MV33	1.0000	1.0000	1.0000	.1771	.0000
SciAtt	Inward	Endogen	1 MVs		
MV23	1.0000	1.0000	1.0000	.2944	.0000
Achieve	Outward	Endogen	3 MVs		
MV37	.3818	.8234	.6779	.3343	.3774
MV38	.4158	.8450	.7140	.3521	.3920
MV39	.4022	.8311	.6908	.3406	.3730
ExpEd	Unity	Endogen	1 MVs		
MV9	1.0000	1.0000	1.0000	.2861	.0000
ExpOcc	Unity	Endogen	1 MVs		
MV6	1.0000	1.0000	1.0000	.2471	.0000
FutEnrol	Unity	Endogen	1 MVs		
MV19	1.0000	1.0000	1.0000	.1005	.0000

Inner Model Diagnostics and Summary

Variable	N Pred	R-square	Det(Pred)	MinBeta	MaxBeta	MinDlta	MaxDlta	MaxTol
Aptitude	1	.178	1.0000	.421	.421	.178	.178	.000
SciValue	2	.039	.9973	-.120	.165	.014	.027	.003
Effort	3	.177	.8198	.135	.290	.018	.069	.180
SciAtt	1	.294	1.0000	.543	.543	.294	.294	.000
Achieve	3	.493	.8198	.122	.630	.012	.325	.180
ExpEd	4	.286	.6679	.169	.229	.028	.039	.250
ExpOcc	2	.247	.8286	.131	.428	.014	.152	.171
FutEnrol	4	.100	.5399	.096	.135	.006	.013	.317

Inner Model Equations

```
Aptitude= HomeBack
SciValue= Aptitude Gender
Effort   = HomeBack Aptitude Gender
SciAtt   = SciValue
Achieve  = HomeBack Aptitude Gender
ExpEd    = HomeBack Aptitude SciValue Effort
ExpOcc   = Aptitude ExpEd
FutEnrol= Aptitude SciValue SciAtt   ExpEd
```

Direct Inner Model Effects

	HomeBack	Aptitude	Gender	SciValue	Effort	SciAtt	Achieve	ExpEd	ExpOcc	FutEnrol
HomeBack	.000	.000	.000	.000	.000	.000	.000	.000	.000	.000
Aptitude	.421	.000	.000	.000	.000	.000	.000	.000	.000	.000
Gender	.000	.000	.000	.000	.000	.000	.000	.000	.000	.000
SciValue	.000	.165	-.120	.000	.000	.000	.000	.000	.000	.000
Effort	.170	.290	.135	.000	.000	.000	.000	.000	.000	.000
SciAtt	.000	.000	.000	.543	.000	.000	.000	.000	.000	.000
Achieve	.122	.630	-.166	.000	.000	.000	.000	.000	.000	.000
ExpEd	.206	.229	.000	.169	.193	.000	.000	.000	.000	.000
ExpOcc	.000	.131	.000	.000	.000	.000	.000	.428	.000	.000
FutEnrol	.000	.117	.000	.135	.000	.096	.000	.116	.000	.000

Total Inner Model Effects

	HomeBack	Aptitude	Gender	SciValue	Effort	SciAtt	Achieve	ExpEd	ExpOcc	FutEnrol
HomeBack	1.000	.000	.000	.000	.000	.000	.000	.000	.000	.000
Aptitude	.421	1.000	.000	.000	.000	.000	.000	.000	.000	.000
Gender	.000	.000	1.000	.000	.000	.000	.000	.000	.000	.000
SciValue	.069	.165	-.120	1.000	.000	.000	.000	.000	.000	.000
Effort	.292	.290	.135	.000	1.000	.000	.000	.000	.000	.000
SciAtt	.038	.089	-.065	.543	.000	1.000	.000	.000	.000	.000
Achieve	.387	.630	-.166	.000	.000	.000	1.000	.000	.000	.000
ExpEd	.371	.313	.006	.169	.193	.000	.000	1.000	.000	.000
ExpOcc	.214	.265	.002	.072	.082	.000	.000	.428	1.000	.000
FutEnrol	.105	.184	-.022	.207	.022	.096	.000	.116	.000	1.000

Appendix 7.1 PLSPATH Output

Country: Australia (Population 2)
Input correlation matrix: between schools
Outcome variable: Achieve

LV-Correlations

	HomeBack	Gender	TchBack	TchGend	SchFacil	TchEmph	Aptitude	SciValue	Effort	SciAtt	Achieve
HomeBack	1.000										
Gender	-.086	1.000									
TchBack	.217	-.142	1.000								
TchGend	.152	.404	-.373	1.000							
SchFacil	.221	-.183	.286	-.229	1.000						
TchEmph	.179	.128	.153	.035	.330	1.000					
Aptitude	.714	.070	.245	.075	.137	.209	1.000				
SciValue	.260	-.045	.067	.026	.058	.092	.246	1.000			
Effort	.503	.194	-.058	.257	-.064	-.077	.572	.152	1.000		
SciAtt	.238	-.175	.043	.022	.030	.081	.257	.610	.251	1.000	
Achieve	.691	-.179	.331	-.116	.114	.133	.823	.203	.471	.231	1.000

Outer Model Summary

Variable	Mode	Status	N of MVs	AvgRed	AvgCom	MinCom	MaxCom
HomeBack	Inward	Exogen	3	-	.810	.787	.841
Gender	Unity	Exogen	1	-	1.000	1.000	1.000
TchBack	Inward	Exogen	2	-	.833	.676	.991
TchGend	Unity	Exogen	1	-	1.000	1.000	1.000
SchFacil	Unity	Endogen	1	.141	1.000	1.000	1.000
TchEmph	Inward	Endogen	1	.159	1.000	1.000	1.000
Aptitude	Inward	Endogen	2	.484	.901	.879	.924
SciValue	Inward	Endogen	2	.037	.610	.266	.953
Effort	Inward	Endogen	1	.458	1.000	1.000	1.000
SciAtt	Inward	Endogen	2	.299	.693	.446	.941
Achieve	Outward	Endogen	3	.650	.856	.840	.881

Outer Model Diagnostics

Variable	Mode	N of MVs	Status	Det(Block)	Min R**2	Max R**2
HomeBack	Inward	3	Exogen	.1933	.5601	.6293
Gender	Unity	1	Exogen	1.0000	.0000	.0000
TchBack	Inward	2	Exogen	.4168	.5832	.5832
TchGend	Unity	1	Exogen	1.0000	.0000	.0000
SchFacil	Unity	1	Endogen	1.0000	.0000	.0000
TchEmph	Inward	1	Endogen	1.0000	.0000	.0000
Aptitude	Inward	2	Endogen	.3519	.6481	.6481
SciValue	Inward	2	Endogen	.8986	.1014	.1014
Effort	Inward	1	Endogen	1.0000	.0000	.0000
SciAtt	Inward	2	Endogen	.7822	.2178	.2178
Achieve	Outward	3	Endogen	.1171	.6599	.7257

List of Manifest Variables (MVs)

MV2	No. of Science Laboratories	MV47	Beneficial Aspects of Science
MV12	Sex of Teacher	MV48	Interest in Science
MV13	Age of Teacher	MV50	Non-Harmful Aspects of Science
MV16	Years of Teaching Experience	MV52	Ease of Learning Science
MV21	Proportion Teaching Science	MV58	Hours Homework
MV34	Father's Education	MV62	Biology test score
MV37	Books in Home	MV63	Chemistry test score
MV40	Sex of Student	MV64	Physics test score
MV44	Word Knowledge test score	MV70	Father's Occupation
MV45	Mathematics test score		

Outer Model Equations

```
HomeBack= MV34  MV37  MV70          Aptitude= MV44  MV45
Gender  = MV40                      SciValue= MV47  MV50
TchBack = MV13  MV16                Effort  = MV58
TchGend = MV12                      SciAtt  = MV48  MV52
SchFacil= MV2                       Achieve = MV62  MV63  MV64
TchEmph = MV21
```

Outer Model Results

Variable	Weight	Loading	Communality	Redundancy	Tolerance
HomeBack	Inward	Exogen	3 MVs		
MV34	.4068	.9171	.8411	.0000	.6293
MV37	.3911	.8947	.8006	.0000	.5601
MV70	.3121	.8873	.7874	.0000	.6134
Gender	Unity	Exogen	1 MVs		
MV40	1.0000	1.0000	1.0000	.0000	.0000
TchBack	Inward	Exogen	2 MVs		
MV13	.1489	.8222	.6760	.0000	.5832
MV16	.8816	.9954	.9908	.0000	.5832
TchGend	Unity	Exogen	1 MVs		
MV12	1.0000	1.0000	1.0000	.0000	.0000
SchFacil	Unity	Endogen	1 MVs		
MV2	1.0000	1.0000	1.0000	.1408	.0000
TchEmph	Inward	Endogen	1 MVs		
MV21	1.0000	1.0000	1.0000	.1594	.0000
Aptitude	Inward	Endogen	2 MVs		
MV44	.5862	.9611	.9237	.4962	.6481
MV45	.4657	.9376	.8791	.4723	.6481
SciValue	Inward	Endogen	2 MVs		
MV47	.9036	.9763	.9532	.0575	.1014
MV50	.2283	.5160	.2663	.0161	.1014
Effort	Inward	Endogen	1 MVs		
MV58	1.0000	1.0000	1.0000	.4582	.0000
SciAtt	Inward	Endogen	2 MVs		
MV48	.8417	.9700	.9409	.4059	.2178
MV52	.2748	.6677	.4458	.1923	.2178
Achieve	Outward	Endogen	3 MVs		
MV62	.3237	.9167	.8404	.6379	.6939
MV63	.3828	.9388	.8813	.6689	.7257
MV64	.3739	.9198	.8460	.6421	.6599

Inner Model Diagnostics and Summary

Variable	N Pred	R-square	Det(Pred)	MinBeta	MaxBeta	MinDlta	MaxDlta	MaxTol
SchFacil	3	.141	.7661	.164	.216	.021	.042	.216
TchEmph	3	.159	.9173	.121	.340	.014	.107	.076
Aptitude	3	.537	.7954	-.111	.746	.010	.530	.199
SciValue	1	.060	1.0000	.246	.246	.060	.060	.000
Effort	5	.458	.3960	-.167	.449	.025	.092	.542
SciAtt	3	.431	.9344	-.188	.571	.034	.316	.064
Achieve	4	.759	.3681	-.157	.707	.014	.231	.557

Inner Model Equations

SchFacil=	HomeBack	TchBack	TchGend		
TchEmph =	HomeBack	Gender	SchFacil		
Aptitude=	HomeBack	Gender	TchGend		
SciValue=	Aptitude				
Effort =	HomeBack	Gender	TchBack	TchEmph	Aptitude
SciAtt =	Gender	SciValue	Effort		
Achieve =	HomeBack	Gender	TchGend	Aptitude	

Direct Inner Model Effects

	HomeBack	Gender	TchBack	TchGend	SchFacil	TchEmph	Aptitude	SciValue	Effort	SciAtt	Achieve
HomeBack	.000	.000	.000	.000	.000	.000	.000	.000	.000	.000	.000
Gender	.000	.000	.000	.000	.000	.000	.000	.000	.000	.000	.000
TchBack	.000	.000	.000	.000	.000	.000	.000	.000	.000	.000	.000
TchGend	.000	.000	.000	.000	.000	.000	.000	.000	.000	.000	.000
SchFacil	.216	.000	.164	-.201	.000	.000	.000	.000	.000	.000	.000
TchEmph	.121	.200	.000	.000	.340	.000	.000	.000	.000	.000	.000
Aptitude	.746	.179	.000	-.111	.000	.000	.000	.000	.000	.000	.000
SciValue	.000	.000	.000	.000	.000	.000	.246	.000	.000	.000	.000
Effort	.275	.191	-.167	.000	.000	-.218	.449	.000	.000	.000	.000
SciAtt	.000	-.188	.000	.000	.000	.000	.000	.571	.201	.000	.000
Achieve	.193	-.157	.000	-.136	.000	.000	.707	.000	.000	.000	.000

Total Inner Model Effects

	HomeBack	Gender	TchBack	TchGend	SchFacil	TchEmph	Aptitude	SciValue	Effort	SciAtt	Achieve
HomeBack	1.000	.000	.000	.000	.000	.000	.000	.000	.000	.000	.000
Gender	.000	1.000	.000	.000	.000	.000	.000	.000	.000	.000	.000
TchBack	.000	.000	1.000	.000	.000	.000	.000	.000	.000	.000	.000
TchGend	.000	.000	.000	1.000	.000	.000	.000	.000	.000	.000	.000
SchFacil	.216	.000	.164	-.201	1.000	.000	.000	.000	.000	.000	.000
TchEmph	.194	.200	.056	-.068	.340	1.000	.000	.000	.000	.000	.000
Aptitude	.746	.179	.000	-.111	.000	.000	1.000	.000	.000	.000	.000
SciValue	.183	.044	.000	-.027	.000	.000	.246	1.000	.000	.000	.000
Effort	.567	.227	-.180	-.035	-.074	-.218	.449	.000	1.000	.000	.000
SciAtt	.219	-.118	-.036	-.022	-.015	-.044	.230	.571	.201	1.000	.000
Achieve	.721	-.030	.000	-.214	.000	.000	.707	.000	.000	.000	1.000

Appendix 7.2 PLSPATH Output

Country: Australia (Population 2)
Input correlation matrix: between schools
Outcome variable: FutEnrol

LV-Correlations

	HomeBack	Gender	TchBack	TchGend	SchFacil	TchEmph	Aptitude	SciValue	Effort	SciAtt	Achieve	ExpEd	ExpOcc	FutEnrol
HomeBack	1.000													
Gender	-.095	1.000												
TchBack	.213	-.142	1.000											
TchGend	.162	.404	-.373	1.000										
SchFacil	.222	-.183	.286	-.229	1.000									
TchEmph	.175	.128	.153	.035	.330	1.000								
Aptitude	.710	.074	.242	.078	.139	.208	1.000							
SciValue	.258	-.105	.049	.016	.055	.075	.220	1.000						
Effort	.501	.194	-.058	.257	-.064	-.077	.573	.133	1.000					
SciAtt	.242	-.177	.046	.019	.031	.081	.254	.613	.250	1.000				
Achieve	.686	-.179	.331	-.116	.114	.133	.821	.190	.471	.235	1.000			
ExpEd	.690	-.046	.084	.180	.194	.052	.625	.240	.623	.294	.534	1.000		
ExpOcc	.602	.095	.030	.216	.104	.120	.602	.206	.506	.261	.476	.655	1.000	
FutEnrol	.336	-.094	.133	.125	.187	.299	.349	.228	.137	.255	.287	.287	.283	1.000

Outer Model Summary

Variable	Mode	Status	N of MVs	AvgRed	AvgCom	MinCom	MaxCom
HomeBack	Inward	Exogen	3	-	.807	.725	.852
Gender	Unity	Exogen	1	-	1.000	1.000	1.000
TchBack	Inward	Exogen	2	-	.833	.676	.991
TchGend	Unity	Exogen	1	-	1.000	1.000	1.000
SchFacil	Unity	Endogen	1	.143	1.000	1.000	1.000
TchEmph	Inward	Endogen	1	.159	1.000	1.000	1.000
Aptitude	Inward	Endogen	2	.483	.903	.899	.906
SciValue	Inward	Exogen	1	-	1.000	1.000	1.000
Effort	Inward	Endogen	1	.458	1.000	1.000	1.000
SciAtt	Inward	Endogen	2	.299	.699	.466	.931
Achieve	Outward	Endogen	3	.647	.856	.840	.881
ExpEd	Unity	Endogen	1	.597	1.000	1.000	1.000
ExpOcc	Unity	Endogen	1	.500	1.000	1.000	1.000
FutEnrol	Unity	Endogen	1	.241	1.000	1.000	1.000

Outer Model Diagnostics

Variable	Mode	N of MVs	Status	Det(Block)	Min R**2	Max R**2
HomeBack	Inward	3	Exogen	.1933	.5601	.6293
Gender	Unity	1	Exogen	1.0000	.0000	.0000
TchBack	Inward	2	Exogen	.4168	.5832	.5832
TchGend	Unity	1	Exogen	1.0000	.0000	.0000
SchFacil	Unity	1	Endogen	1.0000	.0000	.0000
TchEmph	Inward	1	Endogen	1.0000	.0000	.0000
Aptitude	Inward	2	Endogen	.3519	.6481	.6481
SciValue	Inward	1	Exogen	1.0000	.0000	.0000
Effort	Inward	1	Endogen	1.0000	.0000	.0000
SciAtt	Inward	2	Endogen	.7822	.2178	.2178
Achieve	Outward	3	Endogen	.1171	.6599	.7257
ExpEd	Unity	1	Endogen	1.0000	.0000	.0000
ExpOcc	Unity	1	Endogen	1.0000	.0000	.0000
FutEnrol	Unity	1	Endogen	1.0000	.0000	.0000

List of Manifest Variables (MVs)

MV2	No. of Science Laboratories	MV47	Beneficial Aspects of Science
MV12	Sex of Teacher	MV48	Interest in Science
MV13	Age of Teacher	MV52	Ease of Learning Science
MV16	Years of Teaching Experience	MV58	Hours Homework
MV21	Proportion Teaching Science	MV62	Biology test score
MV34	Father's Education	MV63	Chemistry test score
MV37	Books in Home	MV64	Physics test score
MV40	Sex of Student	MV70	Father's Occupation
MV43	Post-Science Education	MV72	Expected Occupation
MV44	Word Knowledge test score	MV73	Expected Education
MV45	Mathematics test score		

Outer Model Equations

```
HomeBack= MV34  MV37  MV70
Gender  = MV40
TchBack = MV13  MV16
TchGend = MV12
SchFacil= MV2
TchEmph = MV21
Aptitude= MV44  MV45
SciValue= MV47
Effort  = MV58
SciAtt  = MV48  MV52
Achieve = MV62  MV63  MV64
ExpEd   = MV73
ExpOcc  = MV72
FutEnrol= MV43
```

Outer Model Results

Variable	Weight	Loading	Communality	Redundancy	Tolerance
HomeBack	Inward	Exogen	3 MVs		
MV34	.4231	.9232	.8522	.0000	.6293
MV37	.2599	.8513	.7248	.0000	.5601
MV70	.4223	.9190	.8445	.0000	.6134
Gender	Unity	Exogen	1 MVs		
MV40	1.0000	1.0000	1.0000	.0000	.0000
TchBack	Inward	Exogen	2 MVs		
MV13	.1489	.8222	.6760	.0000	.5832
MV16	.8816	.9954	.9908	.0000	.5832
TchGend	Unity	Exogen	1 MVs		
MV12	1.0000	1.0000	1.0000	.0000	.0000
SchFacil	Unity	Endogen	1 MVs		
MV2	1.0000	1.0000	1.0000	.1425	.0000
TchEmph	Inward	Endogen	1 MVs		
MV21	1.0000	1.0000	1.0000	.1588	.0000
Aptitude	Inward	Endogen	2 MVs		
MV44	.5348	.9517	.9056	.4850	.6481
MV45	.5178	.9483	.8994	.4816	.6481
SciValue	Inward	Exogen	1 MVs		
MV47	1.0000	1.0000	1.0000	.0000	.0000
Effort	Inward	Endogen	1 MVs		
MV58	1.0000	1.0000	1.0000	.4582	.0000
SciAtt	Inward	Endogen	2 MVs		
MV48	.8262	.9649	.9310	.3983	.2178
MV52	.2970	.6827	.4660	.1994	.2178
Achieve	Outward	Endogen	3 MVs		
MV62	.3228	.9165	.8401	.6348	.6939
MV63	.3832	.9388	.8814	.6660	.7257
MV64	.3743	.9199	.8462	.6394	.6599
ExpEd	Unity	Endogen	1 MVs		
MV73	1.0000	1.0000	1.0000	.5973	.0000
ExpOcc	Unity	Endogen	1 MVs		
MV72	1.0000	1.0000	1.0000	.4998	.0000
FutEnrol	Unity	Endogen	1 MVs		
MV43	1.0000	1.0000	1.0000	.2405	.0000

Inner Model Diagnostics and Summary

Variable	N Pred	R-square	Det(Pred)	MinBeta	MaxBeta	MinDlta	MaxDlta	MaxTol
SchFacil	3	.143	.7639	.163	.221	.021	.043	.215
TchEmph	3	.159	.9160	.118	.341	.013	.107	.076
Aptitude	3	.536	.7889	-.122	.748	.012	.527	.204
Effort	5	.458	.3994	-.165	.451	.024	.094	.539
SciAtt	3	.428	.9281	-.156	.570	.023	.313	.062
Achieve	4	.756	.3664	-.156	.707	.015	.232	.558
ExpEd	4	.597	.2943	.112	.401	.006	.082	.568
ExpOcc	3	.500	.2416	.156	.394	.010	.076	.604
FutEnrol	5	.241	.6941	-.192	.264	.015	.062	.219

Inner Model Equations

```
================================================================================
SchFacil= HomeBack TchBack   TchGend
TchEmph  = HomeBack Gender   SchFacil
Aptitude= HomeBack Gender   TchGend
Effort   = HomeBack Gender   TchBack  TchEmph  Aptitude
SciAtt   = Gender   SciValue Effort
Achieve  = HomeBack Gender   TchGend  Aptitude
ExpEd    = HomeBack SchFacil Aptitude Effort
ExpOcc   = HomeBack Aptitude ExpEd
FutEnrol= Gender   TchGend  TchEmph  Aptitude SciAtt
================================================================================
```

Direct Inner Model Effects

	HomeBack	Gender	TchBack	TchGend	SchFacil	TchEmph	Aptitude	SciValue	Effort	SciAtt	Achieve	ExpEd	ExpOcc	FutEnrol
HomeBack	.000	.000	.000	.000	.000	.000	.000	.000	.000	.000	.000	.000	.000	.000
Gender	.000	.000	.000	.000	.000	.000	.000	.000	.000	.000	.000	.000	.000	.000
TchBack	.000	.000	.000	.000	.000	.000	.000	.000	.000	.000	.000	.000	.000	.000
TchGend	.000	.000	.000	.000	.000	.000	.000	.000	.000	.000	.000	.000	.000	.000
SchFacil	.221	.000	.163	-.205	.000	.000	.000	.000	.000	.000	.000	.000	.000	.000
TchEmph	.118	.201	.000	.000	.341	.000	.000	.000	.000	.000	.000	.000	.000	.000
Aptitude	.748	.194	.000	-.122	.000	.000	.000	.000	.000	.000	.000	.000	.000	.000
SciValue	.000	.000	.000	.000	.000	.000	.000	.000	.000	.000	.000	.000	.000	.000
Effort	.272	.191	-.165	.000	.000	-.217	.451	.000	.000	.000	.000	.000	.000	.000
SciAtt	.000	-.156	.000	.000	.000	.000	.000	.570	.204	.000	.000	.000	.000	.000
Achieve	.192	-.156	.000	-.139	.000	.000	.707	.000	.000	.000	.000	.000	.000	.000
ExpEd	.401	.000	.000	.000	.112	.000	.117	.000	.363	.000	.000	.000	.000	.000
ExpOcc	.156	.000	.000	.000	.000	.000	.246	.000	.000	.000	.000	.394	.000	.000
FutEnrol	.000	-.192	.000	.171	.000	.252	.264	.000	.000	.130	.000	.000	.000	.000

Total Inner Model Effects

	HomeBack	Gender	TchBack	TchGend	SchFacil	TchEmph	Aptitude	SciValue	Effort	SciAtt	Achieve	ExpEd	ExpOcc	FutEnrol
HomeBack	1.000	.000	.000	.000	.000	.000	.000	.000	.000	.000	.000	.000	.000	.000
Gender	.000	1.000	.000	.000	.000	.000	.000	.000	.000	.000	.000	.000	.000	.000
TchBack	.000	.000	1.000	.000	.000	.000	.000	.000	.000	.000	.000	.000	.000	.000
TchGend	.000	.000	.000	1.000	.000	.000	.000	.000	.000	.000	.000	.000	.000	.000
SchFacil	.221	.000	.163	-.205	1.000	.000	.000	.000	.000	.000	.000	.000	.000	.000
TchEmph	.193	.201	.055	-.070	.341	1.000	.000	.000	.000	.000	.000	.000	.000	.000
Aptitude	.748	.194	.000	-.122	.000	.000	1.000	.000	.000	.000	.000	.000	.000	.000
SciValue	.000	.000	.000	.000	.000	.000	.000	1.000	.000	.000	.000	.000	.000	.000
Effort	.567	.235	-.177	-.040	-.074	-.217	.451	.000	1.000	.000	.000	.000	.000	.000
SciAtt	.116	-.108	-.036	-.008	-.015	-.044	.092	.570	.204	1.000	.000	.000	.000	.000
Achieve	.720	-.019	.000	-.225	.000	.000	.707	.000	.000	.000	1.000	.000	.000	.000
ExpEd	.719	.108	-.046	-.052	.085	-.079	.280	.000	.363	.000	.000	1.000	.000	.000
ExpOcc	.623	.090	-.018	-.050	.033	-.031	.356	.000	.143	.000	.000	.394	1.000	.000
FutEnrol	.261	-.104	.009	.120	.084	.247	.276	.074	.027	.130	.000	.000	.000	1.000

Appendix 8.1 HLM-2 Output

Country:	Australia (Population 2)
Sub-population	Coeducational Schools
Outcome variable:	Achieve

```
  SPECIFICATIONS FOR THIS MIXED-MODEL  RUN OF HLM OF Wed Apr 03 08:28:56 1991
 -----------------------------------------------------------------------------

  PROBLEM TITLE: AUSTRALIA COED RUN13: ACHIEVE

  THE DATA SOURCE FOR THIS RUN =                          aus2ssm
  OUTPUT FILE NAME IS          = AUSMIX13
  THE MAXIMUM NUMBER OF UNITS =  188
  THE MAXIMUM NUMBER OF ITERATIONS =    5


  THE HLM MODEL SPECIFIED FOR THE FIXED EFFECTS WAS:
 _________________________________________________

      WITHIN-UNIT     BETWEEN-UNIT
       VARIABLES        VARIABLES
      _________      ___________
            BASE              BASE
                          HOMEBACK
                          REGION
                          APTITUDE
                          EFFORT
       *HOMEBACK              BASE
       *APTITUDE              BASE
       *GENDER                BASE
       *SCIVALUE              BASE
       *EFFORT                BASE
       *SCIATT                BASE
  '*' - THIS WITHIN-UNIT VARIABLE HAS BEEN CENTERED AROUND ITS UNIT MEAN

 THE HLM MODEL SPECIFIED FOR THE COVARIANCE COMPONENTS WAS:
 __________________________________________________________

         SIGMA SQUARED (CONSTANT ACROSS UNITS)

         TAU DIMENSIONS
                   BASE
               HOMEBACK
               APTITUDE
               GENDER
               SCIVALUE
               EFFORT
               SCIATT


 PARAMETER      RELIABILITY ESTIMATES:
 _________    _____________________________
      BASE    TAU(1,1) / D_BAR(1,1) =  .605
  HOMEBACK    TAU(2,2) / D_BAR(2,2) =  .067
  APTITUDE    TAU(3,3) / D_BAR(3,3) =  .046
  GENDER      TAU(4,4) / D_BAR(4,4) =  .151
  SCIVALUE    TAU(5,5) / D_BAR(5,5) =  .003
  EFFORT      TAU(6,6) / D_BAR(6,6) =  .143
  SCIATT      TAU(7,7) / D_BAR(7,7) =  .102

 NOTE:    THESE VALUES ARE BASED ON ONLY 186 OF 188 UNITS THAT
          HAD SUFFICIENT DATA FOR COMPUTATION
```

THE GAMMA(*)-STANDARD ERROR-T STATISTIC TABLE:

	GAMMA(*)	STANDARD ERROR	T STATISTIC	p-VALUE
FOR BASE COEF.				
BASE	.011765	.018802	.626	.531
HOMEBACK	.124909	.024692	5.059	.000
REGION	.071215	.018794	3.789	.000
APTITUDE	.231657	.024066	9.626	.000
EFFORT	.072580	.027091	2.679	.008
FOR HOMEBACK SLOPE				
BASE	.101894	.013343	7.637	.000
FOR APTITUDE SLOPE				
BASE	.519316	.013796	37.641	.000
FOR GENDER SLOPE				
BASE	-.143525	.013666	-10.502	.000
FOR SCIVALUE SLOPE				
BASE	.094078	.013081	7.192	.000
FOR EFFORT SLOPE				
BASE	.092906	.016330	5.689	.000
FOR SCIATT SLOPE				
BASE	.071366	.014633	4.877	.000

THE CHI SQUARE TABLE:

PARAMETER	ESTIMATED PARAMETER VARIANCE	DEGREES OF FREEDOM	CHI SQUARE	P-VALUE
BASE COEF.	.03602	181	456.84	.000
HOMEBACK SLOPE	.00334	185	214.65	.067
APTITUDE SLOPE	.00288	185	208.16	.117
GENDER SLOPE	.00765	185	194.03	.310
SCIVALUE SLOPE	.00016	185	196.66	.265
EFFORT SLOPE	.01195	185	193.54	.318
SCIATT SLOPE	.00638	185	196.91	.261

NOTE: THESE VALUES ARE BASED ON ONLY 186 OF 188 UNITS THAT HAD SUFFICIENT DATA FOR COMPUTATION

STATISTICS FOR CURRENT VARIANCE COMPONENTS MODEL

DEVIANCE = .867772E+04
NUMBER OF ESTIMATED PARAMETERS = 29

******** END OF OUTPUT ********

Zusammenfassung

Die vorliegende Dissertation befaßt sich in einem internationalen Vergleich von 10 Ländern mit den geschlechtsspezifischen Unterschieden in den naturwissenschaftlichen Leistungen von Schülern verschiedener Altersstufen.

Grundlage der Untersuchung bilden Daten, die die *International Association for the Evaluation for Educational Achievement (IEA)* im Rahmen zweier internationaler Studien, der *First IEA Science Study (FISS)* und *Second IEA Science Study (SISS)*, weltweit erhoben hat. Die Daten der FISS-Studie wurden in den Jahren 1970/71 und die der SISS-Studie in den Jahren 1983/84 gesammelt.

Die Auswahl der in diese Arbeit einbezogenen Länder ergibt sich daraus, daß seinerzeit nur 10 Länder an beiden IEA-Untersuchungen teilgenommen haben. Dies sind:

- ☞ Australien,
- ☞ Finnland,
- ☞ Japan,
- ☞ Schweden,
- ☞ Ungarn und
- ☞ England,
- ☞ Italien,
- ☞ Niederlande,
- ☞ Thailand,
- ☞ die USA.

Um auch einen Längsschnittvergleich der naturwissenschaftlichen Leistungen von Bildungssystemen zu erhalten, was Schlußfolgerungen über curriculare bzw. bildungspolitische Maßnahmen zuläßt, bleibt die Datenanalyse in der vorliegenden Dissertation auf die genannten 10 Länder beschränkt.

Nähere Angaben zur Planung, zur Durchführung und zu den allgemeinen Ergebnissen dieser beiden IEA-Forschungsprojekte finden sich in folgenden Monographien:

FISS: Comber, L. C. and Keeves, J. P. (1973) Science Education in Nineteen Countries. International Studies in Evaluation I. Almqvist and Wiksell, Stockholm.

SISS - Volume 1: Rosier, M. J. and Keeves, J. P. (1991) Science Education and Curricula in Twenty-Three Countries. Pergamon Press, Oxford.

SISS - Volume 2: Postlethwaite, T. N. and Wiley, D. E. (1991) Science Achievement in Twenty-Three Countries. Pergamon Press, Oxford.

SISS - Volume 3: Keeves, J. P. (1991) Changes in Science Education and Achievement, 1970-1984. Pergamon Press, Oxford.

Nach einer grundlegenden Einführung in die aktuelle Problematik geschlechtsspezifischer Unterschiede in naturwissenschaftlichen Leistungen (**Kapitel 1**) wird ein Abriß des gegenwärtigen internationalen Forschungsstandes in einer ausführlichen Analyse der vorhandenen Literatur gegeben (**Kapitel 2**).

Es wurde für notwendig erachtet, in kurzen Zügen die Konzeptualisierung der beiden IEA-Studien aus den Jahren 1970/71 (FISS) und 1983/84 (SISS) zu beschreiben, da die beiden internationalen, naturwissenschaftlichen Studien die Datenbasis dieser Arbeit bilden (**Kapitel 3**). Dargestellt werden u.a. das verwendete Testinstrumentarium bzw. Probleme der Stichprobenziehung, Stichprobenfehlerberechnung und Gewichtung.

Die methodologischen Schwierigkeiten bei der Untersuchung geschlechtsspezifischer Unterschiede in naturwissenschaftlichen Leistungen von Schülern verschiedener Altersstufen werden in **Kapitel 4** erörtert. Hier werden auch die komplexen multivariaten Verfahren, *Partial Least Squares (PLS)* und *Hierarchical Linear Modelling (HLM)*, die maßgeblich zur Analyse der Daten eingesetzt wurden, eingehend erläutert und ihre statistischen und methodologischen Rahmenbedigungen mit anderen Analysetechniken verglichen.

Die Ergebnisse der Analysen werden in den folgenden vier Kapiteln dargestellt (Kapitel 5 bis 8). **Kapitel 5** gibt einen umfassenden Überblick der geschlechtsspezifischen Unterschiede in bezug auf die naturwissenschaftlichen Leistungen für alle drei Altersgruppen in allen 10 Ländern getrennt für die Fächer Biologie, Chemie, und Physik sowie die beiden Meßzeitpunkte 1970/71 und 1983/84.

Im ersten Teil von **Kapitel 6** werden Ergebnisse zu den Einstellungen und Wertvorstellungen von Schülern im Hinblick auf die Naturwissenschaften innerhalb und außerhalb der Schule präsentiert. Gleichzeitig werden naturwissenschaftliche Einstellungen in Relation zu den gezeigten Leistungen gesetzt. Im Anschluß daran wird ein Pfadmodell zur Erklärung der naturwissenschaftlichen Leistungen bei vierzehnjährigen Schülern entworfen und getestet, das die relevanten Einflußfaktoren auf Schülerebene in allen 10 Ländern aufzeigt. Nach der Entwicklung eines über alle Länder 'gemittelten' Pfadmodells (average path model) für die naturwissenschaftliche Leistung auf Schülerebene wird dieses Modell erweitert, um die Faktoren zu bestimmen, die das Interesse von Jungen und Mädchen beeinflussen, auch in der Sekundarstufe II bei entsprechender Wahlfreiheit am naturwissenschaftlichen Unterricht teilzunehmen.

In **Kapitel 7** werden die Ergebnisse von PLS-Analysen beschrieben, die zusätzlich zu den Schüler-relevanten Faktoren auch Einflüsse seitens der Lehrer (u.a. Geschlecht, Ausbildungsniveau, Unterrichtsmethode) und der Schule (u.a. Anzahl der vorhandenen Laboratorien oder Praktikumsräume) auf die Ausprägung von geschlechtsspezifischen Unterschieden in Betracht ziehen. Diese auf Schulebene durchgeführten Pfadanalysen führen zu interessanten Ergebnissen, die jeweils mit einem kurzen, zusammenfassenden Kommentar für jedes Land abgeschlossen werden. In Anlehnung an das erweiterte Pfadmodell, das Faktoren für das von Jungen und Mädchen geäußerte Interesse an weiteren naturwissenschaftlichen Kursen in der Sekundarstufe II bereits in Kapitel 6 aufgezeigt hat, werden nun auch Lehrer- und Schulvariablen in ein erweitertes

Pfadmodell für Vergleiche zwischen den Schulen eines Landes eingearbeitet und analysiert.

Die Analysen werden mit einem Vergleich von Jungen- und Mädchenschulen zu Schulen mit Koedukation in Australien und England beendet (**Kapitel 8**). Dieses waren die beiden einzigen Länder, in denen es während der Datenerhebung der SISS-Studie (1983/84) noch eine nennenswerte Anzahl von Sekundarschulen gab, die nur von Jungen oder nur von Mädchen besucht wurden. Der Vergleich der drei Schulformen - Koedukationsschulen, Schulen für Jungen, Schulen für Mädchen - mit Blick auf geschlechtsspezifische Unterschiede in naturwissenschaftlichen Leistungen wurde mit Hilfe eines in der Bundesrepublik Deutschland bisher noch nicht eingesetzten multivariaten Mehrebenenverfahrens, HLM, durchgeführt. Diese neuartige Analysetechnik gestattet das simultane Abschätzen von Einflußfaktoren auf Schüler- **und** Schulebene, was zur Ausschaltung ungewollter Aggregations- oder Disaggregationseffekte führte.

Die vorliegende Dissertation endet mit einer Zusammenfassung der wichtigsten Forschungsergebnisse und ordnet diese in den Zusammenhang des aktuellen internationalen Standes der Forschung im Hinblick auf geschlechtsspezifische Unterschiede in naturwissenschaftlichen Leistungen auf internationaler Ebene ein (**Kapitel 9**). Abschließend wird ein Ausblick auf zukünftige Forschungsschwerpunkte in diesem Bereich gegeben.

Im folgenden wird ein kurzer Überblick zu den wichtigsten Ergebnissen der Untersuchung geliefert.

Die vorliegende Arbeit bietet eine Reihe von neuen und interessanten Ergebnissen zur Problematik von Einflußfaktoren auf die geschlechtsspezifischen Unterschiede von Schülern verschiedener Altersstufen in bezug auf naturwissenschaftliche Leistungen. Dies wurde nicht zuletzt dadurch ermöglicht, daß mit Hilfe komplexer multivariater Verfahren Daten zweier großer internationaler Studien, die von der IEA in den Jahren 1970/71 und 1983/84 durchgeführt wurden, miteinander verglichen werden konnten.

Zum ersten Mal wurden die vorhandenen Daten von Schülern, Lehrern und Schulen mit dem multivariaten Verfahren PLS und dem Mehrebenenverfahren HLM daraufhin untersucht, welche Faktoren maßgeblichen Einfluß auf die Ausprägung von geschlechtsspezifischen Unterschieden bei naturwissenschaftlichen Leistungen in den 10 Ländern haben und welche Faktoren das Interesse von Jungen und Mädchen bestimmen, im Laufe ihrer Schulausbildung naturwissenschaftliche Kurse, sofern dies möglich ist, zu wählen.

Die Ergebnisse zeigen deutlich, daß geschlechtsspezifische Unterschiede in allen untersuchten Ländern nachweisbar waren. Dennoch gab es markante Unterschiede zwischen...

☞ den Analyseebenen (Schülerebene, Schulebene);
☞ den drei untersuchten Altersstufen;
☞ den Fächern Biologie, Chemie und Physik;
☞ den verschiedenen Schulformen; und
☞ den 10 Ländern, die in die Untersuchung einbezogen waren.

Ein weiteres entscheidendes Ergebnis war, daß geschlechtsspezifische Leistungsunterschiede in aller Regel zugunsten von Jungen oder zugunsten von Klassen, in denen die Jungen zahlenmäßig überwogen, ausfielen. Dies bestätigt frühere Resultate von Comber und Keeves (1973) in ihrer Analyse der FISS-Daten.

Ein anderer interessanter Aspekt, der ebenfalls bereits von Comber und Keeves beschrieben wurde, war, daß Leistungsunterschiede zwischen Jungen und Mädchen mit zunehmendem Alter immer gravierender wurden. Die vorliegenden Ergebnisse dokumentieren dies in aller Deutlichkeit.

Weiterhin sprechen die Resultate dafür, daß die geschlechtsspezifischen Unterschiede in Physik größer als in Chemie sind. Im Vergleich aller 10 Länder ergaben sich die geringsten Unterschiede in Biologie. Darüber hinaus wurde deutlich, wie sehr sich diese leistungsbezogenen Geschlechtsunterschiede in den Jahren seit der ersten Datenerhebung in 1970/71 bis 1983/84 verändert haben. Im allgemeinen konnte ein deutlicher Rückgang der geschlechtsspezifischen Unterschiede in allen Altersstufen festgestellt werden. Diese positiven Veränderungen waren für alle drei Fächer Biologie, Chemie und Physik gültig.

Die Ergebnisse der pfadanalytischen Modelle auf Schüler- und Schulebene ermöglichten, die Bedeutung von Einstellungen und Wertvorstellungen im Hinblick auf Naturwissenschaften in ihrem Effekt auf die gezeigte Leistung zu erkennen. In der Tat scheinen diese geschlechtstypischen Unterschiede in den Einstellungen von Jungen und Mädchen die Leistung der Schüler durchweg bei allen Altersgruppen und in allen Ländern zu bestimmen. Diese Beobachtung erweitert Erkenntnisse einer durch Duncan (1989) in Botswana durchgeführten Studie und zeigt die Anwendbarkeit auch auf industrialisierte Länder wie isnbesondere Finnland, die Niederlande, Schweden oder die USA.

Trotz der Tatsache, daß in den meisten Ländern eine ganze Reihe von latenten Variablen gefunden werden konnten, die einen signifikanten Effekt auf das Leistungsniveau von Jungen und Mädchen im naturwissenschaftlichen Unterricht haben, zeigte sich, daß es der sozioökonomische Hintergrund des Schülers, sein Können und sein Geschlecht waren - und zwar in genau dieser Reihenfolge -, die am meisten zu geschlechtsspezifischen Unterschieden bei Vierzehnjährigen beitrugen. Interessant hierbei war der überraschend starke Einfluß des Geschlechts eines Schülers auf seine naturwissenschaftlichen Leistungen. Es wäre allerdings unwissenschaftlich daraus zu schließen, daß Jungen bzw. Klassen, in denen Jungen dominieren, einen Leistungsvorsprung **aufgrund** ihres Geschlechts haben. Mit anderen Worten, Jungen zeigen keine besseren Leistungen als Mädchen in den naturwissenschaftlichen Fächern, weil sie als Jungen auf die Welt kamen, sondern weil sie als Jungen eine andere geschlechtstypische Sozialisation, Enkulturation und Elementarausbildung erfahren haben als Mädchen (Maccoby and Jacklin; 1974; Kelly, 1978; Marini and Greenberger, 1978; Keeves, 1991; Keeves and Kotte, 1991). Es sind genau diese latenten Variablen, die nicht in den Pfadmodellen erfaßt werden, die das Geschlecht so wichtig für das Leistungsniveau in den Naturwissenschaften in vielen Ländern werden ließen.

Trotz der zum Teil großen landesspezifischen Unterschiede in bezug auf die Einflußgrößen, die bestimmend für die gezeigte Leistung von Jungen und Mädchen ist, wurde deutlich, daß auch zum Zeitpunkt der zweiten naturwissenschaftlichen IEA-Studie 1983/84 die Frage von geschlechtsspezifischen Unterschieden von entscheidender Bedeutung war.

Dies veranlaßte eine eingehendere Analyse Schultyp-spezifischer Effekte in Schulen für Jungen, Schulen für Mädchen und Koedukationsschulen in Australien und England. Obwohl die leistungsbestimmenden Merkmale in den drei Schulformen (nur Jungen; nur Mädchen; Jungen und Mädchen) Parallelen zwischen den Schulsystemen Australiens und Englands zuließen, sind weitere Untersuchungen notwendig, bevor eine weitergehende Generalisierung auch auf andere Schulsysteme möglich ist.

Trotz einiger theoretischer und methodologischer Einschränkungen dürfte die vorliegende Arbeit mit ihren Ergebnissen neue Aspekte und Forschungsgesichtspunkte aufgezeigt haben, die für die Bewertung der Problematik geschlechtsspezifischer Unterschiede in naturwissenschaftlichen Leistungen bei Sekundarstufenschülern, auch in den nun 16 Ländern der Bundesrepublik Deutschland von Relevanz sind.

Auch wenn dies weitere Forschungsarbeiten notwendig macht, um die Mechanismen, die zu leistungsbezogenen Unterschieden zwischen Jungen und Mädchen verschiedener Altersstufen führen, besser zu verstehen, darf es doch als positives Zeichen gewertet werden, daß die vorliegenden Ergebnisse auch deutlich werden lassen, daß sich geschlechtsspezifische Unterschiede zwischen 1970 und 1984 in den untersuchten Ländern im allgemeinen deutlich verringert haben. Dieses ist das wohl ermutigendste Ergebnis eines komplexen internationalen Vergleichs von Schüler-, Lehrer- und Schuldaten.

EMPIRISCHE SCHUL- UND UNTERRICHTSFORSCHUNG

herausgegeben von Manfred Herbig

In dieser Reihe werden empirische Arbeiten veröffentlicht, die inhaltliche Fragestellungen zu Schule und Unterricht verfolgen. Die Bereiche beschränken sich nicht nur auf angewandte und Feldforschung, sondern beziehen auch Grundlagenstudien und Experimente aus den Bereichen Lernen, Entwicklung und gesellschaftliche Randbedingungen mit ein. Außer diesem Band erschien bisher als

Band 1 Manfred Herbig: Anatomie einer Ganztagsschule. Abschlußbericht einer wissenschaftlichen Begleituntersuchung an Eltern, Schülern und Lehrern in Mülheim an der Ruhr. 1984.

Band 2 Michael Kerres: Prüfungsangst und -bewältigung. Eine Untersuchung zu selbstregulativen Aktivitäten im Prüfungsgeschehen. 1988.

Band 3 Helmut Mund: Verständnis chemischer Symbole. Eine Untersuchung von Lehrstrategien, Lernverhalten und funktionalem Denken in Grund- und Erweiterungskursen der 9. Jahrgangsstufe der Integrierten Gesamtschule unter Verwendung der probabilistischen Testtheorie von Rasch. 1990.

Band 4 Mohamed Mahrouse: Effects of Education on Job Levels, Earnings, Vocational Aspirations, and Job Satisfaction of Industrial Workers. An Analytical-Empirical Study on Economic Benefits of Education. 1990.

Band 5 Martina Fach-Overhoff: Computerunterstützte interaktive Lernprogramme für den naturwissenschaftlichen Unterricht. Zur Konzeption und Wirkung der Einführung eines phänomennahen Begriffs "Atombindung". 1990.

Band 6 Martin K. W. Schweer: Bewältigungsstrategien in problematischen Interaktionssituationen. 1991.

Band 7 Wolfgang Sonntag: "Induktiv denken lernen". Experimentelle Untersuchungen einer neuen Lehrtheorie und einer älteren volitionalen Handlungstheorie bei lernbehinderten Sonderschülern. 1991.

Band 8 Tobina Brinker: Dialogvideo im Führungskräfte-Training. Eine Studie zur Effektivität und Akzeptanz. 1991.

Band 9 Dieter Kotte: Gender Differences in Science Achievement in 10 Countries: 1970/71 to 1983/84. 1992.